THE GEOMETRY OF QUANTUM POTENTIAL

Entropic Information of the Vacuum

THE GEOMETRY OF QUANTUM POTENTIAL

Entropic Information of the Vacuum

Davide Fiscaletti

SpaceLife Institute, Italy

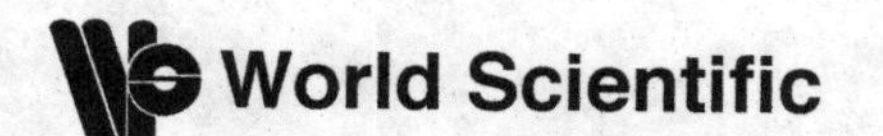

NEW JERSEY · LONDON · SINGAPORE · BEIJING · SHANGHAI · HONG KONG · TAIPEI · CHENNAI · TOKYO

Published by

World Scientific Publishing Co. Pte. Ltd.
5 Toh Tuck Link, Singapore 596224
USA office: 27 Warren Street, Suite 401-402, Hackensack, NJ 07601
UK office: 57 Shelton Street, Covent Garden, London WC2H 9HE

British Library Cataloguing-in-Publication Data
A catalogue record for this book is available from the British Library.

THE GEOMETRY OF QUANTUM POTENTIAL
Entropic Information of the Vacuum

Copyright © 2018 by World Scientific Publishing Co. Pte. Ltd.

All rights reserved. This book, or parts thereof, may not be reproduced in any form or by any means, electronic or mechanical, including photocopying, recording or any information storage and retrieval system now known or to be invented, without written permission from the publisher.

For photocopying of material in this volume, please pay a copying fee through the Copyright Clearance Center, Inc., 222 Rosewood Drive, Danvers, MA 01923, USA. In this case permission to photocopy is not required from the publisher.

ISBN 978-981-3227-97-2

For any available supplementary material, please visit
http://www.worldscientific.com/worldscibooks/10.1142/10653#t=suppl

Desk Editor: Christopher Teo

Typeset by Stallion Press
Email: enquiries@stallionpress.com

Printed in Singapore

Contents

Introduction

Phenomena studied by quantum physics are far outside the range of parameters of our daily experience. The behaviour and the evolution of quantum systems do not seem to be compatible with a description in visualizable terms. Quantum physics turns out to appear extremely exotic, unnatural and mysterious to us because of our incapability to visualize a subatomic particle (such as an electron) and its motion or the uncertainty principle or the processes of creation and annihilation of quanta in the quantum vacuum.

If in classical physics systems can be analysed in terms of elements of information represented by a certain number of well-defined states, which are determined by the values of a certain number of physical quantities (such as position and speed) and our ignorance about the knowledge of the state of the system is owed to the loss of the ability to follow information, in quantum mechanics the physical variables of a system do not seem to have a well defined value at each instant and thus the ignorance seems to be intrinsic to the nature of objects. In the study of the microscopic world it seems that there is no way to avoid the introduction of the probability and that, when a property of an object is measured, its state turns out to be modified dramatically by virtue of the irreversible interaction with the measurement apparatus. Quantum mechanics leads to develop an entirely new kind of programme in order to study physical processes — with respect to classical physics — which takes account of the fact that each system can be found in a state belonging to its Hilbert space with a certain probability distribution. For example, if we prepare an electron with the spin oriented along a vertical direction by creating a large external magnetic field oriented along this direction, it turns out that the electron spin has only two possible configurations, parallel or antiparallel, to the external magnetic field. By repeating the same experiment many times in an identical way, in each experiment we discover the same result regarding the spin of the electron, namely: sometimes the spin is "up" and sometimes the spin is "down". In the light of all experimental results, we can conclude that the states "spin up" and "spin down" are the two possible states of the spin and that the real, actual state of the spin of the electron may be known only after a measurement is performed, in other words probability seems to play an important role as regards the behaviour and the evolution of the spin of the electron.

Quantum mechanics is perhaps the fundamental theory of 20th century which has determined the most profound changes in the understanding of the physical world. Despite its extraordinary successes from the predictive point of view (as regards, in particular, the study of the microscopic world, from atomic physics to nuclear physics and subnuclear physics and so on), it is plagued by several problems of interpretation as regards what it says about the world. After 80 years since its birth, the meaning of quantum theory remains as controversial as ever. Nevertheless, quantum theory has introduced much more spread perspectives and scenarios inside theoretical physics than those offered by every previous theory.

The debate about the meaning of quantum mechanics can be traced back to 1927 when the participants at the fifth Solvay conference in Brussels distinctly failed to arrive at a consensus (this emerges in a clear way from the actual proceedings of the conference) [1]. In order to put in evidence the relevant disagreement existing among the participants at this congress, Paul Ehrenfest made an emblematic gesture, during one of the discussions, by writing the following quotation from the book of Genesis on the blackboard: "And they said one to another: 'Go to, let us build us a tower, whose top may reach unto heaven; and let us make us a name'.

And the Lord said: 'Go to, let us go down, and there confound their language, that they may not understand one another's speech"'.

If the builders of the Tower of Babel could not realize one another's speech, in analogous way in the fifth Solvay Conference in Brussels it seemed as if the eminent physicists gathered in it could no longer understand one another's speech. In particular, at the crucial 1927 Solvay conference, three quite distinct theories of quantum physics were presented and discussed on an equal footing: de Broglie's pilot-wave theory, Schrödinger's wave mechanics, and Bohr's and Heisenberg's quantum mechanics. According to de Broglie's theory, subatomic particles (such as electrons) can be seen as point-like objects moving with continuous trajectories 'guided' or choreographed by the wavefunction. In Schrödinger's wave mechanics, particles can be seen as localized wave packets moving in space, that emerge out of the wavefunction, intended as the fundamental reality. Instead, according to Born's and Heisenberg's view, in the study of microscopic processes the interference between measured system and observer has to be taken into consideration: the idea of definite states of reality at the quantum level cannot be maintained in a way that is independent of human observation.

In order to describe the atomic and subatomic world, quantum theory, as presented in classical textbooks, implies to consider a human observer performing experiments with microscopic quantum systems by making use of a macroscopic classical apparatus. As physicists know from the reading of classical textbooks on these topics, in quantum mechanics the state of a physical system is described by introducing the concept of wavefunction, a mathematical object that is used to calculate probabilities but which provides no clear description of the state of reality of a single system. Instead, the observer and the apparatus are described classically and are assumed to have definite states of reality. For example, a pointer on a measuring apparatus always indicates a particular reading, corresponding to the measured value of the quantity in consideration. Quantum systems seem to inhabit a fuzzy, indefinite realm, while our everyday macroscopic world — although is ultimately built from the quantum laws — does not.

Quantum theory is formulated in such a way that causes a sharply defined boundary between the quantum and classical domains. Even today Newton's laws of classical mechanics are seen as an approximation of the more general quantum mechanics and it is implicitly assumed that they must be considered valid as long as the dimensions of the system under study are big. However, the correspondence principle, which expresses this peculiar characteristic of Newton's laws to appear only at macroscopic level, indeed does not turn out to be demonstrable, and thus one can say that, strictly speaking, the classical domain practically does not exist. The following questions become thus natural. How does everyday reality emerge from the fuzzy and exotic quantum domain? What does it happen to real macroscopic states as we move to smaller scales? In particular, when and in which situations does physical reality generate microscopic fuzziness? What are really the phenomena occurring inside an atom? Despite the astonishing progress made in high-energy physics and in cosmology since the Second World War, today there is no precise and satisfactory answer to these simple questions.

The Founding Fathers of quantum theory, namely the exponents of Copenhagen and Gottinga schools (Bohr, Heisenberg, Born, etc...) handed to us a view known as standard (or Copenhagen) interpretation of quantum mechanics. According to the Copenhagen interpretation, a causal description of atomic subatomic processes in agreement with the motion dogma cannot be provided: the wavefunction carries only the information about possible outcomes of a measurement process [2]. In the original formulation of the theory, a measurement derives from the interaction of a quantum microsystem S with a classical macrosystem O [3, 4]. After the microsystem S under consideration interacts with its surroundings, the microsystem and its surroundings then become entangled and they are in a quantum mechanical superposition. If the macrosystem O interacts with the variable q of the microsystem S, and S is in a superposition of states of different values of q, then the macrosystem O measures only one of the values of q, and the interaction modifies the state of S by projecting it into a state with that value: in every measurement the wavefunction of a microsystem collapses into the state specified by the outcome of the measurement. Until a measurement is not performed, a microscopic system may be in an undefined state, which describes only a "potentiality" of the physical system into consideration, namely contains the information on a set of possible values, each with a corresponding probability to become real and objective when the measurement takes place. This means that, according to the standard interpretation of quantum mechanics, the evolution of a given physical microscopic system materializes only after a measurement is performed and an interaction between the microscopic system and a macroscopic apparatus occurs. Therefore, whenever a measurement is performed the wavefunction of the physical system ceases to evolve according to the Schrödinger equation (the evolution law of every isolated microscopic system) and collapses into one of its eigenstates. The final state is not predictable with certainty but only probabilistically and, according to the standard version, the reduction of the wavefunction is provoked by the observer, in other words happens at the moment in which the information about the interaction between the microsystem and the macrosystem arrives in the observer's brain: it is the observer who forces nature to reveal itself in one of its possible states [5]. The absolute square of the scalar product of the wavefunction with its eigenfunctions are the probabilities (or probability densities) of the occurrence of these particular eigenvalues in the measurement process [3, 6]. Inside the Copenhagen interpretation, the postulate of the reduction of the wavefunction can be also considered as a sort of device, of a way out which is introduced in order to reproduce the non-linear complex nature of the quantum phenomena associated with the problem of the wave-particle duality.

There was a general movement in theoretical physics in the 1920s against the idea that individual atomic events could be visualized as parts of causally connected sequences of spacetime processes. In the paper where he developed the probability interpretation of the wavefunction, Born wrote: "I myself am inclined to give up determinism in the world of atoms. But that is a philosophical question for which physical arguments alone are not decisive" [7]. The standard interpretation of quantum mechanics seemed to consider as almost axiomatic that the trajectory concept of classical mechanics is incompatible with atomic and subatomic processes. As regards the philosophy of standard quantum mechanics, the following Bohr's sentences are significant too: "A phenomenon is not a phenomenon until it is observed" and "It is a mistake to think that physics' aim is to find how nature is. Physics concerns what we can say about nature" [8]. Bohr thought that in quantum mechanics the observer does not reveal the phenomenon but somehow fixes and defines it. According to Bohr's view, the observer plays an active role in the study of the behaviour of elementary particles, in other words the observer influences in a decisive way the experimental results regarding atomic and subatomic processes: the real and objective features of a physical system are ultimately created by the interaction of the observer and its macroscopic measuring apparatus with the measured system.

The central claim made by standard quantum mechanics is that the wavefunction provides a complete description of a quantum system: all we can know about a physical system is contained in its wavefunction. This would seem to imply that quantum mechanics is, fundamentally, a theory about wavefunctions. However, if one accepts such a proposition an inescapable problem is derived, which was eloquently formulated by Schrödinger within the context of his famous cat paradox: namely, the question of what it actually means for an object to literally exist in a superposition of eigenstates of a measurement operator [9]. In the light of its formal structure and its conceptual foundations as regards the study of the measurement processes, the standard interpretation of quantum mechanics can be also considered as the "minimal" interpretation of quantum theory: it provides us only the mathematical structure and the minimum interpretative propositions needed to define the relation between the mathematical structure itself and the experience.

On the other hand, it must be emphasized that several authors introduced the possibility to understand quantum processes on the basis of a geometric interpretation. According to these proposals, quantum mechanics might not be a fundamental theory but rather emerge from an underlying geometry. In particular, in 1984–1885

Santamato first found a similarity between Weyl geometry and the structure of quantum mechanical equations [10, 11]. In 1990 J.T. Wheeler suggested a geometric picture of quantum theory in which the quantum measurement and the related uncertainty emerge directly from Weyl geometry [12]. In 1995 Wood and Papini developed a modified Weyl–Dirac theory which unifies the particle aspects of matter and Weyl symmetry breaking [13]. More recently, studies in this direction focused on the non-commutative non-integrable geometry [14] or on the Ricci flow [15–20] or on a geometric reduction of the dimensionality of space-time [21, 22].

According to the results of other authors, the formalism of quantum mechanics may be derived from the covering space of the symplectic group (see, for example, the works of Guillemin and Sternberg and of De Gosson [23–25]). These research imply that it is possible to derive the Schrödinger equation rigorously from classical symplectomorphisms by lifting the classical phase space behaviour onto a covering space, if one assumes that the Hamiltonian is up to quadratic in position and momentum. In this picture, the wave and particle aspects of the quantum entities may be described inside a unifying formalism: the particle properties are described on the underlying phase space while the wave properties appear at the level of the covering space. Moreover, as shown by de Gosson in the recent paper *The symplectic camel and quantum universal invariants: the angel of geometry versus the demon of algebra*, the orthogonal projection operator of the phase space R^{2n} of a continuous variable system with n degrees of freedom onto an arbitrary symplectic subspace F_{2k} is an ellipse of area not inferior to $\frac{1}{2}h$ (h being Planck's constant), which is a geometric version of Heisenberg's uncertainty principle [26].

On the other hand, although the standard interpretation of quantum mechanics functions perfectly from the predictive point of view, it is nevertheless characterized by inner contradictions and paradoxes. Besides the impossibility to describe quantum processes as events happening in a space-time picture, other crucial problems which plague the Copenhagen interpretation regard the treatment of measurement processes and, in particular, the definition of the boundary between the microscopic world, governed by the superposition principle, and the macroscopic world, in which one has well-defined perceptions as regards the properties of physical systems. The failure of the standard quantum theory to offer any sort of coherent resolution to these problems is largely the reason for which it has continually remained so ambiguous and obscure. As a consequence, the ontology which derives from the standard quantum mechanics cannot be considered a completely satisfactory starting-point in order to develop a coherent geometrodynamic picture of the quantum world.

Since the birth of quantum mechanics in the second half of the 20s of last century, Einstein claimed that the description of quantum phenomena proposed by Copenhagen and Gottinga schools, despite functioned well in the prediction of experimental results, was not complete, did not constitute a perfect theory and that it could be made causal by introducing some additional parameters. Einstein's point of view was thus that the behaviour of subatomic particles is independent from the observer, that it depends on some "hidden variables" besides their wavefunctions. In this way a strong well-known debate between Einstein and Bohr as regards the meaning of quantum theory occurred. Despite the majority of physicists think that the debate on the foundations of quantum mechanics between Einstein and Bohr, lasted many years, was won by Bohr and that the Copenhagen interpretation should simply be accepted, over the last 30 years the consensus about the interpretation of quantum processes has evaporated and physicists find themselves faced with a set of different alternative interpretations of quantum mechanics. In particular, today we can affirm with certainty that what Einstein wished and Bohr thought impossible exists. A quantum mechanics without the observer exists, in which the measurement processes can be analysed in terms of more fundamental concepts.

Today we have got some significant versions of quantum mechanics which do not ascribe a special role to the observer: they demonstrate that Bohr's interpretation of quantum processes cannot be considered convincing and satisfactory, both from the physical and from the philosophical point of view. In this regard, as we have said before, one of the theories presented at the 1927 Solvay Conference was Louis de Broglie's pilot-wave dynamics according to which subatomic particles are point-like objects with continuous trajectories "guided" or choreographed by the wavefunction [27–29]. De Broglie's pilot wave theory can be considered as the first significant geometrodynamic non-linear way of thinking for solving the enigma posed by the wave-corpuscle duality. Moreover, de Broglie provided another development of this line of research, regarding how the pilot wave would actually guide the particle, known as "double solution theory": in this picture, for each quantum particle the fundamental non-linear nature of the interaction determines a singular solution, representing the particle, which fits smoothly onto a weak background field, obeying Schrödinger's equation as a linear approximation.

The concepts of pilot wave and double solution theory introduced by de Broglie lead to important developments towards the construction of a clear ontology and geometrodynamics of quantum processes (which allows us to avoid the spectrum of the observer). On one hand, de Broglie's pilot-wave theory was resurrected in 1952 when Bohm used it to describe a general quantum measurement (for example of the energy of an atom). Bohm showed that the statistical results obtained would be the same as in conventional quantum theory — if one assumes that the initial positions of all the particles involved (making up both "system" and "apparatus") have a Born-rule distribution, that is, a distribution proportional to the squared-amplitude of the wavefunction (as appears in conventional quantum theory). In his classic works of 1952 Bohm developed a mathematical treatment of de Broglie's pilot-wave concept, realizing that this theory provided a foundation for a clear causal ontology of non-relativistic quantum mechanics. Bohm's approach reproduces all the empirical results of quantum theory and at the same time has the merit to explain the quantum behaviour of matter remaining faithful to the principle of causality and the motion dogma, thus avoiding to ascribe a special role to the observer and recovering some causality also in the microscopic world [2, 30–34]. On the other hand, also de Broglie's double solution theory has resurrected today as one of the basic starting-point of the non-linear quantum physics which emerges, for examples, in some works of Croca [35–38]. Croca's non-linear approach suggests that a quantum particle is composed of an extended region, the theta wave, and inside it there is a kind of a very small localized structure, the acron, which satisfies a fundamental principle, the principle of eurhythmy, which states that the acron moves in a stochastic way preferentially to the regions where the intensity of the theta wave field is greater. In the picture of Croca's non-linear quantum physics the classical domain can be obtained directly when there is a complete independence between corpuscles and waves, in other words quantum physics can be seen as a real generalization of classical physics inside an unifying formalism. As a consequence, Croca's non-linear approach can be considered another interesting starting-point in order to build a fundamental clear ontology and geometrodynamics which is at the basis of quantum phenomena. In particular, it allows us to derive a new set of more general uncertainty relations which may be tested by the current super-resolution optical microscopes.

As regards the perspective of a geometric description of quantum processes, according to the author of this book, above all it is important to emphasize that in recent years it is possible to remark a growing interest in De Broglie-Bohm realist interpretation of quantum mechanics. Among the merits of de Broglie-Bohm approach, one of the most significant regards indeed the doors that it opens towards the construction of a purely geometrodynamic picture of the quantum world. De Broglie-Bohm interpretation of quantum mechanics, thanks to its most important element, namely the quantum potential, introduces interesting and relevant perspectives towards the interpretation of quantum phenomena as a modification of the geometrical properties of the physical space, in other words towards the interpretation of quantum mechanics as a corresponding "deformation" of the space-time background. In Bohm's approach, the quantum potential emerges indeed as the crucial physical entity which allows us to construct a satisfactory geometrodynamic picture of the quantum world, as a fundamental reality which underlines the geometry of the quantum processes. The aim of this book is to make a comprehensive state-of-the-art review of some of the most significant elements and results about the geometrodynamic picture determined by the quantum potential in various contexts. In particular, we will explore the most fundamental arena of quantum processes that is subtended by the quantum potential, in other words the link between the geometry associated to the quantum potential and a fundamental quantum vacuum.

This book is structured in the following manner. In chapter 1 we will analyse the geometrodynamic features of the quantum potential, the geometry subtended by the quantum potential in non relativistic de Broglie-Bohm theory, in relativistic quantum mechanics, in relativistic quantum field theory, in quantum gravity and in quantum cosmology as well as the link between Bohm's quantum potential and Weyl geometries, in the light of the current state-of-the-art. In chapter 2, in order to introduce novel perspectives as regards the deep, fundamental arena which is associated with the geometry subtended by the quantum potential we will analyse a recent interpretation of Bohm's quantum potential in terms of a more fundamental entity called quantum entropy (and we will extend this approach from the non-relativistic domain to the relativistic domain, the relativistic quantum field theory, the quantum gravity and the quantum cosmology in the picture of the models analysed in chapter 1). In chapter 3 we will analyse a recent development of the geometrodynamic approach of Bohm's quantum potential represented by a symmetrized quantum potential which implies a three-dimensional timeless background as fundamental arena of quantum processes (and we will extend this approach to the relativistic curved space-time, the relativistic quantum field theory and the quantum cosmology). Finally, in chapter 4 we will make some consideration about the link between the quantum potential and the quantum vacuum, intended as the fundamental arena of physical processes.

Acknowledgments

The author thanks Ignazio Licata and Amrit Sorli for fruitful discussions.

Chapter 1

The Geometry of the Quantum Potential in Different Contexts

1.1 Bohm's Original 1952 Approach on the Quantum Potential

The probabilistic interpretation of the wavefunction developed by the physicists of the Copenhagen and Gottinga schools seems to be in agreement with experimental facts regarding the microscopic world. However, indeed it is not forced by these experimental results: the outcomes of experiments on atomic and subatomic processes seem to indicate only that it is consistent to consider the wavefunction as a parameter which contains information on probability but do not exclude the possibility that the wavefunction may possess other properties. On the other hand, as even Heisenberg observed at the dawning of quantum theory, the standard interpretation is radically a-causal and atomic processes cannot therefore be incorporated into a space-time picture. Since in the standard interpretation of quantum mechanics quantum phenomena cannot be explained as events happening in space-time, namely the dogma of formulation of physics in terms of motion in space–time must be abandoned, the standard quantum mechanics cannot be considered satisfactory if one wants to develop a coherent geometrodynamic picture of the quantum world.

The purely probabilistic interpretation of the wavefunction characteristic of standard quantum mechanics can be considered coherent only if one wishes to reduce physics to a kind of algorithm which is efficient to correlate the statistical results of experiments. If one wishes to do more, and attempts to understand the experimental results regarding the microscopic world in terms of a causally connected series of individual processes and thus to develop a real geometrodynamic picture of the quantum world, then it becomes natural and plausible to search for possible further significances of the wavefunction (beyond its probabilistic aspect), and to introduce other elements in addition to the wavefunction.

This is what Bohm's version of quantum mechanics allows us to realize: to suggest a formulation of quantum mechanics where probability only enters as a subsidiary condition on a causal theory of the motion of individual events. In Bohmian quantum mechanics the wavefunction turns out to have a direct physical significance in each individual process, its statistical meaning is only a secondary property. Besides the wavefunction, this approach introduces an additional element in order to obtain a geometrodynamic description of subatomic processes, namely a particle, conceived in the classical sense as pursuing a definite continuous trajectory in space and time.

Bohmian quantum mechanics, known in the literature also as de Broglie-Bohm pilot wave theory, can be considered as the most significant and satisfactory hidden variables theory predictably equivalent to quantum mechanics, able to give a causal completion to quantum mechanics. It can be inserted inside that important research stream directed to complete standard quantum theory in a deterministic sense.

This approach was originally proposed by Louis de Broglie in 1927 at the Solvay Conference. As regards the non-relativistic problem, de Broglie proposed that the wavefunction of each one-body physical system is associated with a set of identical particles which have different positions and are distributed in space according to the usual quantum formula, given by $|\psi(\vec{x})|^2$. But he recognized a dual role for the wavefunction: on one hand, it carries information about the probable position of the particle (just like in the standard interpretation), on the other hand, it influences the position by exerting a force on the orbit. According to the de Broglie view, the wavefunction would act like a sort of pilot wave which guides the particles in regions where such wavefunction is more intense [27]. In the context of his proposal, de Broglie applied his guidance formula to compute the orbits for the hydrogen atom stationary states. The approach met however with a general lack of enthusiasm at the Solvay Conference. The unfavourable climate, presumably generated by Heisenberg's discovery of the 'uncertainty' relations (and the interpretation of these relations provided by Heisenberg himself, which implies a crucial role of the observer), eventually led him to abandon this research programme. De Broglie returned to research in this field 25 years later when David Bohm rediscovered the approach and developed it to the level of a fully fledged physical theory.

In 1952 David Bohm published in Physical Review two fundamental papers entitled *A suggested interpretation of the quantum theory in terms of hidden variables* [30, 31]. In these two classic works Bohm was able to extend de Broglie's approach in a coherent way also to many-body systems, and to use this approach in order to describe a general quantum measurement (for example of the energy of an atom). In the light of these two papers, which can be indeed considered the starting point of de Broglie-Bohm theory, a new quantum world has been opened, the so-called Bohm's approach to quantum phenomena. Bohm's version of quantum mechanics is practically the de Broglie pilot-wave theory carried to its logical conclusion. The essential insight of Bohm's work lies in the assumption that quantum mechanics can be interpreted as a theory about particles in motion: in addition to the wavefunction, the description of a quantum system should also include its configuration — namely, the precise positions of all the particles of the system at all times. For this reason, Bohm's approach can be considered a hidden variables theory: it is based on the idea that the description of quantum processes provided by the quantum wavefunction is not complete and must be completed by adding supplementary parameters to the quantum formalism. The hidden variables of the model are just the positions of all the particles constituting the physical system into examination: the physical system is prepared in such a way that, at the initial time $t = 0$, it is associated with a specific wavefunction $\psi(\vec{x}, 0)$ which is assumed to be known perfectly, and moreover is in a point $\vec{x}$ (among those compatible with the wavefunction into examination) that instead we ignore. In this approach, the outcome of a single quantum experiment is in line of principle determined by the precise ('hidden variable') positions of all the particles involved. If the experiment is repeated many times, then the outcomes turn out to have a statistical behaviour caused by the spread in the initial distribution of the particle positions.

As shown by Bohm in [30] and [31], the basic postulates which constitute the core of this approach are the following:

1. An individual physical system comprises a wave propagating in space and time together with a point particle which moves continuously under the guidance of the wave.
2. The wave is mathematically described by $\psi(\vec{x}, t)$, a solution to Schrödinger's equation $i\hbar\frac{\partial\psi}{\partial t} = H\psi$ (1.1) (where H is the Hamiltonian of the physical system and $\hbar$ is Planck's reduced constant).
3. The particle motion is obtained as the solution $\vec{x}(t)$ to the equation $\dot{\vec{x}} = \frac{1}{m}\nabla S(\vec{x}, t)|_{\vec{x}=\vec{x}(t)}$ (1.2) where S is the phase of $\psi(\vec{x}, t)$ and m is the mass of the particle. This equation can be solved by specifying the initial condition $\vec{x}(0) = \vec{x}_0$. This specification constitutes the only extra information introduced by the theory that is not contained in the wavefunction $\psi(\vec{x}, t)$ (the initial velocity is fixed once one knows S). An ensemble of possible motions associated with the same wave is generated by varying $\vec{x}_0$.

These three postulates on their own provide a consistent theory of motion from the physical point of view. In order to obtain a compatibility of the motion of the ensemble of particles with the results of quantum mechanics (which are well-founded), Bohm's approach introduces a further fourth postulate:

4. The probability that a particle in the ensemble lies between the points $\vec{x}$ and $\vec{x} + d\vec{x}$ at time t is given by $R^2(\vec{x}, t)d^3x$

where $R^2 = |\psi|^2$ (namely R is the real amplitude function of the wavefunction $\psi(\vec{x}, t)$).

This fourth postulate has the effect to select among all the possible motions implied by the law (1.2) those that are compatible with an initial distribution $R^2(\vec{x}, 0) = R_0^2(\vec{x})$ of the particles associated with the wavefunction. As regards the postulate 4, it is also important to underline that the modulus squared of the wavefunction $|\psi(\vec{x}, t)|^2$ determines the probability density of finding a particle in the volume d^3x at time t if a suitable measurement is carried out.

In Bohm's view of non-relativistic quantum mechanics, the introduction of the particle is motivated by substituting the expression of the wavefunction in polar form $\psi = Re^{iS/\hbar}$ into Schrödinger's equation. By separating real and imaginary part of Schrödinger's equation, Bohm found the following fundamental result regarding the description of the behaviour of one-body systems: the movement of the corpuscle under the guide of the wave happens in agreement with a law of motion which assumes the form

$$\frac{\partial S}{\partial t} + \frac{|\nabla S|^2}{2m} - \frac{\hbar^2}{2m}\frac{\nabla^2 R}{R} + V = 0 \tag{1.3}$$

(where V is the classical potential). This equation is equal to the classical equation of Hamilton-Jacobi except for the appearance of the additional term

$$Q = -\frac{\hbar^2}{2m}\frac{\nabla^2 R}{R} \tag{1.4}$$

having the dimension of an energy and containing Planck's constant and therefore appropriately defined quantum potential. After some mathematical manipulations, the equation of motion of the particle can be expressed also in the form

$$m\frac{d^2\vec{x}}{dt^2} = -\nabla(V + Q), \tag{1.5}$$

where $\vec{x} = \vec{x}(t)$ is the trajectory of the particle associated with its wavefunction. Equation (1.5) is equal to Newton's second law of classical mechanics, always with the presence of the additional term Q of quantum potential. In the light of equation (1.5), the movement of a subatomic particle, according to Bohm's pilot wave theory, is therefore tied to a total force which is given by the sum of two contributions: a classical force (derived from a classical potential) and a quantum force (derived just from the quantum potential) [2, 32, 33].

Therefore, if one studies the conceptual foundations and the mathematical formalism of the de Broglie-Bohm theory, as was originaly developed in the 1952 Bohm papers, one finds that in this approach the behaviour of each subatomic particle turns out to be completely and univocally specified if one provides its wavefunction (which evolves according to the usual Schrödinger equation (1.1)) as well as its configuration: each particle follows a precise trajectory $\vec{x} = \vec{x}(t)$ in space-time that is originated by the action of a classical potential and a quantum potential (and that evolves according to the equation (1.3) or the equivalent equation (1.5)). Moreover, in the Bohm model, there is another important equation which emerges from the decomposition of the Schrödinger equation (1.1) together with the quantum Hamilton-Jacobi equation (1.3), namely the continuity equation for the probability density $\rho(\vec{x}, t) = R^2(\vec{x}, t) = |\psi(\vec{x}, t)|^2$:

$$-\frac{\partial \rho}{\partial t} = \nabla \cdot \left(\rho \frac{\nabla S}{m} \right). \tag{1.6}$$

The continuity equation (1.6) implies that, in de Broglie-Bohm theory, all individual trajectories of the particles put in evidence a collective behavior like a liquid flux, and maybe a superconductive one.

We have to emphasize here that quantum potential must not be considered a term which is introduced ad hoc inside quantum theory, contrary to the opinion of the supporters of the Copenhagen interpretation. In the formal plant of Bohm's non-relativistic theory, it emerges directly from Schrödinger's equation and without it the total energy of the system into consideration should not be conserved. In fact, in equation (1.3), the quantity $-\frac{\partial S}{\partial t}$ constitutes the total energy of the particle whilst $\frac{|\nabla S|^2}{2m}$ is its kinetic energy. As a consequence, equation (1.3) can be also written in the equivalent convenient way

$$\frac{|\nabla S|^2}{2m} - \frac{\hbar^2}{2m}\frac{\nabla^2 R}{R} + V = -\frac{\partial S}{\partial t}, \tag{1.7}$$

which can be seen as a real energy conservation law in quantum mechanics: here one can easily see that without the quantum potential

$$Q = -\frac{\hbar^2}{2m}\frac{\nabla^2 R}{R} \tag{1.4}$$

energy could not be conserved and this means that quantum potential plays an essential role in quantum formalism. On the other hand, it must be observed, as it was shown recently by Hiley in the reference [39], that quantum potential can be derived also by Heisenberg's formalism by choosing a particular representation for operators and also here such term must be present in order to assure the conservation of the total energy of the system.

The treatment provided by relations (1.3)–(1.7) can be extended in a simple straightforward way to many-body systems. In the case of a system of N particles, one considers a wavefunction $\psi = R(\vec{x}_1, \ldots, \vec{x}_N, t)e^{iS(\vec{x}_1,\ldots,\vec{x}_N,t)/\hbar}$, defined on the configuration space R^{3N} and, by following the same procedure followed in the case of the one-body system, one finds that the movement of this system under the action of the wave ψ happens in agreement to the law of motion

$$\frac{\partial S}{\partial t} + \sum_{i=1}^{N}\frac{|\nabla_i S|^2}{2m_i} + Q + V = 0 \tag{1.8}$$

where

$$Q = \sum_{i=1}^{N} -\frac{\hbar^2}{2m_i}\frac{\nabla_i^2 R}{R} \tag{1.9}$$

is the many-body quantum potential. Equation (1.8) can also be written in the convenient way

$$\sum_{i=1}^{N}\frac{|\nabla_i S|^2}{2m_i} + Q + V = -\frac{\partial S}{\partial t} \tag{1.10}$$

which can be considered as a real energy conservation law in quantum mechanics for a many-body system. The equation of motion of the i-th in particle, in the limit of big separations, can also be written in the following form

$$m_i \frac{\partial^2 \vec{x}_i}{\partial t^2} = -[\nabla_i Q(\vec{x}_1, \vec{x}_2, \ldots, \vec{x}_n) + \nabla_i V_i(\vec{x}_i)] \tag{1.11}$$

which is a quantum Newton law for a many-body system. In the light of equation (1.11), one can say that the contribution to the total force acting on the i-th particle coming from the quantum potential, i.e. $\nabla_i Q$, is a function of the positions of all the other particles and thus in general does not decrease with distance. The continuity equation for the probability density becomes:

$$-\frac{\partial \rho}{\partial t} = \sum_{i=1}^{N} \nabla_i \cdot \left(\rho \frac{\nabla_i S}{m} \right). \tag{1.12}$$

The fundamental element of the mathematical formalism of non-relativistic de Broglie-Bohm theory is the quantum potential. In virtue of its mathematical expression, the quantum potential turns out to have different properties with respect to those expected from a classical potential (such as the electromagnetic potential). Its mathematical definition — which is (1.4) for one-body systems and (1.9) for many-body systems — shows that the quantum potential depends on how the amplitude of the wavefunction varies in space. Because of the presence of the Laplace operator, the action of this potential is like-space, namely creating onto the particle a non-local, instantaneous action. In equations (1.4) and (1.9), the appearance of the amplitude of the wavefunction in the denominator also lets us realize why the quantum potential can produce strong long-range effects that do not necessarily fall off with distance and the typical properties of entangled wavefunctions. Thus even though the wavefunction spreads out, the effects of the quantum potential need not necessarily decrease. This type of behaviour of the quantum potential can understand and explain what happens in EPR-type experiments.

Let us consider, for example, the well known two-slit experiment, where a beam of electrons is sent towards two slits and at the end is revealed on a screen, where an interference pattern is observed. As regards the double-slit experiment, the expression of the quantum potential turns out to depend on the width of the slits, their distance apart and the momentum of the particle. In other words, it has a contextual nature, namely providing global information on the process and its environment. It contains instantaneous information about the overall experimental arrangement, the environment. Moreover, this information can be regarded as being active in the sense that it modifies the behaviour of the particle. In the double-slit experiment, this means that, if one of the two slits is closed, the quantum potential changes, and this information arrives instantaneously to the particle, which behaves accordingly. In this regard, as suggested in 1984 by Bohm and Hiley, one can interpret the quantum potential as a sort of "information potential". According to this reading of the quantum potential, the particles in their movement are guided by the quantum potential just as a ship at automatic pilot can be handled by radar waves of much less energy than ship's energy. In the light of this interpretation, the results of double-slit experiment are explained by considering that the quantum potential contains an active information, for example about the slits, and that this information manifests itself in the particles' motions.

Since the end of the 70s, significant research showed that by means of the quantum potential approach it is possible to understand and explain in a satisfactory way all a series of experimental results and quantum phenomena. In particular, in 1979 Philippidis, Dewdney and Hiley [40] found that, as regards the double-slit experiment of the electrons, Bohm's quantum potential allows an explanation of the experimental results (namely the interference pattern) maintaining the concept of trajectory and particle even after the electrons have passed the slits. On the basis of Philippidis's, Dewdney's and Hiley's results, in the double-slit experiment each particle follows a precise and computable trajectory and, despite this, if the distribution of the arrivals on the two slits is uniform, the action of the quantum potential assures that the probability density of presence of the particles remains equal to the modulus square of the wavefunction in all the next instants. As regards the double-slit experiment, on the other hand, if in some works of the 90s some authors claimed that the corpuscular trajectories have to be considered surreal and devoid of physical meaning [41–43], recently it has been shown clearly that these trajectories are the direct consequence of the entanglement, thus recovering the agreement between the quantum description of the particles' propagation and Bohm's realism [44].

The quantum potential can explain in visualizable terms also the other typical quantum phenomenon represented by tunnelling effect through a square barrier, justifying in a clear way why some particles manage to pass the potential barrier and others not. On the basis of the results obtained by Dewdney and Hiley in 1982, the tunnelling effect occurs as a consequence of a significant modification of the classical potential provoked by the quantum potential: the quantum potential turns out to be negative in the region of the barrier and this implies that a particle can have the sufficient energy in order to pass the barrier [45]. The results of Bohm's approach regarding double-slit interference and tunnelling — as well as the fact that during the 80s Bohm's quantum potential approach provided a convincing explanation of many other experiments (in this regard the reader may find details, for example, in the reference [46]) — show the legitimacy and physical coherence of the de Broglie-Bohm theory.

As regards the interpretation of non-relativistic quantum mechanics proposed by Bohm, a great deal of confusion and misunderstanding exists about what Bohm was saying exactly when he first published his 1952 papers. Many research and discussions about the ideas proposed by Bohm in these two papers would seem to lead to the conclusion that the Bohm approach represented some kind of definitive theory. As shown clearly by Hiley in the recent article *Some remarks on the evolution of Bohm's proposals for an alternative to standard quantum mechanics* [47], the misunderstandings about Bohm's original ideas, grew after the appearance of the term "Bohmian mechanics" in a 1992 paper of Dürr, Goldstein and Zanghì [48]. Instead, it must be remarked that, in his classic works of 1952, Bohm never utilized the term mechanics: in these two papers Bohm's intention was not to find a classical order based on a deterministic mechanics from which the quantum formalism could be obtained. On the other hand, the content of his book *Quantum Theory* published in 1951 [49], which gives an exhaustive account of the orthodox view of the theory, already shows the clues of how radical changes according to Bohm are necessary in order to understand the structure that underlies the quantum formalism. By reading the book *Quantum Theory*, one can easily see in Bohm's ideas the need to go beyond mechanical ideas. In the section of this book titled 'The need for a non-mechanical description', Bohm writes "...the entire universe must, in a very accurate level, be regarded as a single indivisible unit in which separate parts appear as idealisations permissible only on a classical level of accuracy of the description. This means that the view of the world as being analogous to a huge machine, the predominant view from the sixteenth to the nineteenth century, is now shown to be only approximately correct. The underlying structure of matter, however, is not mechanical" [24].

Bohm's 1952 papers let us realize that the mathematical formalism summarized in equations (1.1)–(1.12) does not present a deep and unavoidable breaking with the quantum standard formalism: equations (1.1)–(1.12) merely suggest that an alternative view, that attributes definite properties to individual particles, is possible without radically changing the quantum formalism and altering the predictions. Bohm was not providing these proposals as the final definitive interpretation of the quantum formalism in the non-relativistic regime. On the other hand, a detailed reading of his 1952 papers lets us realize how his approach opens up possibilities of modifying the quantum formalism in ways that could not be possible in the standard interpretation. For example, he suggested that there was a possibility of exploring deeper structures that could lie below 10^{-13} cm. And, as regards these considerations, Bohm never remarked them as being, in some sense, a definitive interpretation of the formalism, but using them as possible keys to go beyond, towards something deeper.

In his recent paper [47], Hiley showed clearly that the fundamental key element which indicates that Bohm's 1952 approach was not a mechanical approach but introduces the possibility to explore deeper structures is the quantum potential. As shown by Bohm in his 1952 papers, the mathematical features of the quantum potential suggest that the whole environment determines the properties of the individual particles and their relationship. In the view of the quantum world introduced by Bohm in 1952, the quantum potential implies a universal interconnection of things that could no longer be questioned. In Bohm's ideas, the notion of the particle cannot be considered more fundamental than the quantum potential. In summary, a detailed analysis of Bohm's 1952 papers leads clearly to the the following fundamental conclusion: the quantum potential allows the global properties of quantum phenomena to be treated in such a way that the 'particle' is not independent of the background and that it is the quantum potential that contains the effect of this background. In other words, the particle and quantum potential form indeed an indivisible whole.

1.2 The Geometrodynamic Information of the Quantum Potential

The fact that the quantum potential produces an active information, a global information on the environment means that it cannot be seen as an external entity in space but as an entity which contains a spatial information, as an entity which represents space. In virtue of its features, the quantum potential can be considered a geometric entity, the information determined by the quantum potential is a type of geometric information "woven" into space-time. Quantum potential has a geometric nature just because has a contextual nature, contains a global information on the environment in which the experiment is performed; and at the same time it is a dynamical entity just because its information about the process and the environment is active, determines the behaviour of the particle.

As the author has underlined in the recent article "*The geometrodynamic nature of the quantum potential*" [50], in this geometrodynamic picture we can say that the quantum potential indicates, contains the geometric properties of space from which the quantum force, and thus the behaviour of quantum particles, derive.

Considering the double-slit experiment, the fact that the quantum potential is linked with the width of the slits, their distance apart and the momentum of the particle, namely that brings a global information on the environment, means just that it describes the geometric properties of the experimental arrangement (and therefore of space) which determine the quantum force and the behaviour of the particle. Moreover, the presence of Laplace operator (and of the absolute value of the wavefunction in the denominator) indicates that the geometric properties contained in quantum potential determine a non-local, instantaneous action onto the particle. One can say therefore that Bohm's approach manages to make manifest this essential feature of quantum mechanics, just by means of the geometric properties of space described and expressed by the quantum potential. As regards the geometrodynamic nature of the quantum potential and the non-local nature of the interactions in physical space, one can also say, by paraphrasing J.A. Wheeler's famous saying about general relativity, that the evolution of the state of a quantum system changes active global information, and this in turn influences the state of the quantum system, redesigning the non-local geometry of the universe.

If Bohm's quantum potential determines a non-local geometry of space, it derives then that the Cartesian order could no longer be used to explain quantum processes: what is needed is a radically new order in which to understand quantum phenomena. In this regard, since 1980 Bohm suggested that the new order to explain quantum phenomena would be based on process and called this new order the implicate order: the quantum potential must be considered an active information source linked with a new, fundamental geometrodynamic background, namely just the implicate order. Taking account of the geometrodynamic nature of the quantum potential, one can also say that the quantum potential expresses the geometric properties of space which determine the behaviour of the particles and are derived just from a more fundamental geometrodynamic background, namely the implicate order.

The intention behind the introduction of the implicate order was simply to develop new physical theories together with the appropriate mathematical formalism that will lead to new insights into the behaviour of matter and ultimately to new experimental tests. In this way Bohm in his last years departed from de Broglie's pilot wave: he suggested the necessity to consider non-locality as a primary fundamental characteristic of space-time, to introduce an intrinsic non-locality of the quantum world. The idea of the implicate order can be collocated just in this context. In Bohm's view the non-locality is a characteristic subtended of space-time and the particles are seen as vibration modes of the global field which is the dynamical expression of the fundamental level, of the deep geometrical structure. One can thus say that the geometrodynamic properties of space determine a non-local global field, the implicate order and that the subatomic particles emerge from this fundamental background.

As regards this research line, in the references [51–53] Hiley suggested that quantum processes evolve not in space-time but in a more general space called pre-space, which is not subjected to the Cartesian division between *res extensa* and *res cogitans*. In this view, the space-time of the classical world would be some statistical approximation and not all quantum processes can be projected into this space without producing the familiar paradoxes, including non-separability and non-locality. According to Hiley's approach, quantum domain is to be regarded as a structure or order evolving in space-time, but space-time is to be regarded as a higher order abstraction arising from this process involving events and abstracted notions of space or space-like points.

In summary, according to Bohm's and Hiley's implicate order research, the geometrodynamic features of the quantum potential lead to understand and explain quantum processes on the basis of approaches different from the ordinary space-time manifold. The non-local geometry determined by the quantum potential introduces a fundamental geometrodynamic background (Bohmian implicate order, or also the analogous Hiley's pre-space and notion of underlying process of quantum phenomena) which redesigns the behaviour of the particles.

On the other hand, since the 90s relevant and suggestive developments have occurred as regards the geometrodynamic features of the quantum potential. For example, in the papers [48, 54–62], Dürr, Goldstein, Tumulka and Zanghì developed a well known re-reading of de Broglie-Bohm theory, called as Bohmian mechanics by its authors, which is based on the idea that well-defined Bohmian particle trajectories can be considered the foundation of quantum mechanics. According to the dynamic approach of Dürr, Goldstein, Tumulka and Zanghì, the primary physical reality of quantum processes is represented by the trajectory of the subatomic particles under consideration, which is governed by a guiding equation intended as the fundamental equation of motion. Instead, the quantum potential — and its geometrodynamic features — can be seen here as a secondary element which has the purpose, together with the guiding equation, to find the wavefunction determining the real, actual trajectories of the particles.

More recently, other interesting approaches to the quantum potential appeared in the literature. In particular, in the paper *A quasi-Newtonian approach to Bohmian quantum mechanics* [63], Atiq, Karamian and Golshani developed a quasi-Newtonian approach to Bohm's quantum potential in which quantum processes are studied in a similar way to classical mechanics and the Bohmian formulation of quantum mechanics — and thus the geometrodynamic features of the quantum potential — can be obtained without the necessity to start from the Schrödinger equation. In Atiq's, Karamian's and Golshani's model, the form of quantum potential (1.4), and consequently its geometrodynamic features, are results which derive by minimizing the total energy of the ensemble of particles in the experiment into consideration rather than being a consequence of the Schrödinger equation. On the basis of Atiq's, Karamian's and Golshani's mathematical formalism, the quantum potential and its geometrodynamic features are more fundamental representations of quantum processes than the Schrödinger equation and the concept of the wavefunction. While in Dürr's, Goldstein's and Zanghì's dynamic approach, the quantum potential and its geometrodynamic features emerge as a secondary physical reality with respect to the particles' trajectories, in the picture proposed by Atiq, Karamian and Golshani Bohm's quantum potential and its geometrodynamic features can be considered as the basis of the Schrödinger equation rather than being a consequence of it. Moreover, in the paper *The path integral approach in the frame work of causal interpretation* Abolhasani and Golshani [64] remarked that as regards the propagation of the wavefunction in the picture of the non-relativistic de Broglie-Bohm theory, a Bohmian path integral — depending on the combined action of the classical potential and the quantum potential — can be associated with the Bohmian trajectory of the particle under consideration.

Even more interesting, according to the author, are the following recent re-readings provided about the quantum potential: Grössing's thermodynamic approach in which the action of the quantum potential derives from a thermal flow of a more fundamental arena, and Sbitnev's approach of the quantum potential as an information channel emerging from the zero-point fluctuations of a physical vacuum. These two models introduce the interesting perspective of a connection of the geometrodynamic action of the quantum potential with the geometry of a more fundamental arena. And it is on these models that now we want to focus our attention. Here, the analysis of these approaches follows in part the treatment made by the author in the recent articles *About the different approaches to Bohm's quantum potential in non-relativistic quantum mechanics* [65] and *Perspectives of Bohm's quantum potential towards a geometrodynamic interpretation of quantum physics. A critical survey* [66].

1.2.1 *Grössing's thermodynamic approach to the quantum potential*

In the recent articles *The vacuum fluctuation theorem: Exact Schrödinger equation via non-equilibrium thermodynamics* [67] and *On the thermodynamic origin of the quantum potential* [68] Grössing suggested that quantum phenomena have a fundamental thermodynamic origin and that the quantum potential and its geometric, contextual and active information about the process and the environment arise from the presence of a subtle thermal vacuum energy distributed across the whole domain of an experimental arrangement. This treatment is based on the use of a classical language and the additional features of classical physics which provide a more complete understanding of how the quantum potential influences particles' motion are those of classical diffusion waves (which are characterized by the peculiarity that the time derivative in their defining equation is only of first order and arise when the classical diffusion equation is coupled to an oscillatory force function). In this interpretation, the form of the quantum potential turns out to be exactly identical to the heat distribution derived from the defining equation for classical diffusion-wave fields.

By following Grossing's treatment in the papers [67] and [68], quantum particles are assumed to be surrounded by a reservoir that is very large compared to the small dissipative system and are described by the following equation regarding the detection probability density provided by the environment:

$$\frac{P(\vec{x},t)}{P(\vec{x},0)} = e^{-\frac{\Delta Q_{hf}}{kT}} \tag{1.13}$$

with k being Boltzmann's constant, T the reservoir temperature and ΔQ_{hf} the heat that is exchanged between the particle and its environment. Here quantum particles are assumed to be actually dissipative systems maintained in a non-equilibrium steady-state by a permanent throughput of energy, or heat flow, respectively, and the detection probability density provided by the particle's environment is considered to coincide with a classical wave's intensity according to relation

$$P(\vec{x},t) = R^2(\vec{x},t) \tag{1.14}$$

(with $R(x,t)$ being the wave's real-valued amplitude). In the light of these assumptions, in Grössing's view, equation (1.13) allows us to obtain Schrödinger's equation from classical mechanics with only two supplementary well-known observations. The first is represented by a relation between heat and action of the form

$$\Delta Q_{hf} = 2\omega[\delta S(t) - \delta S(0)], \tag{1.15}$$

the second consists in the requirement that the average kinetic energy of the thermostat turns out to be equal to the average kinetic energy of the oscillator, for each degree of freedom, namely

$$\frac{kT}{2} = \frac{\hbar\omega}{2}. \tag{1.16}$$

Combining equations (1.13), (1.15) and (1.16), one obtains

$$P(\vec{x}, t) = P(\vec{x}, 0)e^{-\frac{2}{\hbar}[\delta S(\vec{x},t) - \delta S(\vec{x},0)]}. \tag{1.17}$$

As a consequence, taking account that, in virtue of equation (1.17), the expression of the momentum fluctuation $\delta\vec{p}$ determined by the heat flow is

$$\delta P(\vec{x}, t) = -\frac{\hbar}{2}\frac{\nabla P(\vec{x}, t)}{P(\vec{x}, t)}, \tag{1.18}$$

the action integral for a system of n particles turns out to be

$$A = \int P\left[\frac{\partial S}{\partial t} + \sum_{i=1}^{n} \frac{1}{2m_i}\nabla_i S \nabla_i S + \sum_{i=1}^{n} \frac{1}{2m_i}\left(\frac{\hbar}{2}\frac{\nabla_i P}{P}\right)^2 + V\right] d^n x dt \tag{1.19}$$

where $P = P(\vec{x}_1, \vec{x}_2, \ldots, \vec{x}_n, t)$. Here, introducing the "Madelung transformation"

$$\psi = Re^{iS/\hbar} \tag{1.20}$$

where $R = \sqrt{P}$ as in (1.14), one obtains

$$A = \int d^n x dt \left[|\psi|^2 \left(\frac{\partial S}{\partial t} + V\right) + \sum_{i=1}^{n} \frac{\hbar^2}{2m_i}|\nabla_i \psi|^2\right]. \tag{1.21}$$

The action (1.21), if one utilizes the identity $|\psi|^2 \frac{\partial S}{\partial t} = -\frac{i\hbar}{2}(\psi^* \psi - \psi \psi^*)$, leads directly to the n-particle Schrödinger equation

$$i\hbar \frac{\partial \psi}{\partial t} = \left(-\sum_{i=1}^{n} \frac{\hbar^2}{2m_i} \nabla_i^2 + V \right) \psi. \tag{1.22}$$

Equation (1.19), by making a variation in P, yields the quantum Hamilton-Jacobi equation of the de Broglie-Bohm theory, i.e.,

$$\frac{\partial S}{\partial t} + \sum_{i=1}^{n} \frac{(\nabla_i S)^2}{2m_i} + V(\vec{x}_1, \ldots, \vec{x}_n, t) + U_q(\vec{x}_1, \ldots, \vec{x}_n, t) = 0 \tag{1.23}$$

where U_q is the "quantum potential" given by relation

$$U_q(\vec{x}_1, \ldots, \vec{x}_n, t) = \sum_{i=1}^{n} \frac{\hbar^2}{4m_i} \left[\frac{1}{2} \left(\frac{\nabla_i P}{P} \right)^2 - \frac{\nabla_i^2 P}{P} \right]$$
$$= -\sum_{i=1}^{n} \frac{\hbar^2}{2m_i} \frac{\nabla_i^2 R}{R}. \tag{1.24}$$

Moreover, with the definitions

$$\vec{u}_i = \frac{\delta p_i}{m_i} = -\frac{\hbar}{2m_i} \frac{\nabla_i P}{P} \quad \text{and} \quad \vec{k}_{\vec{u}i} = \frac{1}{2} \frac{\nabla_i P}{P} = \frac{\nabla_i R}{R} \tag{1.25}$$

one can rewrite the quantum potential (1.24) as

$$U_q(\vec{x}_1, \ldots, \vec{x}_n, t) = \sum_{i=1}^{n} \left[\frac{m_i \vec{u}_i \cdot \vec{u}_i}{2} - \frac{\hbar}{2} (\nabla_i \cdot \vec{u}_i) \right]$$
$$= \sum_{i=1}^{n} \left[\frac{\hbar^2}{2m_i} \left(\vec{k}_{\vec{u}i} \cdot \vec{k}_{\vec{u}i} - \nabla_i \cdot \vec{k}_{\vec{u}i} \right) \right] \tag{1.26}$$

and inserting the dependence of $\vec{u}_i$ on the spatial behaviour of the heat flow expressed by relation

$$\vec{u}_i = \frac{1}{2\omega_i m_i} \nabla_i Q_{hf} \tag{1.27}$$

one obtains the thermodynamic formulation of the quantum potential as

$$U_q(\vec{x}_1, \ldots, \vec{x}_n, t) = \sum_{i=1}^{n} \frac{\hbar^2}{4m_i} \left[\frac{1}{2} \left(\frac{\nabla_i Q_{hf}}{\hbar \omega_i} \right)^2 - \frac{\nabla_i^2 Q_{hf}}{\hbar \omega_i} \right]. \quad (1.28)$$

On the basis of equations (1.13)–(1.28), the following interpretation of the quantum potential becomes natural: the quantum potential (1.28) responsible of the quantum behaviour of a subatomic particle is derived and determined by a heat flow having a spatial behaviour given by equation (1.27), which is in turn produced by the detection probability density provided by the particle's environment (and which coincides with a classical wave's intensity). According to equation (1.28), the heat flow between the particle under consideration and its environment can be considered as the ultimate physical entity which determines the action of the quantum potential of non-relativistic de Broglie-Bohm theory. These results physically mean that there is a deep fundamental arena which is subtended by the action of the quantum potential. In other words, Grössing's thermodynamic formulation of the quantum potential introduces the relevant perspective that the geometrodynamics of quantum processes derived from the action of the quantum potential derives indeed from a more fundamental arena, a more fundamental level of physical reality which, in particular, is associated with a heat flow, a thermal energy produced by the detection probability density provided by the particle's environment in a given experimental arrangement.

This view implies that the energetic scenario of a steady-state oscillator in non-equilibrium thermodynamics corresponds to a throughput of heat, namely a kinetic energy at the subquantum level which allows two important energy quantities to be provided: the energy needed to maintain a constant oscillation frequency ω and some excess kinetic energy which results in a fluctuating momentum contribution $\delta\vec{p}$ to the momentum $\vec{p}$ of the particle. Moreover, the steady-state resonator constituting a "particle" in this thermodynamic environment will not only receive kinetic energy from it, but, in order to balance the stochastic influence of the supplementary momentum fluctuations, it will also dissipate heat into the environment. On the basis of the "vacuum fluctuation theorem" introduced by Grössing in the paper [67], the larger the energy fluctuation of the oscillating "system of interest" is, the higher is the probability that heat will be dissipated into the environment rather than be absorbed. Also, Bohm and Hiley [69] demand in their review of stochastic hidden variable models that, generally, to maintain an equilibrium density distribution like the one given by $P(\vec{x}, t)$ under random processes, the latter must be complemented by a balancing movement. The corresponding balancing velocity is called, referring to the same expression in Einstein's work on Brownian motion, the "osmotic velocity". If one considers the stochastic "forward" movement in this picture, namely $\delta\vec{p}/m = \vec{u}$, or the current $\vec{J} = P\vec{u}$, respectively, this will have to be balanced by the osmotic velocity $-\vec{u}$, or $\vec{J} = -P\vec{u}$, respectively.

By substituting equation (1.25) into the definition of the "forward" diffusive current $\vec{J}$, and taking account of the diffusivity $D = \frac{\hbar}{2m}$, one has

$$\vec{J} = P\vec{u} = -D\nabla P \tag{1.29}$$

which, when is combined with the continuity equation $\dot{P} = -\nabla \cdot \vec{J}$, may be expressed as

$$\frac{\partial P}{\partial t} = D\nabla^2 P. \tag{1.30}$$

Equations (1.29) and (1.30) represent the first and second of Fick's laws of diffusion, respectively, and $\vec{J}$ is the diffusion current.

In the paper [68] Grössing studied the scenarios introduced by the "osmotic" type of dissipation of energy from the particle to its environment in the interpretation of the quantum potential. In this regard, taking account of equation (1.13) and of the strict directionality of any heat flow, equation (1.27) can be rewritten for the case of heat dissipation where $\Delta Q_{hf} = Q_{hf}(t) - Q_{hf}(0) < 0$. If one maintains the heat flow as a positive quantity, namely in the sense of measuring the positive amount of heat dissipated into the environment, one can select the negative part of the above expression,

$-\Delta Q$, and put this into equation (1.27). In this way, the osmotic velocity is obtained

$$\vec{\bar{u}} = D\frac{\nabla P}{P} = -\frac{1}{2\omega m}\nabla Q_{hf}. \tag{1.31}$$

As a consequence, the osmotic current is expressed by relation

$$\vec{\bar{J}} = P\vec{\bar{u}} = D\nabla P = -\frac{P}{2\omega m}\nabla Q_{hf} \tag{1.32}$$

and therefore the corollary to Fick's second law becomes

$$\frac{\partial P}{\partial t} = -\nabla \cdot \vec{\bar{J}} = -D\nabla^2 P = \frac{1}{2\omega m}\left[\nabla P \cdot \nabla Q_{hf} + P\nabla^2 Q_{hf}\right]. \tag{1.33}$$

In the light of equations (1.29)–(1.33), it is possible to understand the thermodynamic meaning of the quantum potential expressed by equation (1.28). In this regard, for simplicity we base our discussion on one-body systems.

Before all, let us now consider the simplest case $U_q = 0$. In this case, from equation (1.28), the thermodynamic consequence of a vanishing quantum potential for the one-body system reads as

$$\nabla^2 Q_{hf} = \frac{1}{2\hbar\omega}(\nabla Q_{hf})^2. \tag{1.34}$$

The osmotic flux conservation (1.112), by using equations (1.13) and (1.16), leads thus to the following relation

$$\frac{\partial P}{\partial t} = \frac{P}{2\omega m}\left[\nabla^2 Q_{hf} - \frac{(\nabla Q_{hf})^2}{\hbar\omega}\right]. \tag{1.35}$$

By inserting equation (1.34) and equation (1.35) into equation (1.28), with $\hbar\omega = $ constant and $\hat{Q}_{hf} = Q_{hf}/\hbar\omega$, one obtains

$$U_q = -\frac{\hbar^2}{4m}\left[\nabla^2 \hat{Q}_{hf} - \frac{1}{D}\frac{\partial \hat{Q}_{hf}}{\partial t}\right], \tag{1.36}$$

or, generally,

$$U_q = -\frac{\hbar^2}{4m}\left[\nabla^2 Q_{hf} - \frac{1}{D}\frac{\partial Q_{hf}}{\partial t}\right]. \tag{1.37}$$

In the case $U_q = 0$, one obtains

$$\left[\nabla^2 Q_{hf} - \frac{1}{D}\frac{\partial Q_{hf}}{\partial t}\right] = 0 \tag{1.38}$$

which is nothing but the classical heat conduction equation. In other words, one can associate a heat dissipation process emanating from the particle even for free particles, both in the classical and in the quantum domain. Therefore, in Grössing's model, a non-vanishing "quantum potential" can be considered as the device that describes the spatial and temporal dependencies of the corresponding thermal flow for interacting particles.

As regards the solutions to equation (1.38), if, for example, one considers a situation such that, on average, the heat flow was constant, one obtains from equation (1.38) a simple Laplace equation,

$$\nabla^2 Q_{hf} = 0 \tag{1.39}$$

whose solutions are harmonic functions. If the temporal behaviour of Q were not periodic, but represented a heat generated only once and then dissipated, namely $Q_{hf} \propto e^{-wt}$, then equation (1.38) leads to the spatial Helmholtz equation

$$(\nabla^2 + k^2) Q_{hf} = 0 \tag{1.40}$$

with periodic solutions $Q_{hf} \propto e^{-i\vec{k}\cdot\vec{r}}$. Extending to the case of the inhomogeneous Helmholtz equation, one can consider the source as given by a delta function, i.e.,

$$(\nabla^2 + k^2) Q_{hf}(\vec{x}) = \delta(\vec{x}). \tag{1.41}$$

In order to have a unique solution one can introduce here the Sommerfeld radiation condition, for example, as a specification of the boundary conditions at infinity. Then $Q_{hf}(\vec{x})$ is well known to be identical to a Green's function $G(\vec{x})$, which is given in three dimensions as

$$Q_{hf}(\vec{x}) = G(\vec{x}) = \frac{e^{ik|\vec{x}|}}{4\pi\,|\vec{x}|}. \tag{1.42}$$

On the ground of equations (1.39)–(1.42), one can say that, even for the case $U_q = 0$, the heat conducting equation (1.38) provides wave-like solutions Q_{hf}, which are Huygens-type waves.

In the general case $U_q \neq 0$ one can introduce an explicitly non-vanishing source term for the quantum potential, namely,

$$\left[\nabla^2 Q_{hf} - \frac{1}{D}\frac{\partial Q_{hf}}{\partial t}\right] = q(x)e^{i\omega t} \neq 0. \tag{1.43}$$

which can be solved via separation of variables. Thus, with the *ansatz*

$$Q_{hf} = X(x)T(t), \quad \text{with} \quad T = e^{i\omega t} \tag{1.44}$$

and

$$q(x) = \alpha(x)X \tag{1.45}$$

division by (XT) then yields the constant

$$\frac{\nabla^2 X}{X} = \frac{\frac{\partial}{\partial t}T}{DT} + \alpha = -\lambda. \tag{1.46}$$

In particular, if one considers, for example, the case of a particle entrapped in a box of length L whose walls are infinitely high, the Dirichlet boundary conditions

$$Q_{hf}(0,t) = Q_{hf}(L,t) = 0 \tag{1.47}$$

can be taken into consideration, which lead to determine the constant λ as

$$\lambda = \frac{n^2\pi^2}{L^2} \equiv k_n^2. \tag{1.48}$$

Equation (1.46) allows us to obtain the solution

$$X = NC_Q \sin\left(\frac{n\pi}{L}x\right) \tag{1.49}$$

where N is a normalization constant and C_Q is a dimensionality preserving constant. Moreover, assuming $T = e^{i\omega_n t}$ one has $\frac{i\omega_n}{D} + \alpha = -k_n^2$, and therefore

$$\alpha = -k_n^2(1+i). \tag{1.50}$$

In the light of equation (1.45), one arrives thus to the following solution for $q(x)$:

$$q(x) = -k_n^2(1+i)NC_Q \sin\left(\frac{n\pi}{L}x\right) \tag{1.51}$$

and, with (1.44), (1.48), and $\tilde{Q}_{hf} = Q_{hf}/C_Q$,

$$\tilde{Q}_{hf}(x,t) = N\sin(k_n x)e^{i\omega_n t}, \tag{1.52}$$

By considering the Dirichlet boundary conditions (1.47), the eigenvalue equation of the Laplacian

$$\nabla^2 e_n = -k_n^2 e_n \tag{1.53}$$

has solutions $e_n = N\sin(k_n x)$ where

$$\langle e_n | e_m \rangle = \int e_n(x)e_m(x)dx = \begin{cases} 0, m \neq n \\ 1, m = n \end{cases}. \tag{1.54}$$

This means that, for $m = n$, (1.54) can be interpreted as a probability density, with

$$\int_0^L P dx = N^2 \int_0^L \sin^2(k_n x)dx = 1. \tag{1.55}$$

From equation (1.55) follows

$$N = \sqrt{\frac{2}{L}}. \tag{1.56}$$

Therefore, the heat distribution in the box is given by

$$\tilde{Q}_{hf}(x,t) = \sqrt{\frac{2}{L}}\sin(k_n x)e^{i\omega_n t} \tag{1.57}$$

with the probability density

$$P = \left|\tilde{Q}(x,t)\right|^2 \tag{1.58}$$

Therefore, the classical state (1.57) turns out to be identical with the quantum mechanical one, and equations (1.58) and (1.24) lead to define a quantum potential expressed by the following relation

$$U_q = \frac{\hbar^2 k_n^2}{2m}. \tag{1.59}$$

The interpretation proposed by Grössing shows a tight connection between the quantum potential and wave-diffusion fields. In fact, introducing the solutions (1.51) into equation (1.43), one obtains a sort of "eigenvalue equation" of the type

$$\left[\nabla^2 \tilde{Q}_{hf} - \frac{1}{D}\frac{\partial \tilde{Q}_{hf}}{\partial t}\right] = q(x)e^{i\omega t} = -(1+i)k_n^2 \tilde{Q}_{hf} \tag{1.60}$$

which, applying a temporal Fourier transformation and introducing the complex diffusion number $\kappa(x,\omega) \equiv \sqrt{\frac{i\omega}{D}}$, becomes a Helmholtz-type pseudo-wave equation

$$\nabla^2 \tilde{Q}_{hf}(x,\omega) - \kappa^2(x,\omega) = \sigma(x,\omega). \tag{1.61}$$

Equation (1.61) is precisely the defining equation for a thermal-wave-field and therefore describes the spatio-temporal behaviour of diffusion waves.

Furthermore, in [68] Grössing demonstrated that by considering both the osmotic current of heat dissipation and the usual forward current, the total quantum potential, integrated over periods of time $t \approx n/\omega$ long enough so that it is characterized by the energy throughput of n total currents, or by equal weights of n "absorption" and n "dissipation" currents, respectively, in the light of equation (1.37) is given by the following relation

$$\begin{aligned} \bar{U}_q &= \frac{1}{n}\left(n\bar{U}_{forward} + n\bar{U}_{osmotic}\right) \\ &= -\frac{\hbar^2}{4m}\frac{1}{\hbar\omega}\left[2\nabla^2 Q_{hf} + \frac{1}{D}\frac{\partial Q_{hf}}{\partial t} - \frac{1}{D}\frac{\partial Q_{hf}}{\partial t}\right] = -\frac{\hbar^2}{2m}\frac{\nabla^2 Q_{hf}}{\hbar\omega}. \end{aligned} \tag{1.62}$$

On the basis of the thermalized version of the quantum potential, given by equation (1.62), the usual quantum Hamilton-Jacobi equation (1.23) of de Broglie-Bohm theory can be formulated as

$$\frac{\partial S}{\partial t} + \sum_{i=1}^{n} \frac{(\nabla_i S)^2}{2m_i} + V - \sum_{i=1}^{n} \left(\frac{L(\omega_i)}{2} \right)^2 \nabla_i^2 Q_{hf} = 0. \quad (1.63)$$

Thus, the equations of quantum motion become of the form

$$m_i \frac{d\vec{v}_i}{dt} = -\nabla_i (V + U_q) = -\nabla_i V + \sum_{i=1}^{n} \left(\frac{L(\omega_i)}{2} \right)^2 \nabla_i (\nabla_i^2 Q_{hf}) = 0. \quad (1.64)$$

Equations (1.63) and (1.64), at least for simple solutions Q_{hf}, allow a simplification in the calculation of quantum trajectories. According to equations (1.62)–(1.64), the heat flow between the system under consideration and its environment can be considered the ultimate element responsible of the quantum processes, by determining an opportune correction term in the total energy of the system. This term depending on the heat flow is fundamental in the sense without this term, the total energy of the system would not be conserved, and it is just this term that determines the motion of the system in appropriate trajectories.

Therefore, in Grössing's approach, the "form" of the quantum potential, as given by $-\frac{\nabla^2 R}{R}$, and its geometrodynamic features, are ultimately connected with a Helmholtz-type dependence $-\nabla^2 Q_{hf}$ of a thermal energy Q_{hf} defining a fundamental vacuum and this thermal energy is associated with wave-diffusion waves and is distributed "non-locally" throughout the environment of the experimental arrangement into consideration. A fundamental feature of the diffusion-wave fields which characterize the thermal vacuum is in fact represented by non-locality. The "propagation speed" of these fields is infinite and, thus, if one considers a prepared neutron source in a reactor, a thermal field in the "vacuum" immediately arises that allows a non-local connection between the neutron oven, the apparatus (including, for example, a Mach-Zehnder interferometer), and the detectors. The particle distributions in the environment of the experiment are described by (typical) Gaussians which thus also contribute in their totality to the form of the heat distribution in the overall system, independently of which particle actually is on its way through the interferometer. As a consequence, all "potential" trajectories are implicitly present throughout the experiment (namely, under constant boundary conditions) in the sense that the corresponding thermal field exists no matter where the particle actually is. The infinite propagation of the diffusion-wave fields of the thermalized vacuum leads therefore a promising perspective for a deeper understanding of the physical origin of quantum non-locality characterizing Bohm's quantum potential.

Because of the parabolic nature of the wave equation corresponding with the Helmholtz-type dependence $-\nabla^2 Q$ one must generally expect the thermal waves' behaviour to be radically different from that of ordinary hyperbolic ones. In fact, Grössing studied several examples of forms of the quantum potential which clearly manifest a departure from ordinary wave behaviour in terms of the appearance of accumulation and depletion zones. Thus, Grössing's research suggests that the "strange" form of the quantum potential as well as the (only seemingly "surrealistic") Bohmian trajectories characterizing quantum phenomena can potentially be fully understood by using the physics of diffusion-wave fields. If one assumes that the quantum potential which appears in the quantum Hamilton-Jacobi equation of the de Broglie-Bohm theory is the projection of more fundamental diffusion-wave fields which characterize a thermalized vacuum, in several classic quantum phenomena (such as the double-slit interference) the Bohmian trajectories, deriving from Bohm's quantum potential, can be understood in terms of the presence of diffusion-wave fields.

In summary, we can conclude that, in Grössing's approach, a fundamental level of physical reality, which is described by a heat flow, a thermal energy distributed non-locally between the particle under consideration and its environment, and connected to wave-diffusion fields, can be considered as the ultimate physical structure which determines the geometrodynamic action of the quantum potential of non-relativistic de Broglie-Bohm theory. In other words, according to this model, the thermal energy distributed non-locally between the particle under consideration and its environment is the primary entity which characterizes the geometrodynamics of quantum processes.

1.2.2 *Quantum potential as an information channel of a special superfluid vacuum*

On the basis of a recent approach proposed by Sbitnev in [70–73], Bohm's quantum potential is an information channel which emerges from the zero-point fluctuations of a physical vacuum acting as a special superfluid medium. In Sbitnev's approach, the physical vacuum consists of an enormous amount of virtual pairs of particles-antiparticles with opposite orientations of spins (thus constituting a Bose ensemble) and is described by the Navier-Stokes equation

$$\rho_M\left(\frac{\partial\vec{v}}{\partial t}+(\vec{v}\cdot\nabla)\vec{v}\right)=\frac{\vec{F}}{\Delta V}-\rho_M\nabla\left(\frac{P}{\rho_M}\right)+\mu(t)\nabla^2\vec{v} \tag{1.65}$$

and the continuity equation

$$\frac{\partial\rho_M}{\partial t}+(\nabla\cdot\vec{v})\rho_M=0. \tag{1.66}$$

Here $\rho_M = M/\Delta V$ is a mass density of the fluid in the volume ΔV, $\vec{v}$ is the flow velocity, $\frac{\vec{F}}{\Delta V}$ is an external force per the volume ΔV, P is the total pressure generated by the collisions of the virtual particles of the medium and μ is a dynamic viscosity term. The last two terms in equation (1.65) are internal forces that are represented by pressure gradients within the fluid and dissipative forces arising due to the fluid viscosity. These two terms are subjected to slight modification in virtue of the following reasons: (a) the dynamic viscosity term μ depends of time (the viscosity in average on time is zero, but its variance is nonzero); (b) the pressure gradient contains an added term describing the entropy gradient multiplied by the pressure, namely $\nabla P \rightarrow \rho_M\nabla(\frac{P}{\rho_M}) = \nabla P - P\nabla\ln(\rho_M)$. Because of these modifications, the Navier-Stokes equation (1.65) can be reduced to the Schrödinger equation describing the behavior of a particle into the vacuum, where the vacuum is a superfluid medium. Here, the first term, ∇P, is the customary pressure gradient appearing in the Navier-Stokes equation. Instead, the second term, $P\nabla\ln(\rho_M)$, is an extra term describing changing of the logarithm of the density along increment of length (the entropy increment) multiplied by P. It may mean that change of the pressure originating from the collisions of the virtual particles of the vacuum, is induced by change of the entropy per length, or else by change of the information flow [74, 75] per length. This term contains clues characteristic of the osmotic pressure, mentioned by Nelson in [76]. It can be interpreted as follows: a semipermeable membrane where the osmotic pressure manifests itself is an instant which separates the past and the future (that is, the 3D brane of our being is the semipermeable membrane in the 4D world).

By following Sbitnev's treatment in [70–73], the pressure P, in more detail, may be expressed as the sum of two pressures P_1 and P_2, which are defined in the following way. Since on the basis of Fick's law the diffusion flux $\vec{J}$ turns out to be proportional to the negative value of the density gradient $\vec{J} = -D\nabla\rho_M$, the pressure P_1 may be defined as

$$P_1 = D\nabla\vec{J} = -\frac{\hbar^2}{4m}\nabla^2\rho_M, \tag{1.67}$$

m being the mass of the subparticles populating the volume ΔV into consideration and

$$D = \frac{\hbar}{2m} \tag{1.68}$$

is the diffusion coefficient of the motion of these subparticles. Moreover, taking into account that the kinetic energy of the diffusion flux is $(m/2)(J/\rho_M)^2$, one finds that one more pressure exists in the form of the average momentum transfer per unit area per unit time:

$$P_2 = \frac{\rho_M}{2}\left(\frac{\vec{J}}{\rho_M}\right)^2 = \frac{\hbar^2}{8m^2}\frac{(\nabla\rho_M)^2}{\rho_M}. \tag{1.69}$$

Thus, by writing the mass density ρ_M as

$$\rho_M = m\rho \tag{1.70}$$

where

$$\rho = \frac{N}{\Delta V} \tag{1.71}$$

is the density of the subparticles, N being the number of identical subparticles in the volume ΔV, one finds that the sum of the two pressures P_1 and P_2 divided by ρ yields the quantum potential:

$$Q = \frac{P_1 + P_2}{\rho} = \frac{\hbar^2}{8m}\left(\frac{\nabla\rho}{\rho}\right)^2 - \frac{\hbar^2}{4m}\frac{\nabla^2\rho}{\rho}. \tag{1.72}$$

In the light of equation (1.72), one can say that if in Grössing's approach the gradient of the quantum potential describes a completely thermalized fluctuating force field, in Sbitnev's model the fluctuating force is expressed via the gradient of the pressure divided by the density distribution of particles (particle-antiparticle pairs) chaotically moving in the environment. This means that, in Sbitnev's approach, the pressure originating from the collisions of the virtual particles of the vacuum can be considered the ultimate factor which determines the action of the quantum potential and thus the geometrodynamics of quantum processes.

As a consequence of the link of the quantum potential with the pressures originating from the collisions of the virtual particles of the vacuum, the Navier-Stokes equation (1.65) divided by ρ becomes

$$m\left(\frac{\partial\vec{v}}{\partial t} + (\vec{v}\cdot\nabla)\vec{v}\right) = \frac{\vec{F}}{N} - \nabla Q + \nu(t)\nabla^2 m\vec{v}. \tag{1.73}$$

Here $\frac{\vec{F}}{N}$ is the external force per the subparticle, while $\nu(t) = \mu(t)/\rho_M$ is the kinetic having the dimension *length*^{2}/*time* and fluctuating about zero. Now, decomposing the current velocity into the irrotational and solenoidal components

$$\vec{v} = \vec{v}_R + \vec{v}_S, \tag{1.74}$$

taking account that $\vec{v}_S = \nabla S/m$ and that

$$(\vec{v}\cdot\nabla)\vec{v} = \nabla v^2/2 + \vec{\omega}\times\vec{v}, \tag{1.75}$$

where $\vec{\omega} = \nabla \times \vec{v}$ is the vorticity, the Navier-Stokes equation (1.73) may be rewritten as

$$\frac{\partial}{\partial t}(\nabla S + m\vec{v}_R) + \left\{\frac{1}{2m}\nabla((\nabla S)^2 + m^2 v_R^2) + m[\vec{\omega} \times \vec{v}_R]\right\}$$
$$= -\nabla U - \nabla Q + \nu(t)\nabla^2(\nabla S + m\vec{v}_R), \tag{1.76}$$

where U is the potential energy relating to the single subparticle and the last term on the right side describes kinetic losses in the fluid because of the viscosity. By regrouping the terms, equation (1.76) becomes

$$\nabla\left(\frac{\partial}{\partial t}S + \frac{1}{2m}(\nabla S)^2 + \frac{m}{2}v_R^2 + U + Q - \nu(t)\nabla^2 S\right)$$
$$= -m\frac{\partial}{\partial t}v_R - m[\vec{\omega} \times \vec{v}] + \nu(t)m\nabla^2\vec{v}_R. \tag{1.77}$$

By multiplying (1.77) by the curl one finds

$$\nabla \times \nabla\left(\frac{\partial}{\partial t}S + \frac{1}{2m}(\nabla S)^2 + \frac{m}{2}v_R^2 + U + Q - \nu(t)\nabla^2 S\right) = 0 \tag{1.78}$$

and, thus, as a result one arrives at the following modified Hamilton-Jacobi equation

$$\frac{\partial}{\partial t}S + \frac{1}{2m}(\nabla S)^2 + \frac{m}{2}v_R^2 + U + Q - \nu(t)\nabla^2 S = C, \tag{1.79}$$

C being an arbitrary integration constant. The modification of this equation is due to the presence of the quantum potential Q, given by equation (1.72). Moreover, it must be remarked the term (1.75) entering in the Navier-Stokes equation, and therefore the terms on the right side of equation (1.77), are responsible for emergence of vortex structures. In particular, the third term in equation (1.77) is the energy of the vortex.

As regards the vorticity $\vec{\omega}$, it satisfies the general equation

$$\frac{\partial \vec{\omega}}{\partial t} + (\vec{\omega} \cdot \nabla)\vec{v} = \nu(t)\nabla^2 \vec{\omega} \tag{1.80}$$

which, in the cross-section of the vortex, has the following solution

$$\omega(r,t) = \frac{\Gamma}{4\pi(\nu/\Omega)(sen(\Omega t + \phi) + n)} \cdot \exp\left(-\frac{r^2}{4\pi(\nu/\Omega)(sen(\Omega t + \phi) + n)}\right) \tag{1.81}$$

where Ω is an oscillation frequency, ϕ is the uncertain phase, Γ is the integration constant having dimension m^2/s and $n > 1$ is an additional number that prevents appearance of singularity in the cases when $\sin(\Omega t + \phi)$ tends to -1. The velocity of the fluid matter around the vortex results from the integration of the vorticity function

$$v(r,t) = \frac{\Gamma}{2\pi r}\left(1 - \exp\left\{-\frac{r^2}{4\pi(\nu/\Omega)(sen\,(\Omega t + \phi) + n)}\right\}\right). \tag{1.82}$$

In the light of the solutions (1.81) and (1.82), the functions ω(r,t) and v(r,t) show typical behavior for the vortices. The both functions do not decay with time but exhibit pulsations on the frequency Ω. Larger the value of the parameter n, smaller the amplitude of the pulsations. At n tending to infinity the amplitude of the pulsations tends to zero and correspondently the vortex disappears entirely.

In addition to the modified Hamilton-Jacobi equation, from (1.66) taking account of (1.69) and (1.70), a continuity equation can be obtained as follows

$$\frac{\partial \rho}{\partial t} + (\nabla \cdot \vec{v})\rho = \frac{\partial \rho}{\partial t} + (\vec{v} \cdot \nabla)\rho + \rho(\nabla \cdot \vec{v}) = \frac{d\rho}{dt} + \rho(\nabla \cdot \vec{v}) = 0. \tag{1.83}$$

Here, the both real-valued equations (1.79) and (1.83), which are crucial in Bohm's approach to quantum mechanics, give sufficient conditions for transition to the complex-valued Schrödinger equation.

By writing the term $\nabla^2 S$ in the following convenient form

$$\nabla^2 S = m(\nabla \cdot \vec{v}) = -m\frac{d}{dt}\ln\rho = -mf(\rho), \tag{1.84}$$

equation (1.79) may be rewritten as

$$\frac{\partial}{\partial t}S + \frac{1}{2m}(\nabla S)^2 + \frac{m}{2}v_R^2 + U + Q - \nu(t)mf(\rho) = C. \tag{1.85}$$

Hence, by expanding f into Taylor series as a polynomial in $|\psi|^2 = \rho$ with real coefficients, one comes to the Gross-Pitaevskii equation

$$i\hbar\frac{\partial\psi}{\partial t} = \frac{1}{2m}(-i\hbar\nabla + m\vec{v}_R)^2\psi + \nu(t)mf(\rho)\psi + U(r)\psi - C\psi \tag{1.86}$$

which is the nonlinear Schrödinger equation, because of the viscosity term. Here, the kinetic momentum operator $(-i\hbar\nabla + m\vec{v}_R)$ contains the term $m\vec{v}_R$ describing a contribution of the vortex motion. This term is analogous to the vector potential multiplied by the ratio of the charge to the light speed, which appears in quantum electrodynamics [77]. The appearance of this term in this equation is constrained by the Helmholtz theorem, which reads: (i) if fluid particles form, in any moment of the time, a vortex line, then the same particles support the vortex line both in the past and in the future; (ii) ensemble of the vortex lines traced through a closed contour forms a vortex tube. Intensity of the vortex tube is constant along its length and does not change in time. The vortex tube is then characterized by one of these features: (a) either goes to infinity by both endings; (b) or these endings lean on walls of bath containing the fluid; (c) or these endings are locked to each on other forming a vortex ring.

Now, by substituting the wavefunction ψ represented in polar form

$$\psi = \sqrt{\rho}\exp(iS/\hbar). \tag{1.87}$$

into equation (1.86) and separating on real and imaginary parts we come to equations (1.79) and (1.80). Therefore, the Navier-Stokes equation (1.145), when it is expressed in terms of the slightly expanded pressure gradient term, can be reduced to the Schrödinger equation if we take into consideration also the continuity equation.

In the light of Sbitnev's treatment provided in [70–73], the Schrödinger equation (1.83) describes a flow of a physical vacuum, a superfluid medium containing pairs of particle-antiparticles that give origin to a Bose-Einstein condensate. Therefore, one can say that, in Sbitnev's approach, Bohm's quantum potential and its geometrodynamic features emerge as an information channel into the behaviour of quantum particles as a result of more fundamental processes of collisions between the virtual particles of a superfluid physical vacuum. The Navier-Stokes equation supplemented by the two slightly modified terms (the dynamic viscosity term and the pressure gradient) constitutes the fundamental equation underlying the non-relativistic Schrödinger equation and thus describing the basic processes occurring in the physical vacuum which generate the geometrodynamic action of the quantum potential. The two modifications appearing in the Navier-Stokes equation regard the inner forces that take place within the superfluid vacuum. The first force is the gradient pressure supplemented by the gradient of the quantum entropy multiplied by the pressure. It turns out that such a modification induces emergence of the quantum potential. The second force is a dissipative force conditioned by viscosity of the superfluid vacuum. The modification of this term, represented by the fact that the viscosity in the average of the time vanishes but its variance is not zero, implies that there is an energy exchange within this medium. Modification of the second force should be such that a final product would be just the quantum potential. For that reason, the pressure gradient is supplemented by a term representing the pressure multiplied by gradient of the quantum entropy (gradient of the logarithm from the density distribution of virtual particle-antiparticle pairs).

In virtue of the role of the quantum entropy as the fundamental parameter which measures the degree of ordering of the virtual pairs in this superfluid physical vacuum, the gradient of the quantum entropy determines a flow of the ordered distribution of these pairs to a region where the ordering is in deficit. In other words, it describes a sort of osmotic pressure arising in the physical vacuum. In summary, in Sbitnev's model, the geometrodynamic action of the quantum potential emerges from the ratio of the pressure (characterizing the processes of the vacuum) to the density distribution. This means that information about a particle state transmitted on enormous distances is achieved due to the osmotic pressure arising in the superfluid medium. A positive element of this approach lies in its agreement with de Broglie's original idea about the wavefunction as a pilot wave guiding the particle along an optimal path [29]. In the picture proposed by Sbitnev, formation of the pilot wave can be seen as the direct consequence of (more fundamental) constructive and destructive interference waves induced by the particle [78] passing through the vacuum and exchanging by quanta with the zero-point vacuum fluctuations. In fact, the particle moving through that superfluid medium perturbs virtual particle-antiparticle pairs which, in turn, create both constructive and destructive interference ahead the particle. The vacuum undergoes fluctuations in the ground state that have ultimately wave-like nature. These wave-like fluctuations are similar to the subcritical Faraday oscillations and by virtue of them one can derive the motion of the droplet along a optimal, Bohmian path.

1.3 About the Link Between Weyl Geometries and the Quantum Potential

One relevant development of Bohm's approach which puts in evidence the geometry subtended by quantum phenomena is represented by the link between the quantum potential and Weyl geometries. In the picture proposed by Weyl geometries interesting perspectives are opened towards a new geometrical reading of quantum mechanics. Instead of imposing *a priori* that quantum mechanics must be constructed over a Euclidean background, from the point of view of Weyl geometries, quantum processes can be seen as a manifestation of a non-Euclidean structure derived from a variational principle. In this alternative picture, the validity of the specific geometrical structure proposed can be checked *a posteriori* comparing it to the usual nonrelativistic quantum mechanics. The link between Bohm's quantum potential and Weyl geometries introduces the possibility to see quantum phenomena as a modification, as a deformation of the geometrical properties of physical space.

1.3.1 *Weyl's conformal geometrodynamics*

In 1918, roughly at the time when his famous book *Raum· Zeit· Materie* was published, Hermann Weyl proposed a generalization of Riemannian geometry by introducing scale freedom of the underlying metric, in order to provide a more fundamental "purely infinitesimal" point of view to geometry which can be defined as "purely infinitesimal geometry" ([79, 80]; see also [81] for a review of these concepts). Weyl extended his idea of scale gauge to a unified theory of the electromagnetic and gravitational fields, and developed it into the now generally accepted U(1)-gauge theory of the electromagnetic field, postulating that the invariance for local transformations of coordinates is extended also to the calibration of physical lengths.

Although many times Weyl's original scale gauge geometry was proclaimed dead, physically misleading or, at least, useless as a physical concept, indeed it had surprising come-backs in various research programs of physics. Today, at the turn of 21st century, it seems well alive. It resurrected as a fledged idea in the last third of the 20th century in several contexts: scalar tensor theories of gravity, foundations of physics (gravity, quantum mechanics), elementary particle physics, and cosmology.

In particular, standard linear quantum theory has recently found interesting and significant keys of readings in Weyl's conformal geometric invariance properties affecting the very structure of all physical laws [11, 79]. This concept was well expressed by P.A.M. Dirac in a 1973 seminal paper [82]: "There is a strong reason in support of Weyl's theory. It appears as one of the fundamental principles of Nature that the equations expressing basic laws should be invariant under the widest possible group of transformations. The confidence that one feels in Einstein [general relativity] theory arises because its equations are invariant under the wide group of transformations of curvilinear coordinates in Riemannian space... The passage to Weyl's geometry is a further step in the direction of widening the group of transformations underlying the physical laws. One has to consider transformations of gauge as well as transformations of curvilinear coordinates and one has to take one's physical laws to be invariant under all these transformations, which impose stringent conditions on them....". As it was shown by De Martini and Santamato in the recent papers "Nonlocality, no-signalling and Bell's theorem investigated by Weyl's conformal differential geometry" and "Solving the non-locality riddle by conformal quantum geometrodynamics" [83, 84], these stringent conditions express the conformal-covariance (gauge-covariance) of all physical laws, including the ones belonging to electromagnetism and to the standard quantum dynamics. According to Weyl's conformal differential geometry, the formal expression of all physical laws can be expressed in different "gauges" which are related by a conformal mapping preserving the angles between vectors.

In order to analyse briefly Weyl's conformal geometrodynamics, let us consider a mechanical system described by n generalized coordinates q^i $(i = 1, \ldots, n)$, which define a metric tensor $g_{ij}(q)$ in the configuration space V_n. We assume an affine transport law given by the connection fields $\Gamma^i_{jk}(q)$ with zero torsion, i.e. $\Gamma^i_{jk} - \Gamma^i_{kj} = 0$. The connection fields $\Gamma^i_{jk}(q)$ and their derivatives define in V_n a curvature tensor R^i_{jkl} and, together with the metric tensor, a scalar curvature field $R(q) = g^{ij} R^k_{ikj}$.

Weyl's conformal geometrodynamics is based on the introduction of the multiple-integral variational principle

$$\delta\left[\int d^n q \sqrt{g}\rho(g^{ij}\partial_i\sigma\partial_j\sigma + R)\right] = 0 \tag{1.88}$$

where $g = |\det(g_{ij})|$, $R(q)$ is the scalar curvature and $\rho(q)$ and $\sigma(q)$ are scalar fields. Variation with respect to $\rho(q)$ and $\sigma(q)$ yields, respectively [11, 83]

$$g^{ij}\partial_i\sigma\partial_j\sigma + R = 0 \tag{1.89}$$

and

$$\frac{1}{\sqrt{g}}\partial_i(\sqrt{g}\rho g^{ij}\partial_j\sigma) = 0. \tag{1.90}$$

Variation of (1.88) with respect to the connections $\Gamma^i_{jk}(q)$ yields the Weyl conformal connections [79, 83]

$$\Gamma^i_{jk} = -\left\{ {i \atop jk} \right\} + \delta^i_j\phi_k + \delta^i_k\phi_j + g_{jk}\phi^i. \tag{1.91}$$

In equation (1.91) $\left\{ {i \atop jk} \right\}$ are the Christoffel symbols out of the metric g_{ij}, $\phi^i = g^{ij}\phi_j$, and ϕ_i is Weyl's vector given by

$$\phi_i = -\frac{1}{n-2}\frac{\partial_i\rho}{\rho}\,[11, 83]. \tag{1.92}$$

The curvature tensor R^i_{jkl} and the scalar curvature R obtained from the connections (1.91) can be called as the Weyl curvature tensor and the Weyl scalar curvature, respectively. Furthermore, in the light of equation (1.92) the Weyl vector ϕ_i is a gradient, and this implies that the Weyl connection (1.91) is integrable with Weyl's scalar potential $\phi_i = -\frac{1}{n-2}\ln\rho$. From equation (1.92) one finds that, in order to compute the affine connections, ρ can be used as Weyl's potential in the place of ϕ. Inserting equation (1.92) into the well-known expression of Weyl's scalar curvature, one obtains the following expression for the scalar curvature

$$R_w = \bar{R} + \left(\frac{n-1}{n-2}\right)\left[\frac{g^{ij}\partial_i\rho\partial_j\rho}{\rho^2} - \frac{2\partial_i\left(\sqrt{g}g^{ij}\partial_j\rho\right)}{\rho\sqrt{g}}\right] \tag{1.93}$$

where $\bar{R}$ is the Riemann curvature of V_n obtained from the Christoffel symbols of the metric g_{ij} [79]. The affine Weyl connections (1.91) are invariant under the Weyl conformal gauge transformations

$$g_{ij} \to \lambda g_{ij} \tag{1.94}$$

$$\phi_i \to \phi_i - \frac{\partial_i\lambda}{2\lambda}\,[79]. \tag{1.95}$$

The fields $T(q)$ behave in a simple way in the sense that under Weyl-gauge transform as $T \to \lambda^{\omega(T)}T$ where the exponent $w(T)$ is also called the Weyl "weight" of T and is a real number characterizing the Weyl action on T. The principle (1.88) is Weyl-gauge invariant provided $\omega(\sigma) = 0$ and $\omega(\rho) = -\frac{n-2}{2}$. The field equations (1.89), (1.90), (1.92), and (1.93) are the fundamental equations of the theory.

Now, the mechanical interpretation of the field theory based on the variational principle (1.88) is direct. In fact, one can easely see that equation (1.89) has the form of the classical Hamilton-Jacobi equation of mechanics for the action function $\sigma(q)$ of a particle subjected to the scalar potential given by the Weyl curvature (1.93). Moreover, equation (1.89) may be alternatively derived by applying the single-integral variational problem $\delta \int L d\tau$ to the homogeneous Lagrangian

$$L(q, \dot{q}) = \sqrt{-R(q) g_{ij}(q) \dot{q}^i \dot{q}^j} \tag{1.96}$$

having the same form of the Lagrangian of a relativistic particle moving in space-time with mass constant replaced by the curvature field $R_\omega(q)$. The Hamilton-Jacobi equation (1.89) and the variational principle (1.96) provide the mechanical features of the theory. These equations have the same form of the relativistic equations for a free particle, but the particle mass is replaced here by Weyl's curvature (1.93) as required for Weyl-gauge invariance. Any solution $\sigma(q)$ of the Hamilton-Jacobi equation fixes a set of (time-like) trajectories in V_n given by $\dot{q}^i = g^{ij} \partial_j \sigma$. When the system is in the dynamical state defined by $\sigma(q)$, these time-like trajectories in V_n can be associated to possible trajectories of the system in the configuration space, when the system is in the dynamical state defined by $\sigma(q)$. Each trajectory of the set satisfies the Euler-Lagrange equations derived from the Lagrangian (1.96), so that along its motion, the system is subjected to a Newtonian force proportional to the gradient of the Weyl curvature $R_\omega(q)$. Once equations (1.89) and (1.90) are solved, the field $\sigma(q)$ defines the dynamics and the field $\rho(q)$ defines the affine connections from equations (1.91) and (1.92), and the curvature from equation (1.93) [83].

On the other hand, the field equation (1.90) may receive a simple and direct mechanical interpretation as a "continuity equation"

($\partial_i j^i = 0$) for the current density

$$j^i = \sqrt{g}\rho g^{ij}\partial_j\sigma \tag{1.97}$$

which has the property $\omega(j^i) = 0$ and is therefore Weyl-gauge invariant. An interesting feature of the approach based on the continuity equation (1.90) lies in the fact that the continuity equation (1.90) allows us to describe the motion of a fluid of density ρ carried along the set of trajectories defined by σ. And, in this regard, it must be emphasized that the continuity equation and the dependence of R_ω on ρ shown in equation (1.93) are in agreement with the hydrodynamical picture of quantum mechanics first proposed by Madelung [85] and then developed by Bohm [30, 31] where the particle trajectories are deterministically ruled by classical mechanics and quantum effects are associated with a "quantum potential" whose gradient produces the action of a quantum force on the behaviour of the particle.

1.3.2 *Santamato's geometric interpretation of quantum mechanics*

In some 80s papers, Santamato developed the de Broglie-Bohm formulation of quantum mechanics by relating the mysterious quantum potential to fundamental geometric properties in the picture of Weyl's geometry [10, 11, 86]. The net result was that the mysterious quantum effects were demonstrated to be related to the geometric structure of space, in particular to the curvature.

By following the treatment provided in these papers, one assumes that the motion of the particle is given by some random process $q^i(t, \omega)$ in a manifold M with a probability density ρ, and satisfying a deterministic equation

$$\dot{q}^i(t, \omega) = \frac{dq^i}{dt}(t, \omega) = v^i(q^i(t, \omega), t) \tag{1.98}$$

with random initial conditions $q^i(t_0, \omega) = q^i_0(\omega)$. Moreover, Santamato postulated a Lagrangian given by

$$L(q, \dot{q}, t) = L_c(q, \dot{q}, t) + \gamma(\hbar^2/m)R \tag{1.99}$$

where $\gamma = \frac{1}{8}(\frac{n-2}{n-1}) = \frac{1}{16}$ for $n = 3$.

He then went on to obtain the Schrödinger equation as well as the equations of motion of Bohm's approach to non-relativistic quantum mechanics by invoking an Averaged Least Action Principle

$$I(t_0, t_1) = E\left\{\int_{t_0}^{t} L^*(q\,(t,\omega)\,,\dot{q}\,(t,\omega)\,,t)dt\right\} = \text{minimum} \qquad (1.100)$$

with respect to the class of all Weyl geometries of space with a fixed metric tensor. By minimizing the action functional (1.100), one obtains the velocity field $v^i(q,t)$ determining a classical motion with probability one.

Here, the Weyl geometry is associated with the covariant components ϕ_k of an arbitrary vector of M, which appear in the transplantation property of the length l of a vector expressed by relation $\delta l = l\phi_k dq^k$. Since the only term containing the gauge vector $\vec{\phi} = \phi_k$ is the curvature term, by minimizing the action functional (1.100), one obtains

$$R = \dot{R} + (n-1)\left[(n-2)\phi_i\phi^i - 2\left(\frac{1}{\sqrt{g}}\partial_i(\sqrt{g}\phi^i)\right)\right] \qquad (1.101)$$

where $\phi^i = g^{ik}\phi_k$ and $\dot{R}$ is the Riemann curvature based on the metric, and the minimum occurs when $\phi_i(q,t) = -\frac{1}{n-2}[\partial_i \log \hat{\rho}]$ and $\hat{\rho}(q,t) = \rho(q,t)/\sqrt{g}$ which transforms as a scalar under coordinate changes.

This approach shows that the transplantation properties of space are determined by the presence of matter and in turn this change in geometry acts on the particle via a "quantum" force $f_i = \gamma(\hbar^2/m)\partial_i R$ depending on the gauge vector $\vec{\phi}$. By substituting this gauge vector in (1.99) one obtains

$$R_\omega = R = \dot{R} + \frac{1}{2\gamma\sqrt{\hat{\rho}}}\left[\frac{1}{\sqrt{g}}\partial_i\left(\sqrt{g}g^{ik}\right)\partial_k\sqrt{\hat{\rho}}\right] \qquad (1.102)$$

together with a Hamilton-Jacobi equation

$$\partial_t S + H_c(q, \nabla S, t) - \gamma(\hbar^2/m)R = 0. \qquad (1.103)$$

Now, for certain Hamiltonians of the form $H_C = \frac{1}{2m}g^{ik}(p_i - A_i)(p_k - A_k) + V$ with arbitrary fields A_k and V one finds that the function $\psi = \sqrt{\rho}\exp(\frac{i}{\hbar}S(q,t))$ satisfies a Schrodinger equation (in curvi-linear coordinates) of the form

$$i\hbar\partial_t\psi = \frac{1}{2m}\left\{\left[\frac{i\hbar}{\sqrt{g}}\partial_i\sqrt{g}A_i\right]g^{ik}(i\hbar\partial_k + A_k)\right\}\psi + \left[V - \gamma(\hbar^2/m)\dot{R}\right]\psi = 0. \quad (1.104)$$

This Hamiltonian is characteristic of a particle in an electromagnetic field and all Hamiltonians arising in non-relativistic applications may be reduced to the above form with a corresponding Hamilton-Jacobi equation which, omitting the A_i, assumes the form:

$$\partial_t S = \frac{1}{2m}g^{ik}\partial_i S\partial_k S + V - \gamma(\hbar^2/m)R = 0. \quad (1.105)$$

Now, on the basis of equation (1.105), if one makes a comparison for example to equations (1.3), (1.4) and (1.6), one finds that

$$Q \approx -\gamma(\hbar^2/m)R \quad (1.106)$$

with R given by (1.102) and $\gamma = \frac{1}{8}\left(\frac{n-2}{n-1}\right) = \frac{1}{16}$ for $n = 3$. Relation (1.106) implies that the quantum potential Q is expressed in terms of the scalar curvature R. Moreover, from $\phi_i(q,t) = -\frac{1}{n-2}[\partial_i \log\rho]$ one obtains finally

$$Q \approx -\frac{\hbar^2}{16m}\left[\dot{R} + 2\left\{\phi_i\phi^i - \frac{2}{\sqrt{g}}\partial_i(\sqrt{g}\phi^i)\right\}\right] \quad (1.107)$$

which shows that the quantum potential depends directly on the Weyl vector.

Therefore, in summary, in the light of equations (1.100)–(1.107), Santamato obtained the Schrodinger equation as well as the equations of motion of Bohm's approach to non-relativistic quantum mechanics by applying the Averaged Least Action Principle (1.100) with respect to the class of all Weyl geometries of space with a fixed metric tensor. In Santamato's approach, Bohm's quantum potential can be considered a geometrodynamic entity which contains a global information on the environment in which the experiment is performed (and which determines the behaviour of quantum particles and their relationship with the environment) in the sense that derives from a minimum condition of the action inside a Weyl geometry of space, which leads to a direct dependence of the quantum potential on the Weyl vector. Finally, as the regards the link between the quantum potential and the curvature R invoked by Santamato, it is also worth to be mentioned that according to some research it can be interpreted as the result of cosmic fluctuations [87–89].

More recently, in the already mentioned papers "Nonlocality, no-signalling and Bell's theorem investigated by Weyl's conformal differential geometry" and "Solving the nonlocality riddle by conformal quantum geometrodynamics" [83, 84], De Martini and Santamato developed further the idea of Weyl's conformal quantum geometrodynamics, which shows that the active quantum potential originates from geometry, as does gravitation, and arises from the space curvature due to the presence of the non trivial affine connections of Weyl's conformal geometry. In this picture, the conformal invariance requires that the Riemann scalar curvature generates a contribution to the potential, which is absent in Bohm's approach.

The particle dynamics (governed by $\sigma(q)$) and the affine connections (governed by $\rho(q)$) turn out to be compatible, by solving equations (1.89) and (1.90) together. By introducing the complex scalar function ψ given by

$$\psi(q) = \sqrt{\rho} e^{iS(q)/\hbar} \tag{1.108}$$

with $S(q) = \xi\hbar\sigma(q)$, and

$$\xi = \sqrt{\frac{n-2}{4(n-1)}} \tag{1.109}$$

one may convert equations (1.89) and (1.90) into the Klein-Gordon type linear differential equation

$$\nabla_k \nabla^k \psi - \xi^2 \hbar^2 R\psi = 0 \tag{1.110}$$

where $\nabla_k \nabla^k$ is the Laplace-Beltrami operator built from the metric tensor g_{ij} and R is the Riemann curvature appearing in equation

(1.93) and thus the mass term replaced by the Riemann curvature term $\xi^2\hbar^2 R(q)$. This is a striking result as it demonstrates that the Hamilton-Jacobi equation, applied to a general dynamical problem can be transformed into a linear wave equation, the main ingredient of the formal structure of quantum mechanics and of the Hilbert space theory. The value of ξ given by equation (1.109) ensures that equation (1.110) is conformally invariant. Whilst the wavefunction is not gauge invariant, because its modulus has Weyl weight $w(|\psi|) = -(n-2)/4$, the phase of ψ is conformally invariant. In particular, in De Martini's and Santamato's approach, the quantum wavefunction receives the precise meaning of a physical quantum "Weyl's gauge field" acting in a curved configurational space, the square modulus of the wavefunction is identified with the Weyl potential and its gradient with the Weyl vector.

Santamato and De Martini also extended the idea of Weyl-gauge invariance to the formulation of quantum mechanics of spin 1/2 particles. No quantization rules appear explicitly and quantum effects are derived from the existence of a force from Weyl's scalar curvature of the configuration space of the system. Here, a scalar wavefunction is introduced as a useful *ansatz* in order to solve the nonlinear equations of the theory. The particle mass is replaced by Weyl's scalar curvature field and a constant mass appears only at the level of wave equation. In Santamato's and De Martini's approach, in the case of spin 1/2 particles a one-to-one correspondence can be established between the results of the theory and the results of the standard quantum mechanics based on Dirac's spinors and on the second-order Dirac equation [90, 91]. Here, the occurrence of EPR paradox, in particular, can be traced back to a geometrical nonlocal interaction among the particles due to Weyl's curvature. In particular, for a the EPR state of two entangled qubits A and B, Santamato's and De Martini's calculations give the following results as regards the action and Weyl's curvature:

$$S = \hbar\left[\frac{\gamma_A+\gamma_B}{2} + \arctan\left(\csc\frac{\beta_A-\beta_B}{2}\, sen\frac{\beta_A+\beta_B}{2}\tan\frac{\alpha_B-\alpha_A}{2}\right)\right.$$
$$\left. + \arg(\omega_\uparrow^{(A)}(\vec{r}_A,t)) + \arg(\omega_\downarrow^{(B)}(\vec{r}_B,t))\right] \qquad (1.111)$$

and

$$R_W = \frac{22}{5a^2(1 - \cos\beta_A \cos\beta_B - \cos\Delta\alpha sen\beta_A sen\beta_B)} + R_W^{(A)}(\vec{r}_A, t) + R_W^{(B)}(\vec{r}_B, t)\ [92]. \tag{1.112}$$

In equations (1.111) and (1.112), (α, β, γ) are angles defining the space rotations, $\{\omega_\uparrow, \omega_\downarrow\}$ are the components of the Pauli spinor of the particle, $R_W^{(A)}(\vec{r}_A, t)$ is the contribution of $\omega^{(A)}$ to Weyl's curvature, $R_W^{(B)}(\vec{r}_B, t)$ is the contribution of $\omega^{(B)}$ to Weyl's curvature. Equations (1.111)–(1.112) imply that, although the particle motions are independent in space-time, they are still coupled by the Weyl curvature in the angular variables and, beside the self-force, the particles exert a force on each other.

1.3.3 *Novello's, Salim's and Falciano's approach of quantum mechanics as a manifestation of the non-euclidean geometry*

As regards the link between Bohm's quantum potential and Weyl geometries, appealing results have been achieved in the recent article *On a geometrical description of quantum mechanics* by Novello, Salim and Falciano [93]. In this paper, the three authors showed that Bohm's quantum potential can be identified with the curvature scalar of the Weyl integrable space and developed a variational principle that reproduces the Bohmian equations of motion. Novello's, Salim's and Falciano's model suggests that one can reinterpret quantum mechanics as a manifestation of the non-Euclidean structure of the three-dimensional space: by starting from a variational principle which defines the non-Euclidean structure of space, they provided a geometrodynamic interpretation to quantum processes in the picture of the Weyl geometry.

In Novello's, Salim's and Falciano's approach, the geometrical structure of space is derived by starting from the action

$$I = \int dt d^3x \sqrt{g}\Omega^2 \left(\lambda^2 R - \frac{\partial S}{\partial t} - H_m\right). \tag{1.113}$$

Here, the independent variables are the connection of the 3-d space Γ^i_{jk}, Hamilton's principal function S and the scalar function Ω. Moreover, in equation (1.113) one has $g = \det g_{ij}$, $R \equiv g^{ij}R_{ij}$, $R_{ij} = \Gamma^m_{mi,j} - \Gamma^m_{ij,m} + \Gamma^l_{mi}\Gamma^m_{jl} - \Gamma^l_{ij}\Gamma^m_{lm}$ is the Ricci curvature tensor, λ^2 is a constant having dimension of energy multiplied for length squared and the term $\frac{\partial S}{\partial t}$ is related to the particle's energy.

In the case of a point-like particle the matter Hamiltonian is $H_m = \frac{1}{2m}\nabla S \cdot \nabla S + V$. Variation of the action (1.113) with respect to the independent variables yields

$$g_{ij;k} = -4(\ln \Omega)_{,k} g_{ij} \tag{1.114}$$

where ";" denotes covariant derivative and "," simple spatial derivative. Equation (1.112) describes the affine properties of the physical space. Variation with respect to Ω gives

$$\lambda^2 R = \frac{\partial S}{\partial t} + \frac{1}{2m}\nabla S \cdot \nabla S + V. \tag{1.115}$$

On the other hand, by setting $\lambda^2 = \frac{\hbar^2}{16m}$ and taking into account the expression of the curvature in terms of the scalar function Ω, $R = 8\frac{\nabla^2\Omega}{\Omega}$, equation (1.115) may be rewritten as

$$\frac{\partial S}{\partial t} + \frac{1}{2m}\nabla S \cdot \nabla S + V - \frac{\hbar^2}{2m}\frac{\nabla^2\Omega}{\Omega} = 0. \tag{1.116}$$

Finally, by varying the Hamilton's principal function S one obtains

$$\frac{\partial \Omega^2}{\partial t} + \nabla \cdot \left(\Omega^2 \frac{\nabla S}{m}\right) = 0. \tag{1.117}$$

By comparing equations (1.116) and (1.117) with equations (1.3) and (1.6) respectively, if one identifies the scalar function Ω with the amplitude of the wavefunction, one finds immediately that equation (1.116) is analogous to the quantum Hamilton-Jacobi equation (1.3) of the de Broglie-Bohm interpretation of quantum mechanics, while equation (1.117) is identical to the continuity equation for the current density. Therefore, the "action" of a point-like particle non-minimally coupled to geometry given by

$$I = \int dt d^3x \sqrt{g} \Omega^2 \left(\frac{\hbar^2}{16m} R - \frac{\partial S}{\partial t} - H_m \right) \tag{1.118}$$

exactly reproduces the Schrödinger equation and thus the quantum behaviour. In the mathematical formalism (1.113)–(1.118), the fundamental result is that, according to Novello's, Salim's and Falciano's approach based on Weyl integrable space, the quantum potential can be practically identified with the curvature scalar which characterizes this geometry. Moreover, in this picture, the inverse square root of the curvature scalar defines a typical length (Weyl length) that can be used to evaluate the strength of quantum effects

$$L_W = \frac{1}{\sqrt{R}}. \tag{1.119}$$

Inside this picture, the classical behaviour is recovered when the length defined by the Weyl curvature scalar is small compared to the typical length scale of the system. Once the Weyl curvature becomes non-negligible the system goes into a quantum regime. It is also interesting to mention that in this approach, if one accepts that quantum mechanics is a manifestation of a non-Euclidean geometry, all theoretical topics related to quantum effects can receive a pure geometrical meaning. In particular, in this picture, the identification of Weyl integrable space's curvature scalar as the ultimate origin, as ultimate key of explanation of quantum effects leads to a geometrical interpretation of Heisenberg's uncertainty principle. This geometrical description considers the uncertainty principle as a breakdown of the classical notion of standard rulers. The uncertainty principle derives from the fact that we are unable to perform a classical measurement to distances smaller than the Weyl curvature length. In other words, the size of a measurement has to be bigger than the Weyl length

$$\Delta L \geq L_W = \frac{1}{\sqrt{R}}. \tag{1.120}$$

The quantum regime is entered when the Weyl curvature term is dominant. One can make a parallelism of Novello's, Salim's and

Falciano's interpretation of Heisenberg's uncertainty principle with Bohr's complementary principle in the sense that this approach implies that in the quantum domain there is the impossibility of applying the classical definitions of measurements. However, while Bohr's complementary principle is based on the uncontrolled interference of a classical apparatus of measurement, Novello, Salim and Falciano claim that the notion of a classical standard ruler breaks down because its meaning is intrinsically linked to the validity of Euclidean geometry, and this provides a significant physical difference between the two views. In Novello's, Salim's and Falciano's model, the fundamental point is that once it becomes necessary to include the Weyl curvature, one can no longer perform a classical measurement of distance (and in this regard it is not necessary to postulate a crucial role of the observer like in Boh's view).

1.4 The Quantum Potential in the Relativistic Domain

The geometry subtended by Bohm's quantum potential in the relativistic domain has been explored by various authors (the reader can find the mathematical details, for example, in Carroll's book *Fluctuations, information, gravity and the quantum potential* [94]). Here our aim is to illustrate in what sense the quantum potential emerges as the ultimate factor which provides geometrodynamics in relativistic quantum mechanics, in the light of the fundamental models existing in the literature. In particular, in the paragraph 1.4.1 we focus our attention on some significant Bohmian models of the Klein-Gordon relativistic quantum mechanics developed by F. Shojai and A. Shojai, Faraggi and Matone and Nikolic. Then, in the paragraph 1.4.2, we examine the Dirac relativistic quantum mechanics, by taking into examination recent developments in a Bohmian picture provided by Hernandez-Zapata, Nikolic, Chavoya-Aceves and Hiley and Callaghan. In both cases we will see the fundamental role of the quantum potential and its geometrodinamic information in determining the processes.

1.4.1 *The quantum potential in Bohm's approach to Klein-Gordon relativistic quantum mechanics*

The Klein-Gordon equation, which describes the behaviour of relativistic spinless particles and arises as a constraint for the quantization of their world-line, assumes the form

$$\left(\nabla^2 - \frac{1}{c^2}\frac{\partial^2}{\partial t^2}\right)\psi = \frac{m^2c^2}{\hbar^2}\psi \tag{1.121}$$

where m is the mass of the particle, $\hbar$ is Planck's reduced constant and c is the speed of light. It can be obtained by substituting in the relativistic relation between the total energy E and momentum p of a free particle $E^2 = p^2c^2 + m^2c^4$ the quantum mechanical operators of energy $E \to i\hbar\frac{\partial}{\partial t}$ and momentum $p_j \to -i\hbar\frac{\partial}{\partial x^j}$ or by starting from the action

$$A(\psi) = \frac{1}{2}\int d^4x\left(\partial^\mu\partial_\mu\psi - \frac{m^2c^2}{\hbar^2}\psi\right) \tag{1.122}$$

where $x^0 = t$, $x = (\vec{x}, t)$. The Klein-Gordon equation (1.121) can receive a causal interpretation in a Bohmian framework by making the usual polar decomposition of the wavefunction,

$$\psi = R\ e^{iS/\hbar} \tag{1.123}$$

where R, S are real Lorentz scalar functions. If one substitutes (1.123) into the standard Klein-Gordon equation (1.121), equation (1.121) leads to the two following real equations, one for the real part of (1.121) and the other for the imaginary part: a quantum Hamilton-Jacobi equation

$$\partial_\mu S\partial^\mu S = m^2c^2(1+Q) \tag{1.124}$$

and a continuity equation

$$\partial_\mu j^\mu = 0 \tag{1.125}$$

where

$$Q = \frac{\hbar^2}{m^2c^2}\frac{\left(\nabla^2 - \frac{1}{c^2}\frac{\partial^2}{\partial t^2}\right)|\psi|}{|\psi|} \tag{1.126}$$

is the quantum potential and

$$j^\mu = -(R^2/mc)\partial^\mu S \tag{1.127}$$

is the current associated with the wavefunction of the particle under consideration.

In the approach based on equations (1.123)–(1.127), the current satisfying the continuity equation (1.125) defines a congruence of world lines of a set of particles associated with the wave $\psi(\vec{x}, t)$. The tangent to a world line determines the 4-velocity u^μ which can be defined in terms of the 4-momentum p^μ through the relation

$$Mu^\mu = p^\mu = -\partial^\mu S \tag{1.128}$$

where

$$M = m(1+Q)^{1/2} \tag{1.129}$$

is a variable quantum mass. If one solves the differential equation $u^\mu = \frac{dx^\mu}{d\tau}$ where τ is the proper time, it is possible to obtain a trajectory $x^\mu = x^\mu(\tau)$ once the initial position of a particle in the set is specified [32, 95].

Equation (1.129) presents however a physical problem. Since the quantum potential can be a negative number the quantum mass defined by relation (1.129) is not positive definite and this implies the possibility that tachyonic solutions emerge. For this reason, as evidenced by F. Shojai and A. Shojai in the papers [96, 97], the quantum Hamilton-Jacobi equation of the form (1.124) cannot be considered as the correct equation of motion regarding relativistic spinless particles in a Bohmian picture. On the basis of the consideration that a correct relativistic equation of motion should not only be Poincarè invariant but also give the correct non-relativistic limit, in the papers [96, 97] F. Shojai and A. Shojai developed an interesting Bohmian approach to the Klein-Gordon equation (1.121) in which the quantum Hamilton-Jacobi equation which is derived from the decomposition of the wavefunction in its polar form $\psi = |\psi| \exp(\frac{iS}{\hbar})$ has the following form

$$\partial_\mu S \partial^\mu S = m^2c^2 \exp Q. \tag{1.130}$$

Equation (1.130) has the merit to be Poincarè invariant and to provide the correct non-relativistic limit and here the quantum potential is defined as (1.126), while the continuity equation assumes the form

$$\partial_\mu(\rho\partial^\mu S) = 0 \tag{1.131}$$

where ρ is the density of particles in the element of volume d^3x around a point $\vec{x}$ at time t associated with the wavefunction $\psi(\vec{x}, t)$ of the individual physical system under consideration. On the basis of the quantum Hamilton-Jacobi equation (1.130), in the relativistic regime the mass of a particle is determined and generated by the quantum potential. In fact, equation (1.131) leads directly to relation

$$M = m\sqrt{\exp Q}. \tag{1.132}$$

The quantity given by (1.132) can be defined as a variable quantum mass of the relativistic spinless particle under consideration, which is Poincarè invariant and has the correct non-relativistic limit. Moreover, in F. Shojai's and A. Shojai's approach, the particle trajectory emerges by starting from the guidance formula and, if one differentiates equation (1.131), one obtains the particle trajectory expressed by a Newton's-type equation of motion:

$$M\frac{d^2x^\mu}{d\tau^2} = (c^2\eta^{\mu\nu} - u^\mu u^\nu)\partial_\nu M \tag{1.133}$$

where $\eta_{\mu\nu} = diag(1, -1, -1, -1)$ is the signature of the Minkowski space-time metric, $u^\mu = \frac{dx^\mu}{d\tau}$ is the four-velocity, τ being the proper time.

As regards Klein-Gordon's relativistic quantum mechanics in a Bohmian framework, interesting results have been obtained, besides by F. Shojai and A. Shojai, also in the context of the Bertoldi, Faraggi and Matone theory, in a model of Nikolic as well as in Sbitnev's recent approach of physical vacuum as a special superfluid medium.

A detailed analisys of the references [98–124] shows that the Bertoldi-Faraggi-Matone theory, exclusively on the basis of an equivalence principle — which states that all physical systems can be connected by a coordinate transformation to the free situation with vanishing energy — allows us to obtain a quantum stationary Hamilton-Jacobi equation which is a third order nonlinear differential equation providing a trajectory representation of quantum mechanics. By following the treatment of [98–124], in the relativistic domain, this approach starts by considering the relativistic classical Hamilton-Jacobi equation of the following form

$$\frac{1}{2m}\sum_{1}^{D}(\partial_k S^{cl}(q,t))^2 + \mathrm{M}_{rel}(q,t) = 0 \qquad (1.134)$$

where

$$\mathrm{M}_{rel}(q,t) = \frac{1}{2mc^2}\left[m^2c^4 - \left(V(q,t) + \frac{\partial}{\partial t}S^{cl}(q,t)\right)^2\right]. \qquad (1.135)$$

In particular, in the time-dependent case the relativistic classical Hamilton-Jacobi equation (1.134) becomes

$$\frac{1}{2m}\eta^{\mu\nu}\partial_\mu S^{cl}\partial_\nu S^{cl} + \mathrm{M}'_{rel} = 0 \qquad (1.136)$$

where $\eta_{\mu\nu} = diag(-1, 1, \ldots, 1)$ is the Minkowskian metric and

$$\mathrm{M}'_{rel}(q,t) = \frac{1}{2mc^2}[m^2c^4 - V^2(q) - 2cV(q)\partial_0 S^{cl}(q)] \qquad (1.137)$$

with $q = (ct, q_1, \ldots, q_D)$. The equivalence principle is embedded by modifying the classical equation with any additional function that must be determined, namely

$$\frac{1}{2m}(\partial S)^2 + \mathrm{M}_{rel} + Q = 0, \qquad (1.138)$$

where

$$\mathrm{M}_{rel}(q,t) = \frac{1}{2mc^2}[m^2c^4 - V^2(q) - 2cV(q)\partial_0 S(q)], \qquad (1.139)$$

S being the action determined by the presence of the function Q. Implementation of the equivalence principle has the consequence that an arbitrary M^a state satisfies the following relation

$$M^b_{rel}(q^b) = (p^b|p^a)M^a_{rel}(q^a) + (q^a;q^b);$$
$$Q^b(q^b) = (p^b|p^a)Q(q^a) - (q^a;q^b) \tag{1.140}$$

where $(p^b|p) = \eta^{\mu\nu}p^b_\mu p^b_\nu/\eta^{\mu\nu}p_\mu p_\nu$. Hence, after some algebra, if R satisfies the continuity equation $\partial \cdot (R^2\partial S) = 0$, equation (1.137) may be expressed as

$$M_{rel} = \frac{\hbar^2}{2m}\frac{\left(\nabla^2 - \frac{1}{c^2}\frac{\partial^2}{\partial t^2}\right)(R\exp(iS/\hbar))}{R\exp(iS/\hbar)} \tag{1.141}$$

and thus the corresponding quantum potential becomes

$$Q_{rel} = -\frac{\hbar^2}{2m}\frac{\left(\nabla^2 - \frac{1}{c^2}\frac{\partial^2}{\partial t^2}\right)R}{R}. \tag{1.142}$$

In this way, in the time-dependent case the relativistic quantum Hamilton-Jacobi equation becomes

$$\frac{1}{2m}(\partial S)^2 + \mathrm{M}_{rel} - \frac{\hbar^2}{2m}\frac{\left(\nabla^2 - \frac{1}{c^2}\frac{\partial^2}{\partial t^2}\right)R}{R} = 0 \tag{1.143}$$

while the continuity equation is

$$\partial \cdot (R^2\partial S) = 0. \tag{1.144}$$

Therefore, the treatment of Klein-Gordon relativistic quantum mechanics proposed by the Bertoldi-Faraggi-Matone theory suggests that the quantum potential (1.142) derives from the implementing of the equivalence principle. The quantum potential associated with the Klein-Gordon equation provides a geometrodynamic description of the processes in the sense that emerges and tries its justification from a more fundamental equivalence principle. In this picture, all the Klein-Gordon relativistic quantum mechanics in a Bohmian picture can be derived by starting from the equivalence principle, which can be considered as a fundamental principle describing the behaviour of relativistic spinless particles and ruling the geometry of their processes. As regards the Bertoldi-Faraggi-Matone theory, it is also interesting to make a comparison with Atiq's, Karamian's and Golshani's quasi-Newtonian approach of the quantum potential developed in [63]. In Atiq's, Karamian's and Golshani's approach the mathematical form (and consequently the physical features) of the quantum potential in non-relativistic quantum mechanics can be derived, in the context of a quasi-Newtonian picture, by applying a minimum condition on the total energy of ensemble, without appealing to the Schrödinger equation and the wavefunction. In an analogous way, in the Bertoldi-Faraggi-Matone theory, the mathematical form (and consequently the geometrodynamic nature) of the quantum potential is obtained directly by implementing the equivalence principle, without appealing to the Klein-Gordon equation and the wavefunction. Therefore, just like it occurs in Atiq's, Karamian's and Golshani's approach about the Schrödinger equation, in the Bertoldi-Faraggi-Matone theory the quantum potential concerning the behaviour of relativistic spinless particles can be considered, in the light of the application of the fundamental structure represented by the equivalence principle, as the basis of the Klein-Gordon equation rather than being a consequence of it. The results of the Bertoldi-Faraggi-Matone theory show thus that the quantum potential can be considered as a fundamental concept, namely the geometrodynamics of quantum processes associated with the quantum potential deriving from the implementing of the equivalence principle emerges as a crucial concept of quantum processes also for relativistic spinless particles.

Moreover, in analogy to F. Shojai's and A. Shojai's model, also the Bertoldi-Faraggi-Matone theory introduces important perspectives as regards the problem of generation of masses. Starting from equation (1.140), one finds that also in the Bertoldi-Faraggi-Matone approach to the Klein-Gordon equation masses are expressed in terms of the quantum potential on the basis of the following relation

$$\frac{1}{2mc^2} = (p|p^0)Q^0(q^0) - Q(q). \tag{1.145}$$

In this regard, the role of the quantum potential was seen as a sort of intrinsic self energy which resembles in some way the relativistic self energy [123].

Finally, by following the philosophy that underlies Novello's, Salim's and Falciano's model analysed in chapter 1.3.3, in the picture of the Bertoldi-Faraggi-Matone theory one can think to introduce a quantum length associated with the implementing of the equivalence principle

$$L_{quantum} = \frac{1}{\sqrt{\frac{\left(\nabla^2 - \frac{1}{c^2}\frac{\partial^2}{\partial t^2}\right)R}{R}}}. \qquad (1.146)$$

In the Bertoldi-Faraggi-Matone theory, the quantum length (1.146) can be used to evaluate the strength of quantum effects and, therefore, the modification of the geometry in a quantum relativistic regime — corresponding with the implementing of the equivalence principle — with respect to the Euclidean geometry characteristic of classical physics. Once the quantum length (1.146) becomes non-negligible the spinless particle into consideration goes into a quantum regime. And Heisenberg's uncertainty principle derives from the fact that we are unable to perform a classical measurement to distances smaller than this quantum length (1.146) corresponding with the implementing of the equivalence principle.

As regards the Klein-Gordon equation, other relevant and interesting perspectives have been obtained recently by Nikolic [125–128], who developed a Lorentz-covariant Bohmian interpretation which has at least two important merits. On one hand, in this model the lack of statistical transparency (namely the fact that one cannot compute the statistical distribution of particle positions in a simple way if one only knows the wavefunction, without the knowledge of particle trajectories) appears a virtue of the Bohmian interpretation, in the sense that it opens the possibility of experimentally distinguishing its predictions from the predictions of other possible interpretations. On the other hand, this model shows that the equations for Bohmian particle trajectories, although are non-local, can be naturally formulated in a Lorentz-covariant form without a preferred Lorentz frame.

In Nikolic's approach, the Klein-Gordon equation takes the form

$$(\partial_0^2 - \nabla^2 + m^2c^2)\phi = 0 \qquad (1.147)$$

where, as usual, $\eta_{\mu\nu} = diag(-1, 1, \ldots, 1)$. If $\psi = \phi^+$ and $\psi* = \phi^-$ correspond to positive and negative frequency parts of $\phi = \phi^+ + \phi^-$ the particle current is given by $j_\mu = i\psi^* \overleftrightarrow{\partial}_\mu \psi$ (where $a\overleftrightarrow{\partial}^\mu b = a\partial^\mu b - b\partial^\mu a$) which is conserved ($\partial^\mu j_\mu = 0$) and $N = \int d^3x j_0$ is the particle number. Trajectories assume the form

$$\frac{d\vec{x}}{dt} = \frac{\vec{j}(t, \vec{x})}{j_0(t, \vec{x})} \tag{1.148}$$

namely

$$\frac{dx^\mu}{ds} = \frac{j^\mu}{2m\psi^*\psi} \tag{1.149}$$

where s is an affine parameter along the curves in the 4-dimensional Minkowski space-time. Hence, for $t = x_0$, in a natural system of units ($c = \hbar = 1$), one arrives at two real quantum Hamilton-Jacobi equations of the form

$$\partial^\mu (R^2 \partial_\mu S) = 0 \tag{1.150}$$

and

$$\frac{(\partial^\mu S)(\partial_\mu S)}{2m} - \frac{m}{2} + Q = 0 \tag{1.151}$$

where the quantum potential is

$$Q = -\frac{1}{2m} \frac{\partial^\mu \partial_\mu R}{R}. \tag{1.152}$$

By using equations (1.149), (1.151) and the identity $\frac{d}{ds} = \frac{dx^\mu}{ds}\partial_\mu$ one also finds the equation of motion

$$\frac{d^2 x^\mu}{ds^2} = \partial^\mu Q \tag{1.153}$$

The equations (1.147)–(1.153) for the particle trajectories of Nikolic's approach are non-local, but Lorentz covariant because of the fact that the trajectories in space-time do not depend on the choice of the affine parameter s. Moreover, the quantum potential (1.152) can be considered, together with the particle current, the fundamental physical entity which rules the motion, the trajectories and thus the geometry of relativistic spinless particles in the context of the Klein-Gordon relativistic quantum mechanics.

Also in the context of Nikolic's approach, one can characterize the geometrical properties of the background associated with the quantum potential (1.152) by introducing the quantum length

$$L_{quantum} = \frac{1}{\sqrt{\frac{\partial^{\mu}\partial_{\mu}R}{R}}}. \tag{1.154}$$

The quantum length (1.154) regarding Nikolic's approach can be used to evaluate the strength of quantum effects and, therefore, the modification of the geometry in a quantum relativistic regime — and corresponding with the quantum entropy together with the particle current — with respect to the Euclidean geometry characteristic of classical physics.

Finally, in the picture proposed recently by Sbitnev in [70–73] where the quantum potential is the expression of the zero-point fluctuations of a physical vacuum consisting of an enormous amount of virtual pairs of particles-antiparticles organized in a Bose ensemble acting as a special superfluid medium, the relativistic hydrodynamical equations are examined with the aim of extracting the quantum-mechanical equations (not only the Schrödinger equation in the non-relativistic limit, but also the relativistic Klein-Gordon equation in the general case).

In Sbitnev's model, the equation of the relativistic hydrodynamics, describing motion of the perfect superfluid medium, plus the continuity equation, can lead to the relativistic quantum equation, the Klein-Gordon equation, as soon as the quantum potential is found. Just like in the non-relativistic limit with the Schrödinger equation, the problem of determining the quantum potential goes back to Fick's laws, in this case written down in the relativistic sector.

A main essence of the laws of Sbitnev's approach is that they describe currents induced by gradients of the pressures arising in a physical vacuum acting as a special superfluid medium. Just like in the non-relativistic domain, the quantum potential describes influence through differences of the pressures that arise between ensembles of virtual particles populating the vacuum.

By following Sbitnev's treatment in [70–73], Sbitnev's relativistic hydrodynamics starts from a description of the energy-momentum tensor

$$T^{\mu\nu} = (\varepsilon + p)u^{\mu}u^{\nu} + p\eta^{\mu\nu} \tag{1.155}$$

where ε and p are functions per unit volume expressed in units of pressure and the metric tensor $\eta^{\mu\nu}$ has the spacelike signature $(-,+,+,+)$. As regards the momentum tensor (1.155), one can write the conservation law in the following way:

$$\partial_{\mu}(T^{\mu\nu}/\rho) = 0 \tag{1.156}$$

where ρ is the density distribution of the virtual particles constituting the vacuum medium. The pressure p is generated by the collisions of the virtual particles of the medium and can be expressed as

$$p = p_1 + p_2 \tag{1.157}$$

where

$$p_1 = -\frac{\hbar^2}{4m}\left[\nabla^2\rho - \frac{1}{c^2}\frac{\partial^2}{\partial t^2}\rho\right] \tag{1.158}$$

derives from Fick's law and

$$p_2 = \frac{\hbar^2}{8m\rho}\left[(\nabla\rho)^2 - \frac{1}{c^2}\left(\frac{\partial}{\partial t}\rho\right)^2\right]. \tag{1.159}$$

Equations (1.157)–(1.159) lead to define the quantum potential:

$$Q = \frac{p_1 + p_2}{\rho} = -\frac{\hbar^2}{4m\rho}\left[\nabla^2\rho - \frac{1}{c^2}\frac{\partial^2}{\partial t^2}\rho\right] + \frac{\hbar^2}{8m\rho^2}\left[(\nabla\rho)^2 - \frac{1}{c^2}\left(\frac{\partial}{\partial t}\rho\right)^2\right] = -\frac{\hbar^2}{2m}\frac{\partial_{\mu}\partial^{\mu}R}{R} \tag{1.160}$$

where R is the square ratio of the density distribution ρ of the virtual particles in the vacuum. On the basis of equation (1.160), the quantum potential describes the influence through the pressures that arise between ensembles of virtual particles populating the vacuum.

Now, on the basis of equations (1.156) and (1.160) one has

$$\left(\frac{\varepsilon + p}{p}\gamma\right)\partial^{\nu} u_{\nu} u^{\nu} - \partial^{\nu}\left(\frac{\varepsilon + p}{p}\gamma\right) + \partial^{\nu} Q = 0, \tag{1.161}$$

where

$$\gamma = \left(1 - \frac{v^2}{c^2}\right)^{-1/2}. \tag{1.162}$$

The solution of (1.161) is

$$mv^2 - mc^2 + 2Q = 2C. \tag{1.163}$$

Here, $v_{\mu} = u_{\mu} c$ and C is an integration constant having the dimension of energy. Since the superfluid medium is irrotational, one can express the velocity v_{μ} by introducing the scalar field $S + \hbar\omega\gamma^{-1} t$, where S characterizes a degree of mobility of the virtual particles in the vicinity of the 4-point $(t, \vec{r})$, on the basis of relation

$$v^2 = \frac{1}{m^2}\partial_{\mu} S \partial^{\mu} S + \frac{2E}{\gamma m^2 c}\partial_0 S + \frac{E^2}{\gamma^2 m^2 c^2}. \tag{1.164}$$

Hence, multiplying equation (1.163) by m, one obtains

$$\partial_{\mu} S \partial^{\mu} S + \frac{2E}{c\gamma}\partial_0 S + \frac{E^2}{c^2\gamma^2} - m^2 c^2 - \hbar^2 \frac{\partial_{\mu}\partial^{\mu} R}{R} = 2mC. \tag{1.165}$$

In the relativistic domain one has $v \to c$, $\gamma \to \infty$ and thus $\frac{2E}{c\gamma}\partial_0 S + \frac{E^2}{c^2\gamma^2} \to 0$; in this regime, equation (1.165) becomes therefore

$$\partial_{\mu} S \partial^{\mu} S - m^2 c^2 - \hbar^2 \frac{\partial_{\mu}\partial^{\mu} R}{R} = 2mC \tag{1.166}$$

which is a quantum Hamilton-Jacobi equation extracted, together with the continuity equation

$$\partial_{\mu}(\rho \partial^{\mu} S) = 0 \tag{1.167}$$

from the Klein-Gordon equation

$$\partial_{\mu}\partial^{\mu}\psi + \frac{m^2 c^2}{\hbar^2}\psi + 2\frac{m}{\hbar^2}C\psi = 0 \tag{1.168}$$

if one substitutes in this equation the wavefunction ψ represented in polar form

$$\psi = R\exp(iS/\hbar) \tag{1.169}$$

and separates the solutions into the real and imaginary parts.

In summary, on the basis of equations (1.157)–(1.169), in Sbitnev's approach the introduction of the concept of a physical vacuum acting as a superfluid medium, and consisting of an enormous amount of virtual pairs of particles-antiparticles and which is defined by the energy-momentum tensor (1.155), the quantum potential describes the influence through the pressures that arise between ensembles of virtual particles populating the vacuum and the standard Klein-Gordon equation emerges as a special case, in the relativistic domain, of a more general energy conservation law (1.164) of the physical vacuum. An appropriate Hamilton-Jacobi equation, associated with the Klein-Gordon equation, can be derived as a result of the pressure due to the collisions of the virtual particles of the medium, which is linked with the fluctuations of the quantum vacuum energy density.

1.4.2 *About a Bohmian approach to the Dirac relativistic quantum mechanics*

The Dirac equation, which describes the behaviour of spin 1/2 particles in a relativistic regime, in the case of a free particle assumes the form

$$i\hbar\gamma^{\mu}\partial_{\mu}\psi = mc\psi \tag{1.170}$$

where m is the mass of the particle and the matrices γ_{μ} are defined in terms of the 2×2 Pauli matrices as follows:

$$\gamma_i = \begin{pmatrix} 0 & \sigma_i \\ -\sigma_i & 0 \end{pmatrix}; i = 1,2,3 \text{ where } \sigma_1 = \begin{pmatrix} 0 & 1 \\ 1 & 0 \end{pmatrix}, \quad \sigma_1 = \begin{pmatrix} 0 & -i \\ i & 0 \end{pmatrix},$$

$$\sigma_1 = \begin{pmatrix} 1 & 0 \\ 0 & -1 \end{pmatrix} \quad \text{and} \quad \gamma_4 = i\beta,\ \beta = \begin{pmatrix} 1 & 0 & 0 & 0 \\ 0 & 1 & 0 & 0 \\ 0 & 0 & -1 & 0 \\ 0 & 0 & 0 & -1 \end{pmatrix}.$$

Dirac developed equation (1.170) on the basis of the requirement that in a relativistic treatment time and spatial coordinates must be treated in a symmetrical way and thus the wave equation must be of first order with respect both to time and to space, and furthermore the solutions must be compatible with Klein-Gordon equation, which is directly obtained by the relativistic expression of the energy.

As regards the construction of a Bohmian model of Dirac's relativistic quantum mechanics, several attempts have been restricted to the use of the expression for the Dirac current in order to get trajectories for the Dirac electron. For example, this approach has been used by Bohm and Hiley [69, 129] and Holland [32], who discussed some applications. Gull, Lasenby and Doran [130] also used the Dirac current to calculate trajectories in their investigation of quantum tunnelling in the relativistic domain. However, in none of these cases, a quantum Hamilton-Jacobi equation for the conservation of energy had been derived, and consequently no relativistic expression for the quantum potential was obtained.

As regards Dirac's relativistic quantum mechanics, in this chapter we want to focus our attention, on one hand, on the de Broglie-Bohm like model for the Dirac equation developed by Chavoya-Aceves in the paper [131] and the Bohmian models of Dirac's particles developed by Nikolic [132–134] and Hernández-Zapata [135], and, on the other hand, on a fully relativistic description in a Bohmian picture of a Dirac particle based on Clifford algebras [136].

By following Hernández-Zapata [135], from a Bohmian point of view, the Dirac equation (1.170) can be supplemented with the following Bohmian-like equation that describes how the velocity of the particle is determined by the spinor $\psi = \begin{pmatrix} \psi_1 \\ \psi_2 \\ \psi_3 \\ \psi_4 \end{pmatrix}$:

$$\frac{dX_\mu}{d\sigma} = \frac{c\bar{\psi}\gamma_\mu\psi}{|\bar{\psi}\psi|} \tag{1.171}$$

where the right-hand side of the equation is evaluated on the particle configurational coordinates $\vec{X}$, icT, $d\sigma$ is a parameter with time units — introduced by Nikolic in [61] — used to describe the dynamics (and here, of course, $\bar{\psi} = \psi^+ \beta$, where $\psi^+ = (\psi_1^* \, \psi_2^* \, \psi_3^* \, \psi_4^*)$). By decomposing equation (1.171) on spatial and temporal parts one obtains

$$\frac{d\vec{X}}{dT} = \frac{c\psi^+ \vec{\alpha}\psi}{\psi^+\psi}. \tag{1.172}$$

By considering the right-hand side of equation (1.172) as a function of the configurational coordinates $\vec{X}$ and time T the wave equation deriving from Bohm's version of the Dirac equation can be written as

$$i\hbar \frac{\partial \psi}{\partial T} = (c\vec{\alpha} \cdot \vec{p} + mc^2 \beta)\psi \tag{1.173}$$

while the continuity equation is

$$\frac{\partial}{\partial x_i}(c\psi^+ \alpha_i \psi) + \frac{\partial}{\partial T}(\psi^+\psi) = 0. \tag{1.174}$$

From equations (1.173) and (1.174) it follows that if at time $T = 0$ the particle is distributed with probability $\psi^+\psi$, once normalized the function in space, this will remain so for every T. This property defines a "quantum equilibrium" that is valid only in a preferential frame of reference but not in any frame of reference.

Now, in Nikolic's papers [132–134] it was shown that a general solution of the Dirac equation (1.170) may be written as

$$\psi(x) = \psi^{(P)}(x) + \psi^{(A)}(x) \tag{1.175}$$

where $\psi^{(P)}(x)$ represents the wavefunction associated with particles of spin 1/2 and $\psi^{(A)}(x)$ represents the wavefunction associated with the corresponding antiparticles (and we have denoted $x = (t, \vec{x})$). These two sets of wavefunctions can be expanded as

$$\psi^{(P)}(x) = \sum_k b_k u_k(x), \tag{1.176}$$

$$\psi^{(A)}(x) = \sum_k d_k^* v_k(x) \tag{1.177}$$

respectively, where u_k are positive frequency 4-spinors while v_k are negative frequency 4-spinors. In equations (1.176) and (1.177), the label k is an abbreviation for the set $(\vec{k}, s)$ where $\vec{k}$ is the

3-momentum $\vec{k} = (p_1, p_2, p_3)$ and $s = \pm\frac{1}{2}\hbar$ is the spin label. Writing $\Omega^P(x, x') = \sum u_k(x)u_k^+(x')$ and $\Omega^A(x, x') = \sum v_k(x)v_k^+(x')$, one has

$$\psi^P(x) = \int d^3x'\Omega^P(x, x')\psi(x') \tag{1.178}$$

and

$$\psi^A(x) = \int d^3x'\Omega^A(x, x')\psi(x'). \tag{1.179}$$

By introducing the expression (1.175) for the spinor into equations (1.173) and (1.174), these two latest equations respectively become

$$i\hbar\frac{\partial(\psi^{(P)} + \psi^{(A)})}{\partial T} = (c\vec{\alpha}\cdot\vec{p} + mc^2\beta)(\psi^{(P)} + \psi^{(A)}) \tag{1.180}$$

and

$$\frac{\partial}{\partial x_i}\left[c(\psi^{(P)} + \psi^{(A)})^+\alpha_i(\psi^{(P)} + \psi^{(A)})\right] + \frac{\partial}{\partial T}\left[(\psi^{(P)} + \psi^{(A)})^+(\psi^{(P)} + \psi^{(A)})\right] = 0. \tag{1.181}$$

Now, by following the treatment of Chavoya-Aceves [131], we express the components of the two sets of 4-spinors $\psi^{(P)}(x)$ and $\psi^{(A)}(x)$ in polar form, in terms of their amplitudes and phases, as:

$$\psi_1^{(P)} = \sqrt{\rho_1^{(P)}}e^{\frac{i}{\hbar}S_1^{(P)}} \tag{1.182}$$

$$\psi_2^{(P)} = \sqrt{\rho_2^{(P)}}e^{\frac{i}{\hbar}S_2^{(P)}} \tag{1.183}$$

$$\psi_3^{(P)} = \sqrt{\rho_3^{(P)}}e^{-\frac{i}{\hbar}S_3^{(P)}} \tag{1.184}$$

$$\psi_4^{(P)} = \sqrt{\rho_4^{(P)}}e^{-\frac{i}{\hbar}S_4^{(P)}} \tag{1.185}$$

and

$$\psi_1^{(A)} = \sqrt{\rho_1^{(A)}}e^{\frac{i}{\hbar}S_1^{(A)}} \tag{1.186}$$

$$\psi_2^{(A)} = \sqrt{\rho_2^{(A)}}e^{\frac{i}{\hbar}S_2^{(A)}} \tag{1.187}$$

$$\psi_3^{(A)} = \sqrt{\rho_3^{(A)}}e^{-\frac{i}{\hbar}S_3^{(A)}} \tag{1.188}$$

$$\psi_4^{(A)} = \sqrt{\rho_4^{(A)}}e^{-\frac{i}{\hbar}S_4^{(A)}}. \tag{1.189}$$

Taking account of relations (1.182)–(1.189), the wave equation (1.180) and the continuity equation (1.181) read respectively as:

$$i\hbar\frac{\partial}{\partial T}\begin{pmatrix} \sqrt{\rho_1^{(P)}}e^{\frac{i}{\hbar}S_1^{(P)}} + \sqrt{\rho_1^{(A)}}e^{\frac{i}{\hbar}S_1^{(A)}} \\ \sqrt{\rho_2^{(P)}}e^{\frac{i}{\hbar}S_2^{(P)}} + \sqrt{\rho_2^{(A)}}e^{\frac{i}{\hbar}S_2^{(A)}} \\ \sqrt{\rho_3^{(P)}}e^{\frac{i}{\hbar}S_3^{(P)}} + \sqrt{\rho_3^{(A)}}e^{\frac{i}{\hbar}S_3^{(A)}} \\ \sqrt{\rho_4^{(P)}}e^{\frac{i}{\hbar}S_4^{(P)}} + \sqrt{\rho_4^{(A)}}e^{\frac{i}{\hbar}S_4^{(A)}} \end{pmatrix}$$

$$= (c\vec{\alpha}\cdot\vec{p} + mc^2\beta)\begin{pmatrix} \sqrt{\rho_1^{(P)}}e^{\frac{i}{\hbar}S_1^{(P)}} + \sqrt{\rho_1^{(A)}}e^{\frac{i}{\hbar}S_1^{(A)}} \\ \sqrt{\rho_2^{(P)}}e^{\frac{i}{\hbar}S_2^{(P)}} + \sqrt{\rho_2^{(A)}}e^{\frac{i}{\hbar}S_2^{(A)}} \\ \sqrt{\rho_3^{(P)}}e^{\frac{i}{\hbar}S_3^{(P)}} + \sqrt{\rho_3^{(A)}}e^{\frac{i}{\hbar}S_3^{(A)}} \\ \sqrt{\rho_4^{(P)}}e^{\frac{i}{\hbar}S_4^{(P)}} + \sqrt{\rho_4^{(A)}}e^{\frac{i}{\hbar}S_4^{(A)}} \end{pmatrix} \tag{1.190}$$

and

$$\frac{\partial}{\partial x_i}\left[c\begin{pmatrix} \sqrt{\rho_1^{(P)}}e^{\frac{i}{\hbar}S_1^{(P)}} + \sqrt{\rho_1^{(A)}}e^{\frac{i}{\hbar}S_1^{(A)}} \\ \sqrt{\rho_2^{(P)}}e^{\frac{i}{\hbar}S_2^{(P)}} + \sqrt{\rho_2^{(A)}}e^{\frac{i}{\hbar}S_2^{(A)}} \\ \sqrt{\rho_3^{(P)}}e^{\frac{i}{\hbar}S_3^{(P)}} + \sqrt{\rho_3^{(A)}}e^{\frac{i}{\hbar}S_3^{(A)}} \\ \sqrt{\rho_4^{(P)}}e^{\frac{i}{\hbar}S_4^{(P)}} + \sqrt{\rho_4^{(A)}}e^{\frac{i}{\hbar}S_4^{(A)}} \end{pmatrix}^+\right.$$

$$\times\, \alpha_i \begin{pmatrix} \sqrt{\rho_1^{(P)}}e^{\frac{i}{\hbar}S_1^{(P)}} + \sqrt{\rho_1^{(A)}}e^{\frac{i}{\hbar}S_1^{(A)}} \\ \sqrt{\rho_2^{(P)}}e^{\frac{i}{\hbar}S_2^{(P)}} + \sqrt{\rho_2^{(A)}}e^{\frac{i}{\hbar}S_2^{(A)}} \\ \sqrt{\rho_3^{(P)}}e^{\frac{i}{\hbar}S_3^{(P)}} + \sqrt{\rho_3^{(A)}}e^{\frac{i}{\hbar}S_3^{(A)}} \\ \sqrt{\rho_4^{(P)}}e^{\frac{i}{\hbar}S_4^{(P)}} + \sqrt{\rho_4^{(A)}}e^{\frac{i}{\hbar}S_4^{(A)}} \end{pmatrix} \Bigg]$$

$$+ \frac{\partial}{\partial T}\left[\begin{pmatrix} \sqrt{\rho_1^{(P)}}e^{\frac{i}{\hbar}S_1^{(P)}} + \sqrt{\rho_1^{(A)}}e^{\frac{i}{\hbar}S_1^{(A)}} \\ \sqrt{\rho_2^{(P)}}e^{\frac{i}{\hbar}S_2^{(P)}} + \sqrt{\rho_2^{(A)}}e^{\frac{i}{\hbar}S_2^{(A)}} \\ \sqrt{\rho_3^{(P)}}e^{\frac{i}{\hbar}S_3^{(P)}} + \sqrt{\rho_3^{(A)}}e^{\frac{i}{\hbar}S_3^{(A)}} \\ \sqrt{\rho_4^{(P)}}e^{\frac{i}{\hbar}S_4^{(P)}} + \sqrt{\rho_4^{(A)}}e^{\frac{i}{\hbar}S_4^{(A)}} \end{pmatrix}^+ \right.$$

$$\left. \times \begin{pmatrix} \sqrt{\rho_1^{(P)}}e^{\frac{i}{\hbar}S_1^{(P)}} + \sqrt{\rho_1^{(A)}}e^{\frac{i}{\hbar}S_1^{(A)}} \\ \sqrt{\rho_2^{(P)}}e^{\frac{i}{\hbar}S_2^{(P)}} + \sqrt{\rho_2^{(A)}}e^{\frac{i}{\hbar}S_2^{(A)}} \\ \sqrt{\rho_3^{(P)}}e^{\frac{i}{\hbar}S_3^{(P)}} + \sqrt{\rho_3^{(A)}}e^{\frac{i}{\hbar}S_3^{(A)}} \\ \sqrt{\rho_4^{(P)}}e^{\frac{i}{\hbar}S_4^{(P)}} + \sqrt{\rho_4^{(A)}}e^{\frac{i}{\hbar}S_4^{(A)}} \end{pmatrix}\right] = 0 \tag{1.191}$$

Moreover, always following Chavoya-Aceves [131], the four-velocities of the Dirac particle can be expressed by equations

$$mv_{i,\mu} = -\partial_\mu S_i - \partial_\mu \Phi_i \tag{1.192}$$

where the functions Φ_i are hidden variables. In other words, the four-velocities are explicitly:

$$mv_1^\mu = -\frac{\partial}{\partial X_\mu}(S_1^{(P)} + S_1^{(A)} + \Phi_1) \tag{1.193}$$

$$mv_2^\mu = -\frac{\partial}{\partial X_\mu}(S_2^{(P)} + S_2^{(A)} + \Phi_2) \tag{1.194}$$

$$mv_3^\mu = -\frac{\partial}{\partial X_\mu}(S_3^{(P)} + S_3^{(A)} + \Phi_3) \tag{1.195}$$

$$mv_4^\mu = -\frac{\partial}{\partial X_\mu}(S_4^{(P)} + S_4^{(A)} + \Phi_4). \tag{1.196}$$

where the hidden variables have opportune values in such a way that

$$v_i^\mu v_{i,\mu} = c^2. \tag{1.197}$$

Equations (1.193)–(1.196) mean that, in a Bohmian approach, the velocity of the Dirac relativistic particle satisfying the equation of motion (1.173) is associated with the phases of the different components of the 4-spinors $\psi^{(P)}(x)$ and $\psi^{(A)}(x)$ as well as the hidden variables. On the basis of the four-velocities (1.193)–(1.196), by indicating $(S_i^{(P)} + S_i^{(A)}) = S_i$ one obtains

$$\begin{aligned}\partial^\mu S_i \partial_\mu S_i &= m^2(v_i^\mu + \partial^\mu \Phi_i)(v_{i,\mu} + \partial_\mu \Phi_i)\\ &= m^2c^2 + 2mv_i^\mu \partial_\mu \Phi_i + \partial^\mu \Phi_i \partial_\mu \Phi_i\end{aligned} \tag{1.198}$$

and thus the Dirac equation (3.53) leads to the two following equations

$$\begin{aligned}\sum_{i=1,2} \rho_i \left(v_i^\mu \partial_\mu \Phi_i + \frac{\partial^\mu \Phi_i \partial_\mu \Phi_i}{2m} \right) &= \sum_{i=1,2} \frac{\hbar^2}{2m} (\sqrt{\rho_i} \partial^\mu \partial_\mu \sqrt{\rho_i})\\ &\quad + \frac{\hbar q}{2mc} \begin{pmatrix} \psi_1^{(P)} + \psi_1^{(A)} \\ \psi_2^{(P)} + \psi_2^{(A)} \end{pmatrix}^+\\ &\quad \times \vec{\sigma} \cdot \begin{pmatrix} \psi_1^{(P)} + \psi_1^{(A)} \\ \psi_2^{(P)} + \psi_2^{(A)} \end{pmatrix}\end{aligned} \tag{1.199}$$

and

$$\sum_{i=3,4} \rho_i \left(v_i^{\mu} \partial_\mu \Phi_i + \frac{\partial^\mu \Phi_i \partial_\mu \Phi_i}{2m} \right) = \sum_{i=3,4} \frac{\hbar^2}{2m} (\sqrt{\rho_i} \partial^\mu \partial_\mu \sqrt{\rho_i}) + \frac{\hbar q}{2mc} \begin{pmatrix} \psi_3^{(P)} + \psi_3^{(A)} \\ \psi_4^{(P)} + \psi_4^{(A)} \end{pmatrix}^+ \times \vec{\sigma} \cdot \begin{pmatrix} \psi_3^{(P)} + \psi_3^{(A)} \\ \psi_4^{(P)} + \psi_2^{(A)} \end{pmatrix} \tag{1.200}$$

which can be considered as the counterparts of equations (1.190) and (1.191), expressed in terms of the velocity of the particle. Moreover, the first term on the right side of equations (1.199) and (1.200) can be identified with the quantum potential associated with the first two components and the second two components of the spinors respectively:

$$Q_{1,2} = \sum_{i=1,2} \frac{\hbar^2}{2m} (\sqrt{\rho_i} \partial^\mu \partial_\mu \sqrt{\rho_i}), \tag{1.201}$$

$$Q_{3,4} = \sum_{i=3,4} \frac{\hbar^2}{2m} (\sqrt{\rho_i} \partial^\mu \partial_\mu \sqrt{\rho_i}). \tag{1.202}$$

According to equations (1.201) and (1.202), in a Bohmian approach to Dirac's equation, the quantum potential emerges as a geometric information channel into the behaviour of the particle which is determined by the density distribution corresponding to the different components of the 4-spinors. Starting from equations (1.201) and (1.202), by following the philosophy at the basis of Novello's, Salim's and Falciano's approach analysed in chapter 1.3.3, one can also introduce two appropriate quantum lengths expressing the geometrical properties of the background corresponding to the first two components and the second two components of the spinor respectively:

$$L_{quantum(1,2)} = \frac{1}{\sqrt{\sum_{i=1,2}\left(-\frac{\hbar^2}{2m}\left(\sqrt{\rho_i}\partial^\mu\partial_\mu\sqrt{\rho_i}\right)\right)}} \quad (1.203)$$

$$L_{quantum(3,4)} = \frac{1}{\sqrt{\sum_{i=3,4}\left(-\frac{\hbar^2}{2m}(\sqrt{\rho_i}\partial^\mu\partial_\mu\sqrt{\rho_i})\right)}}. \quad (1.204)$$

The quantum lengths (1.203) and (1.204) can be used to evaluate the modification of the geometry characterizing a relativistic Dirac particle with respect to the Euclidean geometry of classical physics. Once the quantum lengths (1.203) and (1.204) become non-negligible the Dirac particle into consideration goes into a quantum relativistic regime.

On the other hand, recently, Hiley and Callaghan developed a fully relativistic description in a Bohmian picture of a Dirac particle by utilizing an approach based on Clifford algebras [136]. In Hiley's and Callaghan's model, it is possible to derive expressions for the Bohm energy-momentum density, a relativistic quantum Hamilton-Jacobi equation for the conservation of energy which contains an expression for the quantum potential and a relativistic time development equation for the spin vectors of the particle. However, the mathematical formalism of this model appears a little inconvenient and cumbersome with respect to the usual Dirac's formalism of 4-spinors. Hiley's and Callaghan's approach to Dirac's theory of electron based on the using of Clifford algebras tries somewhat to resurrect an original approach already developed in 1930 by prominent physicist and mathematician Fritz Sauter [137] (which indeed did not survive because of its inconvenience and obscure meaning).

By following [136], in order to construct a Dirac theory in a Bohmian picture, Hiley and Callaghan considered the Clifford algebra $C_{1,3}$ which is the algebra generated by $\{1, \gamma_\mu\}$ where $\mu = 0, 1, 2, 3$ and $[\gamma_\mu, \gamma_\nu] = 2g_{\mu\nu}$. In this approach the Clifford algebra is assumed to be as fundamental and a space-time manifold emerges by applying the mapping $\eta : \gamma_\mu \to \hat{e}_\mu$ where $\hat{e}_\mu$ is a set of orthonormal unit vectors in a vector space $V_{1,3}$, the Minkowski space-time for an equivalent class of Lorentz observers. In this way, in the context of this approach, Hiley and Callaghan obtained an energy conservation equation of the form

$$(\partial^\mu \partial_\mu \Phi_L)\Phi_R + \Phi_L(\partial^\mu \partial_\mu \Phi_R) + 2m^2 \Phi_L \Phi_R = 0 \tag{1.205}$$

and the following equation for the time evolution of the spin and its components

$$\Phi_L(\partial^\mu \partial_\mu \Phi_R) + (\partial^\mu \partial_\mu \Phi_L)\Phi_R = 0. \tag{1.206}$$

In equations (1.205) and (1.206), Φ_R and Φ_L are two entities of $C_{1,3}$ respectively called minimal left element and minimal right element and are linked by the Clifford density element

$$\rho_C = \Phi_L \Phi_R = \phi_L \varepsilon_\gamma \phi_R, \tag{1.207}$$

Φ_R is the conjugate to Φ_L,

$$\Phi_L = 2\mathrm{Re} \times \left[\frac{(a - ib)\varepsilon_\gamma + (c - id)\gamma_{23}\varepsilon_\gamma + (h - in)\gamma_{30}\varepsilon_\gamma + (f + ig)\gamma_{01}\varepsilon_\gamma}{4} \right], \tag{1.208}$$

where

$$\varepsilon_\gamma = \frac{(1 + \gamma_0)(1 + i\gamma_{12})}{4}, \tag{1.209}$$

while

$$\phi_L = a + b\gamma_{12} + c\gamma_{23} + d\gamma_{13} + f\gamma_{01} + g\gamma_{02} + h\gamma_{03} + n\gamma_5 \tag{1.210}$$

is the even part of Φ_L, ϕ_R is the even part of Φ_R (ϕ_R is the conjugate of ϕ_L) and a, b, c, d, f, g, h, n are eight real functions that can be used to specify the quantum state of the Dirac particle. The minimal left element (1.208) contains all the information that is normally included in the wavefunction which uses 4 complex numbers. The Clifford density element (1.207) corresponds to $\bar{\psi}\psi$ of the standard Hilbert approach (where $\bar{\psi}$ is the adjoint wavefunction, ψ is the usual wavefunction at four components which are related to (1.208) via relations

$$\psi_1 = a - ib, \quad \psi_2 = -d - ic, \quad \psi_3 = h - in, \quad \psi_4 = f + ig. \tag{1.211}$$

The Clifford density element (1.207) plays a central role inside the Clifford algebra in the sense that it contains all the information needed to describe the state of a system completely.

Here, we concentrate our attention on the energy conservation equation (1.205). By considering the variables

$$2\rho P^{\mu} = [(\partial^{\mu}\phi_L)\gamma_{012}\phi_R - \phi_L\gamma_{012}(\partial^{\mu}\phi_R)] \tag{1.212}$$

and

$$2\rho W^{\mu} = -\partial^{\mu}(\phi_L\gamma_{012}\phi_R) \tag{1.213}$$

which characterize the geometry of the Clifford background of the Dirac particle under consideration and lead to the following relations

$$-\partial^{\mu}\phi_L = [P^{\mu} - W^{\mu}]\phi_L \tag{1.214}$$

$$\partial^{\mu}\phi_R = \gamma_{012}\phi_R[P^{\mu} + W^{\mu}], \tag{1.215}$$

after the application of some algebra the energy conservation equation (1.215) becomes

$$P^2 + W^2 + [J\partial_{\mu}P^{\mu} - \partial_{\mu}P^{\mu}J] + [J\partial_{\mu}W^{\mu} + \partial_{\mu}W^{\mu}J] - m^2 = 0. \tag{1.216}$$

Equation (1.216) can be further simplified by separating it into its Clifford scalar and pseudoscalar parts. The pseudoscalar part of equation (1.216) is

$$[J\partial_{\mu}P^{\mu} - \partial_{\mu}P^{\mu}J] = 0. \tag{1.217}$$

Equation (1.217) provides a constraint on the relation between the spin and the momentum of the particle.

The scalar part of equation (1.216) is

$$P^2 + W^2 + [J\partial_\mu W^\mu + \partial_\mu W^\mu J] - m^2 = 0. \tag{1.218}$$

Equation (1.218) can be compared with the energy equation

$$p_\mu p^\mu - m^2 = 0 \tag{1.219}$$

corresponding with the standard Dirac equation, if one expresses the momentum (1.212) in terms of the Bohm energy-momentum vector:

$$2\rho P_B^\mu = tr[\gamma^0(\phi_L \overleftrightarrow{\partial}_\mu \gamma_{012}\phi_R)]. \tag{1.220}$$

In order to achieve this, the following relations can be used

$$4\rho^2 P^2 = \sum_{i=0}^{3} A_{i\nu}A_i^\nu, \tag{1.221}$$

$$4\rho^2 P^2 = 4\rho^2 P_B^2 + \sum_{i=1}^{3} A_{i\nu}A_i^\nu \tag{1.222}$$

$$4\rho^2 \Pi^2 = \sum_{i=1}^{3} A_{i\nu}A_i^\nu, \tag{1.223}$$

where ρ is defined as

$$\rho(F_{\mu 4}) = \gamma_\mu \tag{1.224}$$

with $\mu = 0, 1, 2, 3$

$$\rho(F_5) = i, \tag{1.225}$$

the A_i are given by

$$\begin{aligned}
A_0^\nu &= -(a\overleftrightarrow{\partial}^\nu b + c\overleftrightarrow{\partial}^\nu d + f\overleftrightarrow{\partial}^\nu g + h\overleftrightarrow{\partial}^\nu n)\\
A_1^\nu &= -(a\overleftrightarrow{\partial}^\nu g + b\overleftrightarrow{\partial}^\nu f + c\overleftrightarrow{\partial}^\nu h + d\overleftrightarrow{\partial}^\nu n)\\
A_1^\nu &= (a\overleftrightarrow{\partial}^\nu f - b\overleftrightarrow{\partial}^\nu g - c\overleftrightarrow{\partial}^\nu n + d\overleftrightarrow{\partial}^\nu h)\\
A_1^\nu &= (a\overleftrightarrow{\partial}^\nu n - b\overleftrightarrow{\partial}^\nu h + c\overleftrightarrow{\partial}^\nu f - d\overleftrightarrow{\partial}^\nu g)
\end{aligned} \tag{1.226}$$

and $\overleftrightarrow{\partial}^{\nu}$ is the operator defining those entities called by Takabayasi [138] the bilinear invariants of the second kind:

$$\psi\overleftrightarrow{\partial}\bar{\psi} = (\partial\psi)\bar{\psi} - \psi(\partial\bar{\psi}) \tag{1.227}$$

and

$$J = \phi_L \gamma^0 \phi_R \tag{1.228}$$

is the axial current.

In this way, equation (1.218) may be formulated as follows

$$P_B^2 + \Pi^2 + W^2 + [J\partial_\mu W^\mu + \partial_\mu W^\mu J] - m^2 = 0. \tag{1.229}$$

Equation (1.229) is the quantum Hamilton-Jacobi equation which regards Dirac's relativistic quantum mechanics inside Hiley's and Callaghan's model and thus can be considered as the fundamental equation of this approach. From equation (1.229) the quantum potential for Dirac's relativistic quantum mechanics can be defined as the quantity

$$Q_D = \Pi^2 + W^2 + [J\partial_\mu W^\mu + \partial_\mu W^\mu J]. \tag{1.230}$$

In the light of the treatment provided by Hiley and Callaghan in [136], the quantum potential is a physical reality of Dirac's relativistic quantum mechanics which is drawn directly from the Clifford algebra $C_{1,3}$ which can be considered as the fundamental arena of quantum processes for Dirac's particles. In this model of Dirac's relativistic quantum mechanics, the Clifford arena $C_{1,3}$ described by the fundamental equations (1.205) and (1.206) can be interpreted as the foreground, the implicate order of quantum processes and the quantum potential may be obtained directly from this arena. This means that the quantum potential (1.230) characterizing Dirac's relativistic quantum mechanics cannot be considered ad hoc but is entirely within the structure of the quantum world and derives from the algebra of process of the Clifford arena $C_{1,3}$ which can be considered as the ultimate reality of Dirac's particles in a relativistic regime. In other words, in this approach the quantum potential and thus its geometrodynamical action can be considered as emerging structures which derive from something more primitive, from a more fundamental background — characterized by a non-commutative algebra — namely the Clifford arena $C_{1,3}$ which underlies the quantum processes.

Finally, in analogy to the other models of Klein-Gordon and Dirac relativistic quantum mechanics, even in Hiley's and Callaghan's approach of Dirac relativistic quantum mechanics based on the quantum potential (1.230), one can introduce an appropriate quantum length expressing the geometrical properties of the background associated with the quantum potential (1.230) in a Clifford arena:

$$L_{quantum(1,2)} = \frac{1}{\sqrt{\Pi^2 + W^2 + [J\partial_\mu W^\mu + \partial_\mu W^\mu J]}} \tag{1.231}$$

The quantum length (1.231) can be used to evaluate the modification of the geometrical properties characterizing a relativistic Dirac particle with respect to the Euclidean geometry of classical physics in the picture of the Clifford algebra $C_{1,3}$.

1.5 The Quantum Potential in Relativistic Quantum Field Theory

Relativistic quantum mechanics in its first quantization formulation, summarized in the Klein-Gordon and Dirac equations, presents some unsatisfactory features. In particular, Klein-Gordon and Dirac equation turn out to possess energy eigenvalues that extends towards $-\infty$ without a lower limit or a definition of a ground state. Such inconsistencies emerge since relativistic wavefunctions have a probabilistic interpretation in position space, but probability conservation is not a relativistically covariant concept. In quantum field theory, instead, position is not an observable, and therefore one does not need the concept of a position space probability density [139].

The quantum field theory is the mature daughter of quantum mechanics and the most general syntax we know to describe forces. Quantum field theory occupies a central position in our description of Nature, providing both our best working description of fundamental physical laws, and a fruitful tool for investigating the behavior of complex systems [140]. It replaces the "hard" naïve particle of classical physics and quantum mechanics with a network of interactions. While in quantum mechanics wavefunctions are acted upon by operators, in quantum field theory wavefunctions are operator valued fields that act on the states which are associated with quantum fields.

On the other hand, quantum field theory greatly modifies the traditional vision of "permanent object". In particular, one can remark the following essential points regarding the formal and conceptual structure of quantum field theory:

a) The physical world is described as a discrete net of interaction vertices where some properties (space-time position, four-impulse, spin, etc.) are destroyed or created. From the point of view of quantum field theory, such properties' measurement is all that we know of the physical world from an operational point of view. Any other construction in Physics — like the continuous space-time notion itself or the evolution operators — has the role to causally connect the measured properties and thus can be considered as emergent structures with respect to the network of events;
b) In quantum field theory waves and particles cannot be considered as primary physical realities, but only entangled modes of the quantized field have a primary physical existence and each physical event may be correlated to the others by means of a fundamental vacuum.
c) In quantum field theory Heisenberg uncertainty principle assumes the more direct form $\Delta n \Delta \phi \geq \hbar$, where n is the number of the quanta and ϕ is the phase of the modes and, therefore, does not indicate the limits of measurement between classical variables, but the applicability limit of the continuous space-time concept [see 141, 142];
d) Motion is no more a continuous phenomenon, but a discontinuous process in the space-time coordinates. There are no more "objects" as exclusive bearers of permanent "qualities". Quantum field theory introduces — by the discontinuity and the relativistic invariance — non-locality as a fundamental ingredient.

Quantum field theory constitutes a sort of very general syntax, which developed in agreement to the quantum-relativistic conditions [143]. It is however affected by a set of conceptual and mathematical difficulties which yield physicists to consider the possibility that this kind of grammar — in its standard interpretation — presents strong intrinsic limits. In this regard, Ignazio Licata writes with a deep insight in the preface to *Vision of oneness*: "Despite the centrality and extraordinary fecondity of quantum field theory in describing the deep level of the universe, [...] this theory is considered more a practical instrument than for its enormous theoretical and cultural possibilities. Till now quantum field theory has not become a global view of the physical world as it happened for relativity and quantum mechanics. Such a problem has also affected the formal language, when one talks about 'first' and 'second' quantization and, in general, the concept of identity of a physical object. The first step to make in order to cover such a conceptual missing point is to put in evidence the non-locality in quantum field theory" [141].

In substance, as already evidenced by Licata, the real crucial core is to develop quantum field theories that are naturally non-local. In this regard, the idea of Bohm's quantum potential is surely the simplest and most general way to introduce non-locality, in which non-locality is present *ab initio* and is not an unexpected host [144]. In this chapter, our purpose is to make some considerations, in the light of the existing literature, as regards the geometrodynamic features of the quantum potential in relativistic quantum field theory, the geometry of the quantum processes derived from the quantum potential in relativistic quantum field theory.

1.5.1 *The quantum potential in bosonic quantum field theory*

We start by reviewing the first attempts to construct a Bohm's picture of relativistic bosonic quantum field theory, referring the reader to the original literature of Bohm, Hiley, Bell and Kaloyerou for full details (e.g. [69, 145, 146]). Let us consider Bose-Einstein fields in the Schrödinger representation:

$$\{\phi(\vec{x})\} = \phi_1(\vec{x}), \phi_2(\vec{x}), \phi_3(\vec{x}), \ldots . \tag{1.232}$$

The Schrödinger equation for the wavefunctional $\Psi(\{\phi(\vec{x})\}, t)$ (here we assume $\hbar = c = 1$) is

$$H\Psi(\{\phi(\vec{x})\}, t) = i\frac{\partial}{\partial t}\Psi(\{\phi(\vec{x})\}, t) \tag{1.233}$$

where

$$H = \sum_k \int d^3x \left[-\frac{1}{2}\frac{\delta^2}{\delta\phi_k^2(\vec{x})} + \frac{1}{2}|\nabla\phi_k(\vec{x})|^2 \right] + V(\{\phi(\vec{x})\}) \tag{1.234}$$

is the Hamiltonian in the Schrödinger representation.

On the basis of the procedure followed in non-relativistic quantum mechanics and also in relativistic quantum mechanics, also here one can expresses the wavefunctional $\Psi(\{\phi(\vec{x})\}, t)$ in the polar representation:

$$\Psi(\{\phi(\vec{x})\}, t) = R(\{\phi(\vec{x})\}, t)e^{iS(\{\phi(\vec{x})\}, t)} \tag{1.235}$$

where R and S are two real wavefunctionals. In this way, if one inserts this expression into the Schrödinger equation (1.233), one obtains two coupled partial differential functional equations:

$$\frac{\partial S}{\partial t} + \frac{1}{2}\sum_k \int d^3x \left[-\frac{1}{2}\frac{\delta^2}{\delta\phi_k^2(\vec{x})} + \frac{1}{2}|\nabla\phi_k(\vec{x})|^2 \right] + V + Q = 0, \tag{1.236}$$

$$\frac{\partial}{\partial t}R^2 + \sum_k \int d^3x \frac{\delta}{\delta\phi_k(\vec{x})} J_k(\{\phi(\vec{x})\}, t) = 0 \tag{1.237}$$

where

$$Q(\{\phi(\vec{x})\}, t) = -\frac{1}{2}\sum_k \int d^3x \frac{1}{R}\frac{\delta^2 R(\{\phi(\vec{x})\}, t)}{\delta\phi_k^2(\vec{x})} \tag{1.238}$$

is the quantum potential and

$$J_k(\{\phi(\vec{x})\}, t) = R^2 \frac{\delta S}{\delta\phi_k(\vec{x})} \tag{1.239}$$

is the generalized current density in the field space.

Equation (1.236) is a quantum Hamilton-Jacobi equation expressing an energy conservation law for quantum fields while equation (1.237) may be interpreted as a continuity equation for the generalized current density in the field space. In the Bohmian interpretation of bosonic quantum field theory based on equations (1.235)–(1.239), the quantum fields have a deterministic time evolution given by the quantum Hamilton-Jacobi equation (1.236) — which can be considered as the fundamental equation of motion — and the quantum potential (1.238) emerges as the ultimate entity which expresses the geometrical properties of the background space associated with quantum fields from which the behaviour of the particles derive. Moreover, the statistical predictions of this deterministic interpretation are equivalent to those of the conventional interpretation. In this deterministic approach, all quantum uncertainties are a consequence of the ignorance of the actual initial field configuration $(\{\phi(\vec{x})\}, t_0)$. This interpretation turns out to be consistent with the result of the standard interpretation in virtue of the fact that the continuity equation (1.237) indicates that the statistical distribution $\rho(\{\phi(\vec{x})\}, t)$ of field configurations $\{\phi(\vec{x})\}$ is given by the quantum distribution $\rho = R^2$ at any time t, provided that $\rho = R^2$ is given by R^2 at some initial time t_0. Although the initial distribution is arbitrary in line of principle, a quantum H-theorem, on the basis of the results of Valentini [147] explains why the quantum distribution is the most probable.

The geometrodynamics of the Bohmian interpretation of bosonic quantum field theory developed by Bohm, Hiley, Bell and Kaloyerou can be characterized by defining an opportune quantum length evaluating the strength of quantum effects in this regime, which is given by relation

$$L_{quantum,Bose-Einstein} = \frac{1}{\sqrt{\frac{1}{2}\sum_k \int d^3x \frac{1}{R} \frac{\delta^2 R(\{\phi(\vec{x}),t\})}{\delta\phi_k^2(\vec{x})}}}. \tag{1.240}$$

The quantum length (1.240) can be used to describe the modification of the geometry — introduced in the background space by a system with Bose-Einstein fields in a quantum relativistic regime — with respect to the Euclidean geometry characteristic of classical fields. Once the quantum length (1.240) becomes non-negligible we have a physical system with Bose-Einstein fields in a quantum relativistic regime. And here Heisenberg's uncertainty principle can be seen as a consequence of the fact that we are unable to perform a classical measurement to distances smaller than this quantum length corresponding with Bose-Einstein fields in quantum-relativistic regime: the size of a measurement has to be bigger than the quantum length (1.240) according to equation

$$\Delta L \geq L_{quantum,Bose-Einstein} = \frac{1}{\sqrt{\frac{1}{2}\sum_k \int d^3x \frac{1}{R} \frac{\delta^2 R(\{\phi(\vec{x}),t\})}{\delta\phi_k^2(\vec{x})}}}. \tag{1.241}$$

As regards relativistic bosonic quantum field theory in a Bohmian approach, in the papers [148–151] Nikolic developed a treatment of Bohmian particle trajectories in relativistic bosonic quantum field theory. Following the papers [148–151], in relativistic bosonic quantum field theory the operator $\hat{\phi}(x)$ which represents the counterpart of a real scalar field $\phi(x)$ satisfies, in the Heisenberg picture, the following equation:

$$(\partial_0^2 - \nabla^2 + m^2c^2)\hat{\phi} = J(\hat{\phi}) \tag{1.242}$$

where J is a nonlinear function describing the interaction. In the Schrödinger picture the time evolution is determined via the Schrödinger equation in the form

$$H\left[\phi, -i\frac{\delta}{\delta\phi}\right]\Psi(\phi,t) = i\frac{\partial}{\partial t}\Psi(\phi,t) \tag{1.243}$$

where Ψ is a functional with respect to $\phi(\vec{x})$ and a function of time t. Equation (1.243) admits normalized solutions that can be expanded as $\Psi(\phi,t) = \sum_{-\infty}^{+\infty} \tilde{\Psi}_n(\phi,t)$ where the $\tilde{\Psi}_n$ are unnormalized n-particle wavefunctionals. In Nikolic's approach the field $\phi(\vec{x})$ has a causal evolution determined by equation

$$(\partial_0^2 - \nabla^2 + m^2c^2)\phi(x) = J(\phi(x)) - \left(\frac{\delta Q[\phi, t]}{\delta\phi(\vec{x})}\right)_{\phi(\vec{x})=\phi(x)} \tag{1.244}$$

where

$$Q = -\frac{1}{2|\Psi|}\int d^3x \frac{\delta^2|\Psi|}{\delta\phi^2(\vec{x})} \tag{1.245}$$

is the quantum potential which expresses the geometrical properties of space associated with the quantum field $\phi(\vec{x})$ and which has an active information on the motion of the particles. Moreover, in this approach, the n particles associated with the wavefunctional Ψ have causal trajectories determined by a generalization of equation $\frac{d\vec{x}}{dt} = \frac{\vec{j}(t,\vec{x})}{j_0(t,\vec{x})}$ as

$$\frac{d\vec{x}_{n,j}}{dt} = \left(\frac{\psi_n^*(x^{(n)})\overleftrightarrow{\nabla}_j\psi_n(x^{(n)})}{\psi_n^*(x^{(n)})\overleftrightarrow{\partial}_{t,j}\psi_n(x^{(n)})}\right)_{t_1=\cdots=t_n=t} . \tag{1.246}$$

where $\psi_n(\vec{x}^{(n)}, t) = \langle 0|\hat{\phi}(t, \vec{x}_1)\cdots\hat{\phi}(t, \vec{x}_n)\Psi|\rangle$ is the n-particle wavefunction. These n-particles have well defined trajectories even when the probability (in the conventional interpretation of quantum field theory) of the experimental detection is equal to zero. In Nikolic's approach of a Bohmian interpretation of bosonic quantum field theory, the description of processes is achieved also by introducing a causally evolving "effectivity" parameter $e_n[\phi, t]$ defined as

$$e_n[\phi, t] = \frac{|\tilde{\Psi}_n[\phi, t]|^2}{\sum_{n'}^{\infty}|\tilde{\Psi}_{n'}[\phi, t]|^2}. \tag{1.247}$$

The evolution of this parameter is determined by the evolution of ϕ expressed by equation (1.244) and by the solution $\Psi(\phi, t) = \sum_{-\infty}^{+\infty}\tilde{\Psi}_n(\phi, t)$ of the Schrödinger equation (1.243). This parameter might be interpreted as a probability that there are n particles in the system at time t if the field is equal (but not measured) to be $\phi(\vec{x})$ at that time. Moreover in Bohm's interpretation, $e_n[\phi, t]$ is assumed to be an actual property of the particles guided by the wavefunction ψ_n. This parameter constitutes a nonlocal hidden variable which is ascribed to the particles and it is introduced in order to provide a deterministic description of the creation and destruction of particles. Creation processes occur when the effectivity changes continuously from 0 to 1, destruction processes when the effectivity changes continuously from 1 to 0.

Moreover, in analogous way to the original models of Bohm, Hiley, Bell and Kaloyerou, even as regards Nikolic's approach to bosonic relativistic quantum field theory, the geometrodynamics can be characterized by introducing a quantum length of the form

$$L_{quantum,Bose-Einstein} = \frac{1}{\sqrt{\int d^3x \frac{1}{2|\Psi|} \frac{\delta^2|\Psi|}{\delta\phi^2(\vec{x})}}}. \tag{1.248}$$

The quantum length (1.248) is a parameter that, in Nikolic's approach, measures the modification of the geometrical properties — introduced in the background space by a system characterized by the presence of bosonic fields in a quantum relativistic regime — with respect to the Euclidean geometry characteristic of classical fields. Once the quantum length (1.248) becomes non-negligible we have a system of bosonic fields in a quantum relativistic regime. And here Heisenberg's uncertainty principle can be seen as a consequence of the fact that we are unable to perform a classical measurement to distances smaller than this quantum length corresponding with bosonic fields in quantum-relativistic regime: the size of a measurement has to be bigger than the quantum entropic length (1.248) according to equation

$$\Delta L \geq L_{quantum,Bose-Einstein} = \frac{1}{\sqrt{\int d^3x \frac{1}{2|\Psi|} \frac{\delta^2|\Psi|}{\delta\phi^2(\vec{x})}}}. \tag{1.249}$$

Finally, another interesting research regarding bosonic quantum field theory in a Bohmian framework is represented by the Nikolic Bohmian covariant interpretation for the many-fingered-time Tomonaga-Schwinger equation for quantum field theory, where there is not a preferred foliation, in the sense that here the quantum state is a functional of an arbitrary timelike hypersurface [152]. Let $x = (x^0, \vec{x})$ be spacetime coordinates. Let $\phi(\vec{x})$ be a dynamical field on a timelike Cauchy hypersurface Σ defined via $x^0 = T(\vec{x})$ where $\vec{x}$ are coordinates on Σ. If we denote with $\hat{\mathrm{H}}(\vec{x})$ the Hamiltonian density operator, the dynamics of the field ϕ is described by the Tomonaga-Schwinger equation

$$\hat{\mathrm{H}}\Psi[\phi, t] = i\frac{\delta\Psi[\phi, t]}{\delta T(\vec{x})} \tag{1.250}$$

where $\delta T(\vec{x})$ represents an infinitesimal change of the hypersurface Σ. The quantity $\rho[\phi, T] = |\Psi[\phi, T]|^2$ may be interpreted as the probability density for the field to have a value ϕ on Σ or equivalently as the probability density for the field to have a value ϕ at time T. For simplicity, one considers a scalar field with the corresponding Hamiltonian operator given by relation

$$\hat{\mathrm{H}}(\vec{x}) = -\frac{1}{2}\frac{\delta^2}{\delta\phi^2(\vec{x})} + \frac{1}{2}[(\nabla\phi(\vec{x}))^2 + m^2\phi^2(\vec{x})]. \tag{1.251}$$

If one writes the wavefunctional in polar form $\Psi = R\exp(iS)$ with R and S real functionals, equation (1.250) is splitted into the two following real equations

$$\frac{1}{2}\left(\frac{\delta S}{\delta\phi(\vec{x})}\right)^2 + \frac{1}{2}[(\nabla\phi(\vec{x}))^2 + m^2\phi^2(\vec{x})] + Q[\vec{x}, \phi, T] + \frac{\delta S}{\delta T(\vec{x})} = 0, \tag{1.252}$$

$$\frac{\delta\rho}{\delta T(\vec{x})} + \frac{\delta}{\delta\phi(\vec{x})}(\rho\frac{\delta S}{\delta T(\vec{x})}) = 0 \tag{1.253}$$

where

$$Q[\vec{x}, \phi, T] = -\frac{1}{2R}\frac{\delta^2 R}{\delta\phi^2(\vec{x})} \tag{1.254}$$

is the quantum potential. Here, the Bohmian interpretation lies in considering a deterministic time dependent hidden variable such that the time evolution of this variable is coherent with the probabilistic interpretation of the probability density ρ. In order to achieve this one introduces a many-fingered-time field $\Phi[\vec{x}, T]$ that satisfies the following many-fingered-time Bohmian equation of motion:

$$\frac{\partial \Phi[\vec{x},t]}{\partial T(\vec{x})} = \left(\frac{\delta S}{\delta \phi(\vec{x})}\right)_{\phi=\Phi}. \tag{1.255}$$

By combining equation (1.254) and the quantum many-fingered-time Hamilton-Jacobi equation (1.252), one obtains:

$$\left(\left(\frac{\partial}{\partial T(\vec{x})}\right)^2 - \nabla_x^2 + m^2\right)\Phi[\vec{x},T] = \left(-\frac{\partial Q[\vec{x},\phi,T]}{\partial \phi(\vec{x})}\right)_{\phi=\Phi}. \tag{1.256}$$

Equation (1.256) can be interpreted as a many-fingered-time Klein-Gordon equation modified with a nonlocal quantum term on the right side. Now, a manifestly covariant theory may be obtained by introducing the parameters (s^1, s^2, s^3) which work as coordinates on the 3-dimensional manifold Σ in space-time with $X^\mu(\vec{s})$ the embedding coordinates. The parameters (s^1, s^2, s^3) determine on Σ an induced metric of the form

$$q_{ij}(\vec{s}) = g_{\mu\nu}(X(\vec{s}))\frac{\partial X^\mu(\vec{s})}{\partial s^i}\frac{\partial X^\nu(\vec{s})}{\partial s^j}. \tag{1.257}$$

If one makes the substitutions

$$\vec{x} \to \vec{s}; \quad \frac{\delta}{\delta T(\vec{x})} \to \frac{g^{\mu\nu}\tilde{n}_\nu}{\sqrt{|g^{\alpha\beta}\tilde{n}_\alpha\tilde{n}_\beta|}}\frac{\delta}{\delta X^\mu(\vec{s})} \tag{1.258}$$

where $\tilde{n}(\vec{s}) = \varepsilon_{\mu\alpha\beta\gamma}\frac{\partial X^\alpha}{\partial s^1}\frac{\partial X^\beta}{\partial s^2}\frac{\partial X^\gamma}{\partial s^3}$ is a normal to the surface Σ, the Tomonaga-Schwinger equation (1.250) may be formulated as

$$\hat{\mathrm{H}}(\vec{s})\Psi[\phi,X] = in^\mu(\vec{s})\frac{\delta\Psi[\phi,X]}{\delta X^\mu(\vec{s})} \tag{1.259}$$

and the Bohmian equations of motion (1.255) become

$$\frac{\partial \Phi[\vec{s},T]}{\partial \tau(\vec{s})} = \left(\frac{1}{|q(\vec{s})|^{1/2}}\frac{\delta S}{\delta \phi(\vec{s})}\right)_{\phi=\Phi} \tag{1.260}$$

where

$$\frac{\partial}{\partial\tau(\vec{s})} = \lim_{\sigma_x\to 0}\int_{\sigma_x} d^3(\vec{s})n^\mu(\vec{s})\frac{\delta}{\delta X^\mu(\vec{s})}. \tag{1.261}$$

By following the same procedure, the quantum many-fingered-time Klein-Gordon equation (1.256) may be expressed as

$$\left(\left(\frac{\partial}{\partial\tau(\vec{s})}\right)^2 - \nabla^i\nabla_i + m^2\right)\Phi[\vec{s},X] = \left(-\frac{1}{|q(\vec{s})|^{1/2}}\frac{\partial Q[\vec{s},\phi,X]}{\partial\phi(\vec{s})}\right)_{\phi=\Phi} \tag{1.262}$$

where ∇_i is the covariant derivative with respect to s^i and

$$Q[\vec{s},\phi,X] = -\frac{1}{|q(\vec{s})|^{1/2}}\frac{1}{2R}\frac{\delta^2 R}{\delta\phi^2(\vec{s})} \tag{1.263}$$

is the quantum potential. In the light of Nikolic's approach, the quantum potential (1.263) emerges therefore as the fundamental geometrodynamic entity which describes quantum processes in the Bohmian covariant interpretation for the many-fingered-time Tomonaga-Schwinger equation for relativistic quantum field theory. It provides a covariant description of relativistic quantum field theory in the sense that it depends on the induced metric (1.257) in the 3-dimensional manifold Σ. Its active information which is responsible for the behaviour of the fields is linked to the induced metric, which thus represents the ultimate properties of the geometry of the background of processes inside this approach.

Moreover, in analogous way to the original models of Bohm, Hiley, Bell and Kaloyerou and in analogy with Nikolic's approach to bosonic relativistic quantum field theory illustrated above, also for Nikolic's covariant Bohmian model to many-fingered-time Tomonaga-Schwinger equation for relativistic quantum field theory one can introduce a quantum length characterizing its geometrodynamics:

$$L_{quantum,MFT-TS} = \frac{1}{\sqrt{\frac{1}{2R|q(\vec{s})|^{1/2}}\frac{\delta^2 R}{\delta\phi^2(\vec{s})}}}. \tag{1.264}$$

The quantum length (1.264) can be used to evaluate the modification of the geometrical properties — introduced in the background space by the fields $\Phi[\vec{s}, X]$ in a three-dimensional many-fingered-time background in a quantum relativistic regime — with respect to the Euclidean geometry characteristic of classical fields. Once the quantum length (1.264) becomes non-negligible we have fields in a quantum relativistic regime of Tomonaga-Schwinger many-fingered-time background.

1.5.2 *The quantum potential in fermionic quantum field theory*

In [153, 154] Nikolic provided an interesting treatment of Bohmian particle trajectories in relativistic fermionic quantum field theory. A generic quantum state may be obtained by considering the action of the following creation operators regarding particles and the corresponding antiparticles

$$b_k^+ = \int d^3x \hat{\psi}^+(\vec{x}) u_k(\vec{x}); \quad d_k^+ = \int d^3x v_k^+(\vec{x}) \hat{\psi}(\vec{x}) \tag{1.265}$$

on the vacuum represented by the state

$$\Psi_0[\eta, \eta^+] = N \exp\left\{ \int d^3x \int d^3x' \eta^+(\vec{x}) \Omega(\vec{x}, \vec{x}') \eta(\vec{x}') \right\}. \tag{1.266}$$

Here, $u_k(\vec{x})$, $v_k(\vec{x})$ are a complete orthonormal set of spinors which are equal to, respectively, the positive and negative frequency 4-spinors solutions to the Dirac equation, η and η^+ are anticommuting Grassmann numbers satisfying $\{\eta_a(\vec{x}), \eta_{a'}(\vec{x}')\} = \{\eta_a^*(\vec{x}), \eta_{a'}^*(\vec{x}')\} = \{\eta_a(\vec{x}), \eta_{a'}^*(\vec{x}')\} = 0$, $\Omega(\vec{x}, \vec{x}') = (\Omega^A - \Omega^P)(\vec{x}, \vec{x}')$, N is a normalization constant such that $\langle \Psi_0 | \Psi_0 \rangle = 1$ and the scalar product is $\langle \Psi | \Psi \rangle = \int D^2\eta \Psi^*[\eta, \eta^+] \Psi'[\eta, \eta^+]$, $D^2 = D\eta D\eta^+$ and Ψ^* is dual to Ψ. A functional Ψ can be written as $\Psi[\eta, \eta^+] = \sum c_K \Psi_K[\eta, \eta^+]$, where $\Psi_K[\eta, \eta^+]$ constitute a complete orthonormal set of Grassmann valued functionals and its time evolution is governed by the functional Schrödinger equation of the form

$$H[\hat{\psi}, \hat{\psi}^+] \Psi[\eta, \eta^+, t] = i \frac{\partial}{\partial t} \Psi[\eta, \eta^+, t]. \tag{1.267}$$

Now, by following the papers [153, 154], a causal interpretation of fermionic quantum field theory is suggested which is based on the introduction of the quantities

$$J = \frac{\partial \rho}{\partial t} + \sum_a \int d^3x \left[\frac{\delta(\rho u_a[\vec{x}])}{\delta \phi_a(\vec{x})} + \frac{\delta(\rho u_a^*[\vec{x}])}{\delta \phi_a^*(\vec{x})} \right] \tag{1.268}$$

where $\rho = \Psi^* \Psi$ and

$$u_a[\vec{x}] = i \frac{\Psi^* \left[\hat{H}_\phi, \phi_a(\vec{x}) \right] \Psi}{\Psi^* \Psi}; \quad u_a[\vec{x}] = i \frac{\Psi^* [\hat{H}_\phi, \phi_a^*(\vec{x})] \Psi}{\Psi^* \Psi} \tag{1.269}$$

and $\phi_a(\vec{x}, t)$ and $\phi_a^*(\vec{x}, t)$ are hidden variables following a causal evolution according to equations

$$\frac{\partial \phi_a(\vec{x}, t)}{\partial t} = v_a[\phi, \phi^+, \vec{x}, t]; \quad \frac{\partial \phi_a^*(\vec{x}, t)}{\partial t} = v_a^*[\phi, \phi^+, \vec{x}, t], \tag{1.270}$$

where the velocities are

$$\begin{aligned} v_a[\vec{x}] &= u_a[\vec{x}] + \rho^{-1}(e_a(t, \vec{x}) + E_a[\vec{x}]); \\ v_a^*[\vec{x}] &= u_a^*[\vec{x}] + \rho^{-1}(e_a^*(t, \vec{x}) + E_a^*[\vec{x}]) \end{aligned} \tag{1.271}$$

where

$$E_a[\vec{x}] = \frac{\delta \Phi}{\delta \phi_a(\vec{x})}; \quad E_a^*[\vec{x}] = \frac{\delta \Phi}{\delta \phi_a^*(\vec{x})}, \tag{1.272}$$

$$e_a(t, \vec{x}) = -\frac{\int D^2\phi E_a[\phi, \phi^+, t, \vec{x}]}{\int D^2\phi}; \quad e_a^*(t, \vec{x}) = -\frac{\int D^2\phi E_a^*[\phi, \phi^+, t, \vec{x}]}{\int D^2\phi}. \tag{1.273}$$

In this way, the wavefunctional may be expressed as

$$\Phi[\phi, \phi^+, t] = \int D^2\chi \int D^2\phi' \frac{e^{2\pi i \chi \cdot (\phi - \phi')}}{4\pi^2 \chi^2} J[\phi', \phi'^+, t]. \tag{1.274}$$

In analogy with bosonic fields, the effectivity parameters guided by the wavefunction $\psi_{n_P, n_A} \equiv \psi_{b_1 \ldots b_{n_P} d_1 \ldots d_{n_A}}(\vec{x}_1, \ldots, \vec{x}_{n_P}, \vec{y}_1, \ldots, \vec{y}_{n_A}, t) = \langle 0 | \hat{\psi}_{b_1}^P(t, \vec{x}_1) \cdots \hat{\psi}_{d_{n_A}}^A(t, y_{n_A}) | \psi \rangle$ (n_P and n_A being the spinor indices of particles and antiparticles respectively), are given by relation

$$e_{n_P,n_A}[\phi,\phi^+,t] = \frac{|\tilde{\Psi}_{n_P,n_A}[\phi,\phi^+,t]|^2}{\sum_{n'_P,n'_A}^{\infty}|\tilde{\Psi}_{n'_P,n'_A}[\phi,\phi^+,t]|^2}. \tag{1.275}$$

Here, by following the philosophy which underlies Chavoya-Aceves' approach analysed in chapter 1.4.2, in the light of (1.268)–(1.274), the Dirac equation for fermionic fields leads to the following equations

$$\rho\left(v_a\partial_\mu\phi_a + \frac{\partial^\mu\phi_a\partial_\mu\phi_a}{2m}\right) = \frac{\hbar^2}{2m}(\sqrt{\rho}\partial^\mu\partial_\mu\sqrt{\rho}) + \frac{\hbar q}{2mc}(\psi_{n_P,n_A})^+\vec{\sigma}\cdot(\psi_{n_P,n_A}) \tag{1.276}$$

and

$$\rho\left(v_a^*\partial_\mu\phi_a + \frac{\partial^\mu\phi_a^*\partial_\mu\phi_a^*}{2m}\right) = \frac{\hbar^2}{2m}(\sqrt{\rho}\partial^\mu\partial_\mu\sqrt{\rho}) + \frac{\hbar q}{2mc}(\psi_{n_P,n_A})^+\vec{\sigma}\cdot(\psi_{n_P,n_A}) \tag{1.277}$$

which can be considered as the counterparts of equations (3.82) and (3.83), for fermionic field theory. Here, the first term on the right side of equations (1.276) and (1.277) can be identified just with the quantum potential for fermionic fields:

$$Q = \frac{\hbar^2}{2m}(\sqrt{\rho_i}\partial^\mu\partial_\mu\sqrt{\rho_i}). \tag{1.278}$$

The geometrical properties of the background determined by the quantum potential (1.278) can be characterized by introducing a quantum length associated with the Clifford background with fermionic fields of the form

$$L_{quantum,F} = \frac{1}{\sqrt{-\frac{\hbar^2}{2m}(\sqrt{\rho_i}\partial^\mu\partial_\mu\sqrt{\rho_i})}}. \tag{1.279}$$

The quantum length (1.279) can be used to evaluate the strength of quantum effects and, therefore, the modification of the geometry — introduced in the background space by a system characterized by the presence of fermionic fields in a quantum relativistic regime — with respect to the Euclidean geometry characteristic of classical physics. Once the quantum length (1.279) becomes non-negligible we have a system with fermionic fields which enter a quantum relativistic regime. And Heisenberg's uncertainty principle can be seen as a property which indicates that we are unable to perform a classical measurement to distances smaller than this quantum length characterizing the geometrical features of the background of processes of systems with fermionic fields in quantum relativistic regime (1.279).

Finally, we conclude this chapter with some considerations about how one could construct a fully relativistic description of fermionic fields by starting from the ideas of Hiley's and Callaghan's model of Dirac's relativistic quantum mechanics in a Clifford background developed in chapter 1.4.2. In this regard, we base our discussion on [155]. Let us consider an operator $\hat{\phi}(x)$ corresponding with a real scalar field $\phi(x)$ which satisfies, in the Heisenberg picture, the Dirac-type equation:

$$(i\gamma^{\mu}\partial_{\mu} - m)\hat{\phi} = I(\hat{\phi}) \tag{1.280}$$

where I is a nonlinear function describing the interaction. In the Schrödinger picture the time evolution of the wavefunctional is described by the Schrödinger equation of the form

$$H\left[\phi, -i\frac{\delta}{\delta\phi}\right]\Psi(\phi,t) = i\frac{\partial}{\partial t}\Psi(\phi,t) \tag{1.281}$$

where Ψ is a functional with respect to $\phi(\vec{x})$ and a function of time t. A normalized solution of this can be expressed as $\Psi(\phi,t) = \sum_{-\infty}^{+\infty}\tilde{\Psi}_n(\phi,t)$ where the $\tilde{\Psi}_n$ are unnormalized n-particle wavefunctionals.

In a Clifford background the Dirac-type equation (1.280) allows an energy conservation equation to be obtained which has the form

$$(\partial^{\mu}\partial_{\mu}\Phi_L)\Phi_R + \Phi_L(\partial^{\mu}\partial_{\mu}\Phi_R) + 2m^2\Phi_L\Phi_R = \mathrm{I} \tag{1.282}$$

where I is the element of Clifford algebra corresponding with the nonlinear function I. Equation (1.282) can be considered as the generalization of equation (1.205) to relativistic fermionic field theory. Like in Hiley's and Callaghan's approach to Dirac's relativistic quantum mechanics analysed in chapter 1.4.2, in equation (1.282) Φ_R and Φ_L are two entities of $C_{1,3}$ linked by the Clifford density element

$$\rho_C = \Phi_L \Phi_R = \phi_L \varepsilon_\gamma \phi_R, \tag{1.283}$$

that here concerns the wavefunction $\Psi[\phi(x), t]$ and corresponds with $|\Psi|^2$ of the standard Hilbert approach.

If one consider (1.212)–(1.215), after some algebraical manipulations the energy conservation equation (1.282) may be expressed in the form

$$P^2 + W^2 + [J\partial_\mu P^\mu + \partial_\mu P^\mu J] + [J\partial_\mu W^\mu + \partial_\mu W^\mu J] - m^2 = \mathrm{I}. \tag{1.284}$$

The scalar part of equation (1.284) leads to equation

$$P^2 + W^2 + [J\partial_\mu W^\mu + \partial_\mu W^\mu J] - m^2 = \mathrm{I} \tag{1.285}$$

which can be written as

$$P_B^2 + \Pi^2 + W^2 + [J\partial_\mu W^\mu + \partial_\mu W^\mu J] - m^2 = \mathrm{I} \tag{1.286}$$

where

$$2\rho P_B^\mu = tr[\gamma^0(\phi_L \overleftrightarrow{\partial}_\mu \gamma_{012} \phi_R)] \tag{1.287}$$

is the Bohm energy-momentum vector,

$$4\rho^2 P^2 = \sum_{i=0}^{3} A_{i\nu} A_i^\nu, \tag{1.288}$$

$$4\rho^2 \Pi^2 = \sum_{i=1}^{3} A_{i\nu} A_i^\nu, \tag{1.289}$$

where ρ, A_i $\overleftrightarrow{\partial}^\nu$ and J are defined like in chapter 1.4.2.

Equation (1.286) constitutes the quantum Hamilton-Jacobi equation for fermionic field theory in a Clifford background and thus can be considered as the fundamental equation of this approach. In the light of equation (1.286), the quantum potential for fermionic relativistic field theory in a Clifford background can be defined as the quantity

$$Q_D = \Pi^2 + W^2 + [J\partial_\mu W^\mu + \partial_\mu W^\mu J] - \mathrm{I}. \tag{1.290}$$

According to the approach to fermionic relativistic field theory here developed, the quantum potential is the fundamental physical entity which describes the active information regarding the behaviour of the fermionic fields and it emerges directly from the Clifford algebra $C_{1,3}$ which can be considered as the fundamental arena of quantum processes for fermionic fields. In analogy with Hiley's and Callaghan's Dirac relativistic quantum mechanics in a Clifford background analysed in chapter 1.4.2, in this approach of fermionic relativistic field theory, the Clifford arena $C_{1,3}$ described by the fundamental energy conservation equation (1.286) can be defined as the foreground, the implicate order of quantum processes and the quantum potential may be obtained directly from this arena.

Moreover, also here the geometrodynamics of processes can be characterized by introducing a quantum length associated with the Clifford background with fermionic fields of the form

$$L_{quantum,F} = 1/\sqrt{-(\Pi^2 + W^2 + [J\partial_\mu W^\mu + \partial_\mu W^\mu J] - I)}. \tag{1.291}$$

The quantum length (1.291) can be used to evaluate the strength of quantum effects and, therefore, the modification of the geometry — introduced in the Clifford background by fermionic fields in a quantum relativistic regime — with respect to the Euclidean geometry characteristic of classical physics. Once the quantum length (1.291) becomes non-negligible the fermionic fields into consideration go into a quantum relativistic regime. And Heisenberg's uncertainty principle can be seen as a property which indicates that we are unable to perform a classical measurement to distances smaller than this quantum length characterizing the geometrical features of the Clifford background for fermionic fields (1.291).

1.6 The Quantum Potential in Bohmian Quantum Gravity

General relativity and quantum theory can be considered among the greatest intellectual achievements of the 20th century. Each of them has profoundly modified the conceptual structure that underlies our understanding of the physical world. Despite their extraordinary successes in describing the physical phenomena in its own domain to an astonishing degree of accuracy, they offer us *strikingly* different pictures of physical reality, they are indeed mutually incompatible. In summary, quantum theory turns out to be very accurate, efficient and successful as far as the quantum phenomena do not interact with spacetime, which plays the role of a background entity. Nevertheless, it breaks down when spacetime is dynamical and interacts with the quantum phenomena. Therefore, in the light of general relativity and quantum theory, our basic understanding of nature is not only incomplete and fragmented — but inconsistent. In order to provide a coherent and satisfactory description of the physical world, one clearly needs a new consistent theory describing the so-called "quantum gravity domain" which allows us to portray a unifying picture of the universe, and this unifying theory must take the principles of both quantum mechanics and general relativity into account. As regards the building of a consistent unifying theory, the road to quantum gravity has been long, spanning about four decades, with many insights, trials, jubilations, triumphs as well as frustrations and tribulations.

The necessity of reconciling general relativity with quantum physics was already recognized by Einstein in 1916 [156] when he wrote: "Nevertheless, due to the inner-atomic movement of electrons, atoms would have to radiate not only electromagnetic but also gravitational energy, if only in tiny amounts. As this is hardly true in Nature, it appears that quantum theory would have to modify not only Maxwellian electrodynamics, but also the new theory of gravitation.".

Yet, almost a century later, we still do not have a satisfactory reconciliation or pacific coexistence between relativity and quantum physics. Why is the problem so difficult? In this regard, perhaps the primary obstacle is that, among the fundamental forces of Nature, gravity is very special in the sense that it is encoded in the very geometry of spacetime as a consequence of the equivalence principle that constitutes the core of the theory (and therefore should be incorporated at a fundamental level in a viable quantum theory if one wants to obtain a satisfactory unification between relativity and quantum physics). As we will see now in this chapter, Bohm's quantum potential introduces interesting unifying perspectives also as regards the treatment of gravitation, by emerging as the ultimate key for the reading and description of processes, which provides unexpected connections between gravity and the quantum.

1.6.1 *Bohm's quantum potential in relativistic curved space-time*

As regards the treatment of gravity and geometry in order to find the connections between quantum mechanics and general relativity, here we begin by focusing our attention on a very interesting recent research of A. Shojai and F. Shojai. These two authors analysed the behaviour of spinless particles in a curved space-time and found that the quantum potential generates a contribution to the curvature that is added to the classic one and exhibits deep and unexpected connections between gravity and the quantum processes [96, 97]. To develop the discussion about this topic we refer to the articles [96, 157–169] by A. Shojai and F. Shojai.

As regards the quantum effects of matter in a Bohmian framework, A. Shojai's and F. Shojai's toy model suggests that the motion of a particle (of spin zero) with quantum effects is equivalent to its motion in a curved space-time. In this approach, the quantum effects of matter as well as the gravitational effects of matter have geometrical nature and a fundamental connection characterizes them: the quantum potential can be seen as the conformal degree of freedom of the space–time metric and its presence is equivalent to the curved space-time. The presence of the quantum force physically corresponds to a curved space–time which is conformally flat where the conformal factor is expressed in terms of the quantum potential. The link between quantum effects and gravitational effects of matter is expressed by an equation of motion of the form

$$\tilde{g}^{\mu\nu}\tilde{\nabla}_\mu S \tilde{\nabla}_\nu S = m^2 c^2 \tag{1.292}$$

where S is the phase of the wavefunction ψ, $\tilde{\nabla}_\mu$ is the covariant differentiation with respect to the metric

$$\tilde{g}_{\mu\nu} = \frac{M^2}{m^2} g_{\mu\nu} \tag{1.293}$$

(which is a conformal metric) where

$$M^2 = m^2 \exp Q, \tag{1.294}$$

M being the quantum mass and

$$Q = \frac{\hbar^2}{m^2 c^2} \frac{\left(\nabla^2 - \frac{1}{c^2}\frac{\partial^2}{\partial t^2}\right)_g |\psi|}{|\psi|} \tag{1.295}$$

is the quantum potential (in (1.295), of course, c is the light speed and $\hbar$ is Planck's reduced constant).

The mathematical formalism of A. Shojai's and F. Shojai's approach begins by writing the wavefunction in its polar form $\psi = |\psi| \exp(\frac{iS}{\hbar})$ and by decomposing the real and imaginary parts of the Klein-Gordon equation (1.121). In this way, one obtains a quantum Hamilton-Jacobi equation that, by imposing that it be Poincarè invariant and have the correct non-relativistic limit, assumes the following form

$$\partial_\mu S \partial^\mu S = m^2 c^2 \exp Q, \tag{1.296}$$

with the quantum potential defined as

$$Q = \frac{\hbar^2}{m^2 c^2} \frac{\left(\nabla^2 - \frac{1}{c^2}\frac{\partial^2}{\partial t^2}\right) |\psi|}{|\psi|}, \tag{1.297}$$

and the continuity equation

$$\partial_\mu(\rho\partial^\mu S) = 0 \tag{1.298}$$

where ρ is the ensemble of particles. The above quantum Hamilton-Jacobi equation (1.296) leads to define the quantum mass as in (1.294).

By considering Bohm's version of Klein-Gordon equation, A. Shojai and F. Shojai demonstrated that it is possible to embed the de Broglie–Bohm quantum theory of motion and gravity and that the key character of de Broglie–Bohm theory, the quantum potential, can be interpreted as the conformal degree of freedom of the space–time metric. Starting from Bohm's version of Klein-Gordon equation, the extension to the case of a particle moving in a curved background can be achieved by changing the ordinary differentiating ∂_μ with the covariant derivative ∇_μ and by changing the Lorentz metric with the curved metric $g_{\mu\nu}$. In this way one obtains that the equations of motion for a particle (of spin 0) in a curved background can be written as:

$$\nabla_\mu(\rho\nabla^\mu S) = 0 \tag{1.299}$$

$$g^{\mu\nu}\nabla_\mu S\nabla_\nu S = m^2c^2 \exp Q \tag{1.300}$$

where

$$Q = \frac{\hbar^2}{m^2c^2}\frac{\left(\nabla^2 - \frac{1}{c^2}\frac{\partial^2}{\partial t^2}\right)_g |\psi|}{|\psi|} \tag{1.301}$$

is the quantum potential. Here, if one utilizes a fruitful observation of de Broglie [170], the quantum Hamilton-Jacobi equation (1.300) may be conveniently expressed as

$$\frac{m^2}{M^2}g^{\mu\nu}\nabla_\mu S\nabla_\nu S = m^2c^2. \tag{1.302}$$

From equation (1.302) one can conclude that the quantum effects are equivalent to the change of the space-time metric from

$g_{\mu\nu}$ to

$$\tilde{g}_{\mu\nu} = \frac{M^2}{m^2} g_{\mu\nu} \tag{1.293}$$

which is a conformal transformation. In this way equation (1.302) may be formulated just as

$$\tilde{g}^{\mu\nu} \tilde{\nabla}_\mu S \tilde{\nabla}_\nu S = m^2 c^2 \tag{1.292}$$

where $\tilde{\nabla}_\mu$ represents the covariant differentiation with respect to the metric $\tilde{g}_{\mu\nu}$. Furthermore, in this new curved space-time the continuity equation will assume the form

$$\tilde{g}^{\mu\nu} \tilde{\nabla}_\mu (\rho \tilde{\nabla}^\mu S) = 0 \tag{1.303}$$

In the light of equations (1.292)–(1.303), the important conclusion one can draw from F. Shojai's and A. Shojai's model is that the presence of the quantum potential is equivalent to a curved space-time with its metric being given by equation (1.293), providing thus a fundamental geometrization of the quantum aspects of matter. On the basis of F. Shojai's and A. Shojai's model, a dual aspect to the role of geometry in physics emerges naturally. The space-time geometry sometimes looks like what we call gravity and sometimes looks like what we understand as quantum behaviours and the quantum potential can be considered the real intermediate between these two aspects. In fact, on the ground of equation (1.293), one can say that the geometric properties which are expressed by the quantum potential and which determine the behaviour of a spinless particle are linked with the curved space-time. In other words, one can say that the particles generate the curvature of space-time and at the same time the space-time metric is strictly tied with the quantum potential which influences the behaviour of the particles. Quantum potential creates itself a curvature, and thus a deformation of the geometry of space-time, which determines deep effects on the classical contribution to the curvature of the space-time.

In substance, F. Shojai's and A. Shojai's model provides a computed image of quantum geometrodynamics which fuses gravitational and quantum aspects of matter, at least as regards the level of macroscopic description of physical processes. In this picture, the particle trajectory is ruled by Newton's equation of motion:

$$M\frac{d^2x^\mu}{d\tau^2} + M\Gamma^\mu_{\nu\kappa}u^\nu u^\kappa = (c^2 g^{\mu\nu} - u^\mu u^\nu)\nabla_\nu M \tag{1.304}$$

which shows that it is determined by the deformation of the geometry of space corresponding with the quantum potential. Equation (1.304) reduces to the standard geodesic equation by applying the above conformal transformation (1.293).

On the other hand, as regards the geometrodynamic features of the quantum potential in the treatment provided by F. Shojai's and A. Shojai's model for the motion of a particle of spin 0 in a curved space-time, in analogy to Novello's, Salim's and Falciano's proposal of a Weyl curvature length, the author of this book in the recent article *Quantum potential and the geometry of space* [171] introduced a quantum length associated with the conformal metric (1.293) given by the following relation

$$L_{quantum} = \sqrt{\frac{2}{mc^2}\frac{\left(\nabla^2 - \frac{1}{c^2}\frac{\partial^2}{\partial t^2}\right)_g |\psi|}{|\psi|}}. \tag{1.305}$$

The quantum length (1.305) provides a parameter to measure the strength of quantum effects and, therefore, the modification of the geometry with respect to the Euclidean geometry characteristic of classical physics in a relativistic curved space-time. Once the quantum length (1.305) becomes non-negligible the particle at spin 0 into consideration goes into a quantum regime where the quantum and gravitational effects are highly related. With the introduction of the quantum length (1.305) regarding the motion of a sinless particle in a curved space-time, one can interpret Heisenberg's uncertainty principle as a principle which derives from the fact that we are unable to perform a classical measurement to distances smaller than this quantum length. In other words, the size of a measurement has to be bigger than the quantum length

$$\Delta L \geq \quad L_{quantum} = \sqrt{\frac{2}{mc^2} \frac{\left(\nabla^2 - \frac{1}{c^2}\frac{\partial^2}{\partial t^2}\right)_g |\psi|}{|\psi|}} \tag{1.306}$$

which means that a quantum regime characterized by a coupling between gravitational and quantum effects is entered when the quantum length (1.306) must be taken under consideration.

In summary, with regards to the geometrodynamic features of the quantum potential for the motion of a spinless particle in a curved background, one can say that, on the basis of F. Shojai's and A. Shojai's results, the geometry subtended by the quantum potential unifies gravitational and quantum aspects of matter and, at the same time, a quantum length can be introduced which provides a measure of the geometrical properties of the fundamental quantum background where there is a connection between gravitational and quantum effects. In virtue of the conformal metric (1.293), the geometric properties which are expressed by the quantum potential indicate that the quantum behaviour of matter is highly coupled with gravitational effects and the condition (1.306) about the quantum length indicates when this coupling between quantum and gravitational effects must be taken into consideration.

1.6.2 *Bohm's quantum potential in quantum gravity*

Also the results obtained in the quantum gravity domain allow us to put in evidence the active role of the quantum potential in redesigning the geometry of physical space in the presence of gravitational interaction. In this regard, before all, despite some problems and weak points (for example, the fact that it is still an open question among the so-called Bohmian community which sense to give — if any — to the wavefunction of the universe), if we take into account some recent research, the Bohmian interpretation of canonical quantum gravity turns out to have several useful aspects and merits [32, 172–175].

Some of them are:

- It leads to time evolution of the dynamical variables whether the wavefunction is dependent on time or not. Therefore, in Bohmian quantum gravity the time problem is not present.
- As Bohm's theory describes a single system, from the Bohmian point of view we have not the conceptual problem of the meaning of the universe's wavefunction in quantum cosmology.
- In the Bohmian approach, one does not have the necessity to normalize the wavefunction for a single system.
- The classical limit has a well-defined meaning. The classical domain is obtained directly when the quantum potential is negligible with respect the classical potential and the quantum force is negligible with respect the classical force.
- There is no need to separate the classical observer and the quantum system in the measurement problem. Although in the Bohmian picture of the measurement process we have two interacting systems, the system and the observer, after the interaction occurs, the wavefunction of the system is subjected to a reduction in a causal way [32, 176].

Until now the Bohmian interpretation of Wheeler-De Witt quantum gravity and cosmology have given some physical results that could be found in the literature:

- In Bohmian quantum cosmology the quantum force can remove the big bang singularity, because it can behave as a repulsive force [177, 178].
- The quantum force may be present in large scales because the quantum effects of the quantum potential are independent of the scale [179].
- Bohmian quantum cosmology allows us to avoid any singularity — which characterizes instead super string cosmology — between inflation and the successive decelerating expansion: there is simply a smooth change from the two different stages [180].
- Real time tunnelling can take place in the classically forbidden regions, through the quantum potential (as regards this effect in a

closed de Sitter universe in 2+1 dimensions see, for example, the reference [181]).

- Finally, and this is the point towards which now we want to focus our attention, in a generalized geometric picture of Bohm's interpretation one can unify the quantum effects and gravity at the fundamental level of quantum gravity [182–187].

As far as this latest point is concerned, F. Shojai and A. Shojai recently developed a toy model of quantum gravity (providing a scalar-tensor picture of the ideas analysed in section 1.6.1) in which the form of the quantum potential, its geometrodynamic features and its relation to the conformal degree of freedom of the space-time metric can be obtained on the basis of the equations of motion. The mathematical formalism of F. Shojai's and A. Shojai's model shows that the quantum gravity equations of motion are the ultimate key elements which make the quantum potential the fundamental entity which expresses the geometrical properties influencing the behaviour of the particles and which is related to the space-time metric. Thus, it allows us to make significant progresses as regards the issue of unifying the gravitational and quantum aspects of matter at the fundamental level of physical reality represented by quantum gravity. For a discussion about this topic we follow the references [182–187].

In F. Shojai's and A. Shojai's model, before all, a general relativistic system consisting of gravity and classical matter can be associated to the action

$$A_{no-quantum} = \frac{1}{2k}\int d^4x\sqrt{-g}R + \int d^4x\sqrt{-g}\frac{\hbar^2}{m}\left(\frac{\rho}{\hbar^2}\partial_\mu S\partial^\mu S - \frac{m^2}{\hbar^2}\rho\right) \quad (1.307)$$

where $\rho = J^0$ is the ensemble density of the particles, $k = 8\pi G$ and hereafter we chose the units in which $c = 1$. On the other hand, as we have mentioned in chapter 1.6.1, the introduction of the quantum effects is equivalent to the modification of the space-time metric from $g_{\mu\nu}$ to $g_{\mu\nu} \rightarrow g^I_{\mu\nu} = \frac{g_{\mu\nu}}{\exp Q}$ which is a conformal transformation. Therefore, in order to introduce quantum effects, one may make this conformal transformation, instead of adding the quantum potential term.

In this regard, one can write the action with quantum effects as:

$$A[\bar{g}_{\mu\nu}, \Omega, S, \rho, \lambda] = \frac{1}{2k}\int d^4x\sqrt{-\bar{g}}(\bar{R}\Omega^2 - 6\bar{\nabla}_\mu\Omega\bar{\nabla}^\mu\Omega)$$
$$+\int d^4x\sqrt{-\bar{g}}(\frac{\rho}{m}\Omega^2\bar{\nabla}_\mu S\bar{\nabla}^\mu S - m\rho\Omega^4)$$
$$+\int d^4x\sqrt{-\bar{g}}\lambda\left(\Omega^2 - \left(1 + \frac{\hbar^2\left(\nabla^2 - \frac{\partial^2}{\partial t^2}\right)\sqrt{\rho}}{m^2\sqrt{\rho}}\right)\right) \tag{1.308}$$

where $\Omega^2 = \exp Q$ is the conformal factor, a bar over any quantity means that it corresponds to no-quantum regime and λ is a Lagrange multiplier introduced in order to identify the conformal factor with its Bohmian value.

By applying a variational approach to the above action with respect to $\bar{g}_{\mu\nu}$, Ω, ρ, S and λ one obtains the following relations as equations of motion:

1. The equation of motion for Ω:

$$\bar{R}\Omega + 6\left(\bar{\nabla}^2 - \frac{\bar{\partial}^2}{\partial t^2}\right)\Omega + 2\frac{k}{m}\rho\Omega(\bar{\nabla}_\mu S\bar{\nabla}^\mu S - 2m^2\Omega^2) + 2k\lambda\Omega = 0 \tag{1.309}$$

2. The continuity equation for the particles:

$$\bar{\nabla}_\mu(\rho\Omega^2\bar{\nabla}^\mu S) = 0 \tag{1.310}$$

3. The equation of motion for the particles:

$$(\bar{\nabla}_\mu S\bar{\nabla}^\mu S - m^2\Omega^2)\Omega^2\sqrt{\rho} + \frac{\hbar^2}{2m}$$
$$\times\left[\left(\bar{\nabla}^2 - \frac{\bar{\partial}^2}{\partial t^2}\right)\left(\frac{\lambda}{\sqrt{\rho}}\right) - \lambda\frac{\left(\bar{\nabla}^2 - \frac{\bar{\partial}^2}{\partial t^2}\right)\sqrt{\rho}}{\rho}\right] = 0 \tag{1.311}$$

4. The modified Einstein equations for $\bar{g}_{\mu\nu}$:

$$\Omega^2\left[\bar{R}_{\mu\nu}-\frac{1}{2}\bar{g}_{\mu\nu}\bar{R}\right]-\left[\bar{g}_{\mu\nu}\left(\bar{\nabla}^2-\frac{\bar{\partial}^2}{\partial t^2}\right)-\bar{\nabla}_\mu\bar{\nabla}_\nu\right]\Omega^2$$
$$-6\bar{\nabla}_\mu\Omega\bar{\nabla}_\nu\Omega+3\bar{g}_{\mu\nu}\bar{\nabla}_\alpha\Omega\bar{\nabla}^\alpha\Omega+\frac{2k}{m}\rho\Omega^2\bar{\nabla}_\mu S\bar{\nabla}_\nu S$$
$$-\frac{k}{m}\rho\Omega^2\bar{g}_{\mu\nu}\bar{\nabla}_\alpha S\bar{\nabla}^\alpha S+km\rho\Omega^4\bar{g}_{\mu\nu}$$
$$+\frac{k\hbar^2}{m^2}\left[\bar{\nabla}_\mu\sqrt{\rho}\bar{\nabla}_\nu\left(\frac{\lambda}{\sqrt{\rho}}\right)+\bar{\nabla}_\nu\sqrt{\rho}\bar{\nabla}_\mu\left(\frac{\lambda}{\sqrt{\rho}}\right)\right]$$
$$-\frac{k\hbar^2}{m^2}\bar{g}_{\mu\nu}\bar{\nabla}_\alpha\left[\lambda\frac{\bar{\nabla}^\alpha\sqrt{\rho}}{\sqrt{\rho}}\right]=0 \tag{1.312}$$

5. The constraint equation:

$$\Omega^2=1+\frac{\hbar^2}{m^2}\frac{\left(\bar{\nabla}^2-\frac{\bar{\partial}^2}{\partial t^2}\right)\sqrt{\rho}}{\sqrt{\rho}}. \tag{1.313}$$

The equations of motion (1.309)–(1.313) evidence how the quantum geometrodynamics at the quantum gravity level emerges, about a general relativistic system characterized by a significant connection between quantum effects and gravity: they suggest that there are back-reaction effects of the quantum factor on the background. On the basis of the high-coupled five equations above listed, one can say that in the quantum gravity domain the conformal factor $\Omega^2=\exp Q$ (associated with the quantum potential) express the geometrodynamic properties which introduce the links (and thus the back-reaction terms) between the quantum effects and the background.

Moreover, in F. Shojai's and A. Shojai's model, by embedding together equations (1.309) and (1.310), one can arrive at a more simple relation instead of (1.309). In fact, by using the trace of equation (1.310) as well as equation (1.311), some mathematical manipulations lead to the following equation regarding the Lagrange multiplier:

$$\lambda = \frac{\hbar^2}{m^2}\bar{\nabla}_\mu \left[\lambda \frac{\bar{\nabla}^\mu \sqrt{\rho}}{\sqrt{\rho}}\right]. \tag{1.314}$$

If one resolves equation (1.314) in perturbative way in terms of the parameter $\alpha = \frac{\hbar^2}{m^2}$ by writing $\lambda = \lambda^{(0)} + \alpha\lambda^{(1)} + \alpha^2\lambda^{(2)} + \cdots$ and $\sqrt{\rho} = \sqrt{\rho}^{(0)} + \alpha\sqrt{\rho}^{(1)} + \alpha^2\sqrt{\rho}^{(2)} + \cdots$ one gets

$$\lambda^{(0)} = \lambda^{(1)} = \lambda^{(2)} = \cdots = 0. \tag{1.315}$$

Thus, if the perturbative solution of (1.314) is $\lambda = 0$, which is its trivial solution, the equations of quantum gravity may be formulated as:

$$\bar{\nabla}_\mu(\rho\Omega^2\bar{\nabla}^\mu S) = 0 \tag{1.316}$$

$$\bar{\nabla}_\mu S\bar{\nabla}^\mu S = m^2\Omega^2 \tag{1.317}$$

$$G_{\mu\nu} = -kT^{(m)}_{\mu\nu} - kT^{(\Omega)}_{\mu\nu} \tag{1.318}$$

where $T^{(m)}_{\mu\nu}$ is the matter energy-momentum tensor and

$$kT^{(\Omega)}_{\mu\nu} = \frac{\left[g_{\mu\nu}\left(\nabla^2 - \frac{\partial^2}{\partial t^2}\right) - \nabla_\mu\nabla_\nu\right]\Omega^2}{\Omega^2} + 6\frac{\nabla_\mu\Omega\nabla_\nu\Omega}{\omega^2} - 3g_{\mu\nu}\frac{\nabla_\alpha\Omega\nabla^\alpha\Omega}{\Omega^2} \tag{1.319}$$

and

$$\Omega^2 = 1 + \alpha\frac{\overline{\left(\nabla^2 - \frac{\partial^2}{\partial t^2}\right)\sqrt{\rho}}}{\sqrt{\rho}}. \tag{1.320}$$

It can be remarked that equation (1.319) is a Bohmian-type equation of motion, and if one expresses it in terms of the physical metric $g_{\mu\nu}$, it reads as

$$\nabla_\mu S\nabla^\mu S = m^2c^2. \tag{1.321}$$

The next step is to make dynamical the conformal factor and the quantum potential. In this regard, one starts from the most general scalar-tensor action

$$A = \int d^4x \left\{ \phi R - \frac{\omega}{\phi} \nabla^\mu \phi \nabla_\mu \phi + 2\Lambda\phi + L_m \right\} \tag{1.322}$$

in which ω is a constant independent of the scalar field ϕ, Λ is the cosmological constant, and L_m is the matter Lagrangian (which is assumed to be in the form

$$L_m = \frac{\rho}{m} \phi^a \nabla^\mu S \nabla_\mu S - m\rho\phi^b - \Lambda(1+Q)^c \tag{1.323}$$

in which a, b, and c are constants). By using a perturbative expansion for the scalar field and the matter distribution density as $\phi = \phi_0 + \alpha\phi_1 + \cdots$

$$\sqrt{\rho} = \sqrt{\rho_0} + \alpha\sqrt{\rho_1} + \cdots$$

(and imposing opportune physical constraints in order to determine the parameters a, b, and c), F. Shojai and A. Shojai obtained the following quantum gravity equations:

$$\phi = 1 + Q - \frac{\alpha}{2}\left(\nabla^2 - \frac{\partial^2}{\partial t^2}\right) Q \tag{1.324}$$

$$\begin{aligned} \nabla^\mu S \nabla_\mu S = {} & m^2\phi - \frac{2\Lambda m}{\rho}(1+Q)(Q-\tilde{Q}) \\ & + \frac{\alpha\Lambda m}{\rho}\left[\left(\nabla^2 - \frac{\partial^2}{\partial t^2}\right) Q - 2\nabla_\mu Q \frac{\nabla^\mu \sqrt{\rho}}{\sqrt{\rho}}\right] \end{aligned} \tag{1.325}$$

$$\nabla_\mu(\rho\nabla^\mu S) = 0 \tag{1.326}$$

$$\begin{aligned} G^{\mu\nu} - \Lambda g^{\mu\nu} = {} & -\frac{1}{\phi} T^{\mu\nu} - \frac{1}{\phi}\left[\nabla^\mu\nabla^\nu - g^{\mu\nu}\left(\nabla^2 - \frac{\partial^2}{\partial t^2}\right)\right]\phi \\ & + \frac{\omega}{\phi^2}\nabla^\mu\phi\nabla^\nu\phi - \frac{1}{2}\frac{\omega}{\phi^2} g^{\mu\nu}\nabla^\alpha\phi\nabla_\alpha\phi \end{aligned} \tag{1.327}$$

where $\tilde{Q} = \alpha\frac{\nabla_\mu\sqrt{\rho}\nabla^\mu\sqrt{\rho}}{\sqrt{\rho}}$ and $T^{\mu\nu} = -\frac{1}{\sqrt{-g}}\frac{\delta}{\delta g_{\mu\nu}}\int d^4x\sqrt{-g}L_m$ is the energy-momentum tensor.

The geometric quantum gravity model suggested by F. Shojai and A. Shojai, and summarized in equations (1.324)–(1.327), allows us to draw some important conclusions:

— In this model equation (1.327) shows that the gravitational effects of matter determine the causal structure of the space-time $g^{\mu\nu}$. On the basis of equation (1.326) quantum effects determine directly the scale factor of space-time.
— The mass field given by the right-hand side of equation (1.325) consists of two parts. The first part, which is proportional to α, is a purely quantum effect, while the second part, which is proportional to $\alpha\Lambda$, is a mixture of the quantum effects and the large scale structure introduced via the cosmological constant.
— In this model, the scalar field produces the quantum force which appears on the right hand and violates the equivalence principle (just like, in Kaluza-Klein theory, the scalar field — dilaton — produces a fifth force leading to the violation of the equivalence principle [188]).

Moreover, according to the geodesic equation (1.324), the appearance of quantum mass justifies Mach's principle which leads to the existence of an interrelation between the global properties of the universe (space–time structure, the large scale structure of the universe) and its local properties (local curvature, motion in a local frame, etc.). In F. Shojai's and A. Shojai's approach analysed in sections 1.6.1 and 1.6.2, it can be easily seen that the space–time geometry is determined by the distribution of matter. A local variation of matter field distribution changes the quantum potential acting on the geometry. Thus the geometry is changed globally (in conformity with Mach's principle). In this sense the Bohmian approach to quantum gravity turns out to be highly non–local as it is forced by the features of the quantum potential. What one calls geometry is only the gravitational and quantum effects of matter in virtue of the presence of the quantum potential which turns out to be the real intermediary between them, the key which determines a unifying picture of them.

To summarize, one of the merits of F. Shojai's and A. Shojai's approach examined in sections 1.6.1 and 1.6.2 (regarding respectively a relativistic Bohmian theory in a curved space-time and a Bohmian scalar-tensor quantum gravity model) on the geometrization of quantum effects is the dual role of the geometry in physics. The gravitational effects determine the causal structure of space–time as long as quantum effects provide its conformal structure. The conformal factor of the metric is a function of the quantum potential and the mass of a relativistic particle is a field produced by quantum corrections to the classical mass. As we have seen in chapter 1.6.1, according to F. Shojai's and A. Shojai's Bohmian model in a relativistic curved space-time, the presence of the quantum potential is equivalent to a conformal mapping of the metric of space-time. Thus in conformally related frames one measures different quantum masses and different curvatures. In particular, one may consider two specific frames. One contains the quantum mass field (appearing in the quantum Hamilton-Jacobi equation) and the classical metric while the other contains the classical mass (appearing in the classical Hamilton-Jacobi equation) and the quantum metric. In other frames both the space–time metric and the mass field are characterized by quantum properties. In the light of this argument, one can conclude that different conformal frames are equivalent pictures of the gravitational and quantum phenomena.

As regards what happens to the quantum force, the conformally related frames are not distinguishable, just like it occurs at different coordinate systems when we consider gravity. Since the conformal transformation changes the length scale locally, we measure different quantum forces in different conformal frames. This is similar to what happens in general relativity where general coordinate transformation modifies the gravitational force at any arbitrary point. Then, the following crucial question becomes natural. Does applying the above correspondence, between quantum and gravitational forces, and between the conformal and general coordinate transformations, mean that the geometrization of quantum effects implies conformal invariance just as gravitational effects imply general coordinate invariance?

If in general relativity we deal with a general coordinate invariance, in a similar way inside the Bohmian approach to quantum gravity developed by A. Shojai and F. Shojai (and illustrated in sections 1.6.1 and 1.6.2), at any point (or even globally) the quantum effects of matter can be removed by a opportune conformal transformation. Thus in that point(s) matter behaves classically. In this way, F. Shojai's and A. Shojai's geometrodynamic approach leads to propose a new quantum equivalence principle, similar to the standard equivalence principle of general relativity, that can be called the conformal equivalence principle. According to this quantum equivalence principle gravitational effects can be removed if one goes to a freely falling frame while quantum effects can be eliminated if one chooses an appropriate scale. If the equivalence principle of general relativity interconnects gravity and general covariance, the conformal equivalence principle has the same function about quantum and conformal covariance. Both these principles state that there is no preferred frame, either coordinate or conformal. And these aspects of the geometry of physical space regarding frames characterized by quantum and conformal covariance turn out to be results which derive just from the quantum potential.

Finally, one can see here that in the quantum gravity domain Weyl geometry provides additional degrees of freedom which can be identified with quantum effects and allows us to construct a unified geometric framework in order to understand both gravitational and quantum forces. This picture is characterized by these features: (i) Quantum effects seem to be independent of any preferred length scale. (ii) The quantum mass of a particle is a field. (iii) In the light of the geometrodynamic action of the quantum potential, the gravitational constant is also a field which depends on the matter distribution (cfr. [97, 182, 183]). (iv) A local variation of matter field distribution modifies the quantum potential acting on the geometry and perturbs it globally; the nonlocal character of processes is determined by the quantum potential (cf. [184]).

In order to analyse the link between the Bohmian toy model of F. Shojai and A. Shojai and Weyl geometry, we follow the reference

[97]. One starts by considering the Weyl-Dirac action

$$\mathcal{J} = \int d^4x\sqrt{-g}(F_{\mu\nu}F^{\mu\nu} - \beta^{2W}R + (\sigma+6)\beta_{;\mu}\beta^{;\mu} + L_m) \quad (1.328)$$

where L_m is the matter Lagrangian, $F_{\mu\nu}$ is the curl of the Weyl 4-vector ϕ_μ, σ is an arbitrary constant and β is a scalar field of weight -1. In equation (1.328), the symbol ";" represents a covariant derivative under general coordinate and conformal transformations (Weyl covariant derivative) defined as $X_{;\mu} = {}^W\nabla_\mu X - N\phi_\mu X$ where $N$is the Weyl weight of X. The Weyl-Dirac action (1.328) leads then to the following equations of motion

$$\begin{aligned}
\Phi^{\mu\nu} &= -\frac{8\pi}{\beta^2}(I^{\mu\nu} + N^{\mu\nu}) + \frac{2}{\beta}(g^{\mu\nu W}\nabla^{\alpha W}\nabla_\alpha\beta - {}^W\nabla^{\mu W}\nabla^\nu\beta) \\
&\quad + \frac{1}{\beta^2}(4\nabla^\mu\beta\nabla^\nu\beta - g^{\mu\nu}\nabla^\alpha\beta\nabla_\alpha\beta) \\
&\quad + \frac{\sigma}{\beta^2}(\beta^{;\mu}\beta^{;\nu} - \frac{1}{2}g^{\mu\nu}\beta^{;\alpha}\beta_{;\alpha}); \\
{}^W\nabla_\mu F^{\mu\nu} &= \frac{1}{2}\sigma(\beta^2\phi^\mu + \beta\nabla^\mu\beta) + 4\pi J^\mu; \qquad (1.329) \\
R &= -(\sigma+6)\frac{{}^W\left(\nabla^2 - \frac{1}{c^2}\frac{\partial^2}{\partial t^2}\right)\beta}{\beta} + \sigma\phi_\alpha\phi^\alpha - \sigma{}^W\nabla^\alpha\phi_\alpha + \frac{\psi}{2\beta}
\end{aligned}$$

where

$$N^{\mu\nu} = \frac{1}{4\pi}\left(\frac{1}{4}g^{\mu\nu}F^{\alpha\beta}F_{\alpha\beta} - F^\mu_\alpha F^{\nu\alpha}\right) \quad (1.330)$$

and

$$8\pi I^{\mu\nu} = \frac{1}{\sqrt{-g}}\frac{\delta\sqrt{-g}L_m}{\delta g_{\mu\nu}}; \quad 16\pi J^\mu = \frac{\delta L_m}{\delta\phi_\mu}; \quad \psi = \frac{\delta L_m}{\delta\beta}. \quad (1.331)$$

As regards the equations of motion of matter and the trace of the electromagnetic tensor, if one uses the invariance of the action under coordinate and gauge transformations, one arrives to the following equations

$$
\begin{aligned}
&{}^{W}\nabla_{\nu}I^{\mu\nu} - I\frac{\nabla^{\mu}\beta}{\beta} = J_{\alpha}\phi^{\alpha\mu} - \left(\phi^{\mu} + \frac{\nabla^{\mu}\beta}{\beta}\right){}^{W}\nabla_{\alpha}J^{\alpha};\\
&16\pi I + 16\pi {}^{W}\nabla_{\mu}J^{\mu} - \beta\psi = 0
\end{aligned}
\tag{1.332}
$$

The first relation of (1.332) is a geometrical identity (Bianchi identity) while the second shows the mutual dependence of the field equations. In this Weyl-Dirac approach the gravity fields $g_{\mu\nu}$ and ϕ_{μ} and the quantum mass field influence the space-time geometry. Here, if one starts from equations (1.328)–(1.332) a Bohmian quantum gravity which is conformally invariant in the framework of Weyl geometry emerges naturally. The Weyl-Dirac action is a general Weyl invariant action and for simplicity one can assume that the matter Lagrangian does not depend on the Weyl vector so that $J_{\mu} = 0$. The equations of motion are then

$$
\begin{aligned}
\Phi^{\mu\nu} &= -\frac{8\pi}{\beta^2}(I^{\mu\nu} + N^{\mu\nu}) + \frac{2}{\beta}(g^{\mu\nu W}\nabla^{\alpha W}\nabla_{\alpha}\beta - {}^{W}\nabla^{\mu W}\nabla^{\nu}\beta)\\
&\quad + \frac{1}{\beta^2}(4\nabla^{\mu}\beta\nabla^{\nu}\beta - g^{\mu\nu}\nabla^{\alpha}\beta\nabla_{\alpha}\beta)\\
&\quad + \frac{\sigma}{\beta^2}\left(\beta^{;\mu}\beta^{;\nu} - \frac{1}{2}g^{\mu\nu}\beta^{;\alpha}\beta_{;\alpha}\right);\\
{}^{W}\nabla_{\nu}F^{\mu\nu} &= \frac{1}{2}\sigma(\beta^2\phi^{\mu} + \beta\nabla^{\mu}\beta);\\
R &= -(\sigma+6)\frac{{}^{W}\left(\nabla^2 - \frac{1}{c^2}\frac{\partial^2}{\partial t^2}\right)\beta}{\beta} + \sigma\phi_{\alpha}\phi^{\alpha} - \sigma {}^{W}\nabla^{\alpha}\phi_{\alpha} + \frac{\psi}{2\beta}
\end{aligned}
\tag{1.333}
$$

The symmetry conditions are

$$
\begin{aligned}
&{}^{W}\nabla_{\nu}I^{\mu\nu} - I\frac{\nabla^{\mu}\beta}{\beta} = 0;\\
&16\pi I - \beta\psi = 0
\end{aligned}
\tag{1.334}
$$

where $I = I^{\mu\nu}_{\mu\nu}$. Now, by introducing a quantum mass field, if one takes into account that it is proportional to the Dirac field, one finds out that this Weyl approach is related to the geometrodynamic features of Bohm's quantum potential. Thus, if one uses equations (1.333) and (1.334), one obtains equation

$$\left(\nabla^2 - \frac{1}{c^2}\frac{\partial^2}{\partial t^2}\right)\beta + \frac{1}{6}\beta R = \frac{4\pi}{3}\frac{I}{\beta} + \sigma\beta\phi_\alpha\phi^\alpha$$
$$+2(\sigma-6)\phi^\gamma\nabla_\gamma\beta + \frac{\sigma}{\beta}\nabla^\mu\beta\nabla_\mu\beta. \quad (1.335)$$

which can be solved in an iterative way by using relation

$$\beta^2 = \frac{8\pi I}{R} - \left\{\frac{1}{[(R/6) - \sigma\phi_\alpha\phi^\alpha]}\right\}\beta\left(\nabla^2 - \frac{1}{c^2}\frac{\partial^2}{\partial t^2}\right)\beta + \cdots \quad (1.336)$$

Now assuming $I^{\mu\nu} = \rho u^\mu u^\nu$ we multiply equation (1.334) by u^μ and sum to get

$$^W\nabla_\nu(\rho u^\nu) - \rho\left(u_\mu\frac{\nabla^\mu\beta}{\beta}\right) = 0. \quad (1.337)$$

Then by substituting equation (1.334) into equation (1.337) one obtains

$$u^{\nu W}\nabla_\nu u^\mu = \left(\frac{1}{\beta}\right)(g^{\mu\nu} - u^\mu u^\nu)\nabla_\nu\beta \quad (1.338)$$

Moreover, from equation (1.336) one gets

$$\beta^{2(1)} = \frac{8\pi I}{R}; \beta^{2(2)} = \frac{8\pi I}{R}$$
$$\times\left(1 - \frac{1}{\left(\frac{R}{6}\right) - \sigma\phi_\alpha\phi^\alpha}\frac{\left(\nabla^2 - \frac{1}{c^2}\frac{\partial^2}{\partial t^2}\right)\sqrt{I}}{\sqrt{I}}\right); \cdots \quad (1.339)$$

Comparing the above formalism with equations (1.295) and (1.304) lets us understand that the correct equations for the Bohmian theory can be derived if one identifies

$$\beta \approx M; \frac{8\pi I}{R} = m^2; \frac{1}{\sigma\phi_\alpha\phi^\alpha - \frac{R}{6}} \approx \alpha = \frac{\hbar^2}{m^2c^2}. \quad (1.340)$$

Thus β emerges as the Bohmian quantum mass field and the coupling constant α (which depends on $\hbar$) is also a field, connected to the geometrical properties of spacetime. The quantum effects and the length scale of the spacetime are linked because there is a relation between the gauge in which the quantum mass is constant (and the quantum force is zero) and the gauge in which the quantum mass is spacetime dependent in virtue of a scale change of the form

$$\beta = \beta_0 \to \beta(x) = \beta_0 \exp(-\Xi(x)); \phi_\mu \to \phi_\mu + \partial_\mu \Xi. \qquad (1.341)$$

In particular, one finds that ϕ_μ in the two gauges differ by $-\nabla_\mu(\beta/\beta_0)$ and, in virtue of the fact that ϕ_μ is a part of Weyl geometry and the Dirac field represents the quantum mass, the geometrical nature of the quantum effects inside this picture is revealed in a clear way (on the other hand, equation (1.333) shows that ϕ_μ is not independent of β so the Weyl vector is directly associated with the quantum mass and thus the geometrical aspects of the manifold are linked to quantum effects).

1.7 The Quantum Potential in Bohmian Quantum Cosmology

Quantum mechanics is a universal and fundamental theory, applicable to any physical system. The universe can be considered, of course, a physical system on its own: there is a theory, standard cosmology, which allows us to describe it in physical terms, and to make predictions which can be confirmed or refuted by observations. In fact, the observations until now seem to confirm the standard cosmological scenario. Hence, supposing the universality of quantum mechanics, the universe itself must be described by quantum theory, from which we could recover standard cosmology.

According to the author's point of view, the Copenhagen interpretation of quantum mechanics does not provide a satisfactory way in order to develop a coherent quantum theory of cosmology. In von Neumann's view, for instance, the necessity of considering a classical domain derives from the fact that it allows the measurement problem to be solved (the reader may see the reference [189] for a good discussion about this topic). If one makes an impulsive measurement of some observable, the wavefunction of the observed system plus the macroscopic apparatus splits into many branches, corresponding to different macroscopic situations, which almost do not overlap (in order to be a good measurement). Each of these branches encodes the information that the observed system is in an eigenstate of the measured observable, and that in correspondence the pointer of the measuring apparatus points to the corresponding eigenvalue. However, at the end of the measurement, the experimenter observes only one of these eigenvalues, and the measurement is robust in the sense that by repeating it immediately after, the experimenter obtains the same result. So, from the analisys of a measurement process, it seems that the wavefunction collapses, and the other branches disappear. The Copenhagen interpretation assumes that this collapse is real. However, a real collapse cannot be described by the unitary Schrödinger evolution. Hence, the Copenhagen interpretation must assume that there is a fundamental process in a measurement which must occur outside the quantum world, in a classical domain. Of course, if we want to quantize the whole universe, the consideration of a classical domain outside it does not seem to have physical sense, and the Copenhagen interpretation cannot be applied consistently.

Nevertheless, there are some alternative solutions to the quantum cosmological dilemma which, together with decoherence, can solve the measurement problem maintaining the universality of quantum theory. In this regard, one can mention the approach of the consistent histories [190], the many-worlds interpretation [191] and, of course, the de Broglie-Bohm interpretation. In particular, the de Broglie-Bohm interpretation, thanks to its fundamental element, the quantum potential, has the merit to build a real geometrodynamic picture of the behaviour of the universe as a whole.

Besides the several relevant aspects treated in the previous chapters, another important topic which puts in evidence the active role of the quantum potential in redesigning the geometry of physical space is thus represented by the results of quantum cosmology. In this regard, we focus our attention on the famous Wheeler-de Witt (WDW) equation (which might also be thought of as an Einstein-Schrödinger equation). The discussion made here follows the references [94, 158, 167–169, 192] which provide a Bohmian interpretation of quantum gravity and takes into consideration also the papers [175, 193–224] for material on WDW equation and quantum gravity.

1.7.1 F. Shojai's and A. Shojai's geometrodynamic approach to Wheeler-de Witt equation

The standard Wheeler-DeWitt (WDW) equation which characterizes the wavefunctional Ψ of the universe can be written in the following regularized form (here we have made the position $\hbar = c = 1$):

$$\left[(8\pi G)G_{abcd}p^{ab}p^{cd} + \frac{1}{16\pi G}\sqrt{q}(2\Lambda - {}^{(3)}R)\right]\Psi = 0. \qquad (1.342)$$

where $G_{abcd} = \frac{1}{2}\sqrt{q}(q_{ac}q_{bd} + q_{ad}q_{bc} - q_{ab}q_{cd})$ is the supermetric (also called the de Witt metric), p^{ab} are the canonical momentum operators related to the 3-metric q_{ab}, Λ is the cosmological constant, G is the gravitational constant.

As regards the mathematical constraints associated with the WDW equation, by following [167] the Lagrangian density for general relativity can be written in the form

$$L = \sqrt{-g}R = \sqrt{q}N({}^{(3)}R + Tr(K^2)) \qquad (1.343)$$

where ${}^{(3)}R$ is the three-dimensional Ricci scalar, K_{ij} is the extrinsic curvature, N is the lapse function, q_{ij} is the induced spatial metric, $q = \det q_{ij}$. The canonical momenta conjugate to the 3-metric are given by

$$p^{ab} = \frac{\partial L}{\partial \dot{q}_{ab}} = \sqrt{q}(K^{ab} + q^{ab}Tr(K)). \qquad (1.344)$$

A straightforward calculation shows that the classical Hamiltonian of general relativity can be expressed in the form

$$H = \int d^3x\sqrt{q}(NC + N_iC^i) \qquad (1.345)$$

where N_i is the shift function,

$$C = {}^{(3)}R + \frac{1}{q}\left(Tr(p^2) - \frac{1}{2}(Tr(p)^2)\right) = -2G_{\mu\nu}n^\mu n^\nu \quad (1.346)$$

$$C^i = -2G_{\mu i}n^\mu, \quad (1.347)$$

n^μ being the normal vector to the spatial hypersurfaces given by $n^\mu = (1/N, -\vec{N}/N)$.

According to equation (1.342), the wavefunctional Ψ of the universe depends only on the configuration space variables and variations with respect to it and is completely independent of any variable one could interpret as time. As a result, solutions to the WDW equation are stationary states. This fact leads to difficulties, for example, in forming an inner product under which evolves unitarily and with which one can define a clear notion of probability. The WDW equation (1.342) has the following important properties:

1. The time parameter which defines the foliation of the space-time, does not appear in it (we have thus the so-called time-problem in quantum gravity).
2. A different ordering of factors yields a different result.
3. In practice, in order to solve the WDW equation, instead of using an infinite-dimensional superspace, we can limit ourselves to a mini-superspace in which some of the degrees of freedom are non-frozen.
4. The wavefunction has to be square-integrable, in order to obtain a probabilistic interpretation for it. But this is not possible for all cases, because a precise definition of the inner product is not known in quantum gravity.

In the Bohm approach, the fundamental object of quantum cosmology is the geometry of three-dimensional spacelike hypersurfaces, which is assumed to exist independently of any observation or measurement, and the same thing is valid for its canonical momentum, the extrinsic curvature of the spacelike hypersurfaces. Its evolution, labelled by some time parameter, is ruled by a quantum evolution that is different from the classical one due to the presence of a "quantum potential for the gravitational field" which appears naturally from the WDW equation and is added to the classical Hamiltonian. One of the important results obtained in the de Broglie-Bohm approach to quantum cosmology lies in the elimination of cosmological singularities, which has been proved at least for some particular but relevant cases. In the Bohm-de Broglie interpretation, the quantum potential introduces relevant perspectives yielding a repulsive quantum force counteracting the gravitational field, avoiding the singularity and yielding inflation.

In this chapter, we want to analyse the Bohmian version of WDW equation (1.342) and its scenarios and results as regards the geometry of space. In a Bohmian approach to WDW equation (1.342), by decomposing the wavefunctional Ψ in polar form $\Psi = Re^{iS/\hbar}$ one obtains a modified Hamilton-Jacobi equation

$$(8\pi G)G_{abcd}\frac{\delta S}{\delta q_{ab}}\frac{\delta S}{\delta q_{cd}} - \frac{1}{16\pi G}\sqrt{q}(2\Lambda - {}^{(3)}R) + Q_G = 0, \quad (1.348)$$

where

$$Q_G = \hbar^2 N q G_{abcd}\frac{1}{R}\frac{\delta^2 R}{\delta q_{ab}\delta q_{cd}} \quad (1.349)$$

can be defined as "quantum potential for the gravitational field". Equation (1.348) suggests that the only difference between classical and quantum universes is the existence of the quantum potential in the latter. This means that a quantum regime may be obtained by modifying the classical constraints via relation

$$C \to C + \frac{Q_G}{\sqrt{q}N};\ C_i \to C_i. \quad (1.350)$$

As regards the constraint algebra, the following integrated forms of the constraints can be utilized

$$C(N) = \int d^3x\sqrt{q}NC; \tilde{C}(\vec{N}) = \int d^3x\sqrt{q}N^iC_i \quad (1.351)$$

which satisfy the following algebra

$$\{\tilde{C}(\vec{N}), \tilde{C}(\vec{N}')\} = \tilde{C}(\vec{N}\cdot\nabla\vec{N}' - \vec{N}'\cdot\nabla\vec{N});$$
$$\{\tilde{C}(\vec{N}), C(\vec{N})\} = C(\vec{N}\cdot\nabla\vec{N}); \{C(N), C(N')\} \approx 0. \quad (1.352)$$

On the basis of equations (1.352), the first 3-diffeomorphism subalgebra and the second 3-diffeomorphism subalgebra do not change with respect to the classical situation; instead, in the third equation the quantum potential changes the Hamiltonian constraint algebra dramatically according to relation

$$\frac{1}{N}\frac{\delta}{\delta q_{ab}}\frac{Q}{\sqrt{q}} = \frac{3}{4\sqrt{q}}q_{cd}p^{ab}p^{cd}\delta(x-z)$$
$$-\frac{\sqrt{q}}{2}q^{ab}({}^{(3)}R - 2\Lambda)\delta(x-z) - \sqrt{q}\frac{\delta^{(3)}R}{\delta q_{ab}} \quad (1.353)$$

yielding for the Poisson bracket a result weakly equal to zero (namely zero when the equations of motion are satisfied).

Here one can remark that the presence of the quantum potential means that the quantum algebra is the 3-diffeomorphism algebra times an Abelian subalgebra and that this resulting algebra is weakly closed. One can see that the algebra (1.352) is a clear projection of the general coordinate transformations to the spatial and temporal diffeomorphisms and in fact the equations of motion are invariant under such transformations. In particular, although the form of the quantum potential will depend on regularization and ordering, in the quantum constraint algebra the form of the quantum potential is not important; the algebra holds independently of the form of the quantum potential. Further it appears that the inclusion of matter terms will not modify anything.

The next step is to provide a formulation of the quantum Einstein equations inside the geometry of space represented by the algebra here mentioned. In this regard, we follow the references [159, 160, 167, 225, 226] of F. Shojai and A. Shojai (see also [155] and [66] for a review of these concepts). Let us consider, before all, the quantum Einstein equations in absence of source of matter-energy. As regards the dynamical parts, if one considers the Hamilton equations, one gets the following result

$$G^{ab} = -\frac{1}{N}\frac{\delta\int d^3x Q_G}{\delta q_{ab}} \quad (1.354)$$

where $G^{ab} = R^{ab} - \frac{1}{2}q^{ab}R$ is, of course, Einstein's tensor. Equation (1.354) physically means that the quantum force modifies the dynamical parts of Einstein's equations. For the non-dynamical parts, by utilizing the constraint relations (1.346) and (1.347), one obtains

$$G^{00} = \frac{Q_G}{2N^3\sqrt{q}}; \quad G^{0i} = -\frac{Q_G}{2N^3\sqrt{q}}N^i. \tag{1.355}$$

Equations (1.355) can also be written in a compact way as

$$G^{0\mu} = \frac{Q_G}{2\sqrt{-g}}g^{0\mu} \tag{1.356}$$

which shows that the non-dynamical parts are modified by the quantum potential. On the basis of equations (1.354)–(1.356), one can say that, in quantum cosmology, in absence of source of matter-energy, the quantum potential for the gravitational field determines a modification of the geometry of the physical space both for the dynamical parts and for the non-dynamical parts. It is also interesting to observe that the modified Einstein's equations (1.354)–(1.356) are covariant under spatial and temporal diffeomorphisms.

Now, by starting from equations (1.354)–(1.356), inclusion of matter is straightforward. If one inserts the matter quantum potential and introduces the energy-momentum tensor in these equations, the following fundamental equations are obtained:

$$G^{ab} = -kT^{ab} - \frac{1}{N}\frac{\delta\int d^3x(Q_G + Q_m)}{\delta q_{ab}};$$

$$G^{0\mu} = -kT^{0\mu} - \frac{1}{N}\frac{(Q_G + Q_m)}{2\sqrt{-q}}g^{0\mu} \tag{1.357}$$

where

$$Q_m = \hbar^2 \frac{N\sqrt{H}}{2}\frac{\delta^2 R}{\delta\phi^2} \tag{1.358}$$

(ϕ being the matter field), is the quantum potential for matter and, of course,

$$Q_G = \hbar^2 NqG_{abcd}\frac{1}{R}\frac{\delta^2 R}{\delta q_{ab}\delta q_{cd}} \tag{1.349}$$

is the quantum potential for the gravitational field. Therefore, one can say that the Bohmian approach to WDW equation based on equation (1.348) leads to general Bohm-Einstein equations of the form (1.357) which can be considered in fact as the quantum versions of Einstein's equations. Here, since regularization here only affects the quantum potential (cfr. also [159, 160, 167, 225, 226]), for any regularization the Bohm-Einstein equations (1.357) are the same. Moreover, the Bohm-Einstein equations (1.357) are invariant under temporal $\otimes$ spatial diffeomorphisms. Consequently, they can be formulated also in the equivalent compact form

$$G^{\mu\nu} = -kT^{\mu\nu} + S^{\mu\nu} \tag{1.359}$$

where

$$S^{0\mu} = \frac{Q_G + Q_m}{2\sqrt{-g}} g^{0\mu}; \quad S^{ab} = -\frac{1}{N} \frac{\delta \int d^3x (Q_G + Q_m)}{\delta g_{ab}} \tag{1.360}$$

In equations (1.359), (1.360) $S^{\mu\nu}$ is the quantum corrector tensor, under the temporal $\otimes$ spatial diffeomorphisms subgroup of the general coordinate transformations. From the analysis we have made here, we can conclude that equations (1.357) (and the equivalent equation (1.359)) describe the geometry of physical space which is derived from WDW equation and that the quantum corrector tensor $S^{\mu\nu}$ (defined by equations (1.360) and determined by the matter quantum potential and by the quantum potential for the gravitational field) encodes the information about the change, the deformation of the geometry of the physical space produced by matter and gravity in WDW equation's regime. The quantum potential, by introducing appropriate corrector terms in Einstein's equations, determines a deformation of the geometry of the physical space with respect to the classical situation also in the context of quantum cosmology.

In summary, according to F. Shoiai's and A. Shojai's model developed in the papers [159, 160, 167, 225, 226], in the Bohmian approach to WDW equation the complete set of mathematical equations which must be solved in order to describe and obtain the geometry of physical space in quantum cosmology is given by the equation (1.359), the WDW equation and the appropriate equation of matter field given by matter Lagrangian. It must be stressed here that solving the above mentioned equations is mathematically equivalent to solving the WDW equation and then using its decomposition to Hamilton–Jacobi equation and continuity equation, and extracting the Bohmian trajectories (cfr. also [32, 172–175] for more details in this regard). It must be remarked that in the papers [227, 228] Vink obtained the quantum Einstein's equations for a special metric (Robertson–Walker metric). However, in these papers neither any attempt have been made to write the equations for a general metric, nor the symmetries have been investigated. Finally, the Bohmian model for WDW equation suggested by F. Shojai and A. Shojai, by considering the divergence of equation (1.359), leads to obtain the following equation:

$$\nabla_\mu T^{\mu\nu} = \frac{1}{k}\nabla_\mu S^{\mu\nu} \tag{1.361}$$

which can be interpreted as an energy conservation law for WDW equation.

Bohm's approach to WDW equation has been indeed applied to many minisuperspace models obtained by the imposition of homogeneity of the spacelike hypersurfaces. In this regard, the reader can find details, for example, in the references [177, 179, 228–231]. Here, the classical limit, the singularity problem, the cosmological constant problem and the time issue have been discussed. For instance, in some of these papers it was shown that in models involving scalar fields or radiation, which are good representatives of the matter content of the early universe, the singularity can be avoided by quantum effects thanks to the action of the quantum potential which produces a repulsive quantum force counteracting the gravitational field.

1.7.2 *Pinto-Neto's results about Bohmian quantum cosmology*

Relevant results about the geometry of space determined by the quantum potential characterizing WDW equation have been also recently obtained by Pinto-Neto. As regards the Bohmian approach to quantum cosmology in the case of homogeneous minisuperspace models, in his papers [192, 232–240] Pinto Neto found the important result that there is no problem of time and that quantum effects can avoid the initial singularity, create inflation, and isotropize the universe. Pinto-Neto's treatment evidences that, if one considers the general case of superspace canonical quantum cosmology, the Bohmian evolution of the 3-geometries, independently of any regularization and factor ordering of the WDW equation, can be derived from a specific Hamiltonian, which is different from the classical one.

By following Pinto-Neto's analysis in [192, 232–240], in a Bohmian interpretation of canonical quantum cosmology one may obtain a quantum geometrodynamical picture where the Bohmian quantum evolution of three geometries may determine, depending on the features of the wavefunctional, a consistent non degenerate four-geometry (which can be Euclidean — for a very special local form of the quantum potential — or hyperbolic), and a consistent but degenerate four-geometry which indicates the presence of special vector fields and the breaking of the space-time structure intended as a single entity (in a wider class of possibilities).

Pinto-Neto's approach starts by writing the WDW equation in the following unregulated form in the coordinate representation

$$\left[-\hbar^2\left(kG_{abcd}\frac{\delta}{\delta q_{ab}}\frac{\delta}{\delta q_{cd}}+\frac{1}{2\sqrt{q}}\frac{\delta^2}{\delta\phi^2}\right)+V\right]\psi=0 \qquad (1.362)$$

where V is the classical potential given by relation

$$V=\sqrt{q}\left[-\frac{1}{k}\left({}^{(3)}R-2\Lambda\right)+\frac{1}{2}q^{ab}\partial_a\phi\partial_b\phi+U(\phi)\right] \qquad (1.363)$$

and there is the constraint $-2q_{ab}\nabla_b\left(\frac{\delta\Psi}{\delta q_{ab}}\right)+\left(\frac{\delta\Psi}{\delta\phi}\right)\partial_a\phi=0$. If one expresses the wavefunctional in polar form $\Psi=Re^{iS/\hbar}$, equation (1.362) allows the following Hamilton-Jacobi type equation to be obtained

$$kG_{abcd}\frac{\delta S}{\delta q_{ab}}\frac{\delta S}{\delta q_{cd}}+\frac{1}{2\sqrt{q}}\left(\frac{\delta S}{\delta\phi}\right)^2+V+Q=0 \qquad (1.364)$$

where the quantum potential is given by

$$Q = -\frac{\hbar^2}{R}\left(kG_{abcd}\frac{\delta^2 R}{\delta q_{ab}\delta q_{cd}} + \frac{1}{2\sqrt{q}}\frac{\delta^2 R}{\delta\phi^2}\right). \tag{1.365}$$

Inside Pinto-Neto's model, a non degenerate four-geometry emerges if the quantum potential (1.365) assumes the specific form

$$Q = -\sqrt{q}\left[(\varepsilon+1)\left(-\frac{1}{k}{}^{(3)}R + \frac{1}{2}q^{ab}\partial_a\phi\partial_b\phi\right)\right.$$
$$\left. + \frac{2}{k}(\varepsilon\bar{\Lambda} + \Lambda) + \varepsilon\bar{U}(\phi) + U(\phi)\right] \tag{1.366}$$

(where ε is a constant which can be ± 1 depending if the four-geometry in which the 3-geometries are embedded is Euclidean or hyperbolic, providing thus the conditions for the existence of spacetime). If $\varepsilon = -1$, the spacetime has a hyperbolic nature and the quantum potential (1.366) becomes

$$Q = -\sqrt{q}\left[\frac{2}{k}(-\bar{\Lambda} + \Lambda) - \bar{U}(\phi) + U(\phi)\right] \tag{1.367}$$

namely acts as a classical potential. An important effect of the quantum potential (1.367) is to determine a renormalization of the cosmological constant and of the classical scalar field potential. In this regime, the quantum geometrodynamics turns out to be indistinguishable from the classical one. If $\varepsilon = +1$, the space-time has Euclidean properties and thus the quantum potential (1.366) becomes

$$Q = -\sqrt{q}\left[2\left(-\frac{1}{k}{}^{(3)}R + \frac{1}{2}q^{ab}\partial_a\phi\partial_b\phi\right)\right.$$
$$\left. + \frac{2}{k}(\bar{\Lambda} + \Lambda) + \bar{U}(\phi) + U(\phi)\right]. \tag{1.368}$$

In this case, the quantum potential, besides to renormalize the cosmological constant and the classical scalar field potential, modifies the signature of space-time. The total potential $V + Q$ which corresponds to $\varepsilon = +1$ allows us to describe some era of the early universe characterized by a Euclidean signature, but not the present era, which is characterized by hyperbolic features. In order to provide a coherent cosmological description, therefore, one has to invoke a transition between these two phases, namely between the era characterized by Euclidean signature and the era characterized by hyperbolic features and, in the Pinto-Neto approach, this transition happens in a hypersurface which is characterized by a classical condition, namely where $Q = 0$. On the basis of the picture provided by Pinto-Neto, one can conclude that if universe is described by a quantum space-time endowed with different features with respect to the classical observed one, then its signature must be Euclidean. Therefore, one can say that here the transition of the signature of the four-geometry of space-time to a Euclidean one is the only relevant quantum effect which conserves the non-degenerate nature of the four-geometry of space-time. The other quantum effects are either irrelevant or break completely the space-time structure.

However, as shown by Pinto-Neto, the evolution of the three geometries under the action of the quantum potential in general does not produce a non degenerate four-geometry, a single space-time with the causal structure of relativity. The most general structures that are formed in the course of the evolution of the three geometries under the influence of the quantum potential, are degenerate four-geometries characterized by alternative causal structures. In particular, in the case of consistent quantum geometrodynamical evolution in which there is a degenerate four-geometry, Pinto-Neto's approach implies that any real solution of the WDW equation, when the evolution is represented by a degenerate four-geometry, yields a structure which is the idealization of the strong gravity limit of general relativity. An example of this situation is obtained, for real solutions of WDW equation, if the phase S is zero and thus $Q = -V$ which implies that the quantum super Hamiltonian contains only the kinetic term. Another example of this situation occurs if the quantum potential satisfies the condition $Q = \gamma V$ where γ is a (non-local) function of the functional S. In this specific case, the non-local features of the quantum potential cause a breaking of space-time. This type of geometry, which was already studied also in the reference [106], might well be the correct quantum geometrodynamical description of the young universe.

Moreover, for non-local quantum potentials, Pinto-Neto demonstrated that apparently inconsistent quantum evolutions turn out to be indeed consistent if they are restricted to the Bohmian trajectories satisfying the guidance relations corresponding to the Bohm-Einstein equations (1.359). In substance, stringent boundary conditions, like the form (1.366) for the quantum potential (which implies a severe restriction on the solutions to the WDW equation), may be obtained if one imposes that quantum geometrodynamics does not break space-time by determining a non degenerate four-geometry.

As regards Pinto-Neto's results about the geometry of the background of processes derived from the WDW equation, one must remark that, although factor ordering and regularization of the WDW equation do not seem to play a fundamental role and a lot of information can be obtained if one uses the quantum potential approach (without appealing to any probabilistic notion), nonetheless these results are limited by many strong assumptions tacitly made. In particular, here it is supposed that a continuous three-geometry exists at the quantum level (quantum effects could also destroy it), or it is assumed the validity of quantization of standard general relativity, forgetting other developments like, for example, string theory. However, also on the basis of Pinto-Neto's model, it seems that the de Broglie-Bohm interpretation may at least be regarded as a suggestive reference picture to be utilized in order to explore the geometry of space in quantum cosmology. Furthermore, if the finer view of the geometrodynamic approach to quantum cosmology based on Bohm's quantum potential can yield useful information in the form of observational effects, then we will have means to decide between different interpretations, something that will be very important not only for quantum cosmology, but for quantum theory itself.

One might event object that the conclusions regarding the geometry of processes in the WDW equation regime obtained on the basis of the de Broglie-Bohm theory have no physical significance, in the sense that they seem abstractions with no observational consequences. However, as shown by Pinto-Neto and Fabris in the recent paper [240], the hypothesis of the objective reality of quantum Bohmian trajectories in configuration space may lead to peculiar observational consequences related to the evolution of quantum cosmological perturbations in quantum cosmological backgrounds, which might be tested (and which are not known how to be obtained in other approaches, such as the many world approach or the consistent histories approach).

If one raises the question regarding the presence of an initial singularity, in the light of Pinto-Neto's and Fabris' results obtained in [240], the de Broglie-Bohm interpretation turn out to present more advantages than the many worlds approach or the consistent histories approach. While in the many worlds picture, despite the mean value is not singular, there may exist worlds in which the scale factor vanishes, one can assert with certainty that within the de Broglie-Bohm theory the initial singularity is eliminated because there is no quantum Bohmian trajectory which is singular. On the other hand, in the framework of the consistent histories, the question about the existence of quantum bounces occurring in family of histories with more than two moments of time has no meaning, one cannot answer whether quantum bounces take place because histories involving any genuine quantum states are inconsistent (in fact, families of histories containing properties defined in one or more moments of time, besides properties defined in the infinity past and in the infinity future, turn out to be no longer consistent, unless one takes semi-classical states, which of course corresponds to histories without a bounce). On the contrary, if one considers the de Broglie-Bohm theory, where trajectories in configuration space are considered to be objectively real, there exist plenty of non-singular bouncing trajectories which go to the classical cosmological trajectories when the volume of the universe is big. In particular, in the paper [240], Pinto-Neto and Fabris consider, in the framework of Bohmian quantum cosmology, two concrete examples of minisuperspace non singular bouncing models, namely prefect fluids and a free massless scalar field, showing that not only singularities are avoided but also quantum isotropization of the universe occurs.

1.7.3 *The quantum potential ... and the cosmological constant*

Undoubtedly one of the most prominent performances of modern cosmology has been to provide observational evidence of the accelerated expansion of the universe. In this regard, the notion of "dark energy" was introduced; it refers to some mysterious form of diffuse (namely not clustering) energy presumably permeating all corners of the universe, possessing negative pressure and thus being capable of boosting the expansion of the universe as a whole.

One of the most intriguing aspect of the dark energy concerns the so-called cosmological constant problem, which arises when we are at the intersection region between general relativity and quantum field theory. In this regard, the vacuum energy density of general relativity ($\langle \rho_{vac} \rangle$) is usually assumed to be equivalent to a contribution to the 'effective' cosmological constant in Einstein equations

$$\Lambda_{eff} = \Lambda_0 + \frac{8\pi G}{c^4} \langle \rho_{vac} \rangle \tag{1.369}$$

where Λ_0 denotes Einstein's own 'bare' cosmological constant which generates a curvature of empty space, namely in absence of matter and radiation. As a consequence of equation (1.369), anything which contributes to the quantum field theory vacuum energy density determines also a contribution to the effective cosmological constant in general relativity, in other words quantum field theory and general relativity can be seen as two distinct aspects of the same physical reality from the point of view of their contribution to the vacuum energy density.

From the quantum field theories which describe the known particles and forces one can derive various contributions to the vacuum energy density. The vacuum energy density associated with these theories, which has experimentally demonstrated consequences and is thus taken to be physically real, determines cosmological implications if certain assumptions are made about the relation between general relativity and quantum field theory. On the basis of the fundamental theories, one can infer that the total vacuum energy density has at least the following three contributions

$$\begin{pmatrix} Vacuum \\ energy \\ density \end{pmatrix} = \begin{pmatrix} VACUUM \\ ZERO-POINT-ENERGY \\ +FLUCTUATIONS \end{pmatrix} + \begin{pmatrix} QCD \\ gluon-and-quark \\ condensates \end{pmatrix} + \begin{pmatrix} The \\ Higgs \\ field \end{pmatrix} + \cdots \tag{1.370}$$

namely the fluctuations characterizing the zero-point field, the fluctuations characterizing the quantum chromodynamic level of subnuclear physics and the fluctuations linked with the Higgs field, and the dots represent contributions from possible existing sources outside the Standard Model (for instance, GUT's, string theories, and every other unknown contributor to the vacuum energy density). The Standard Model presents no structure which indicates any relations between the terms in equation (1.370). This means that one can assume that the total vacuum energy density $\rho_{vac}$is, at least, as large as any of the individual terms appearing in (1.370). The vacuum energy density estimate within the Standard Model may be reconciled with the observational limits on the cosmological constant $|\Lambda| < 10^{-56}cm^{-2}$ by invoking a programme of "fine-tuning": for example, if the vacuum energy is estimated to be at least as large as the contribution from the QED sector then Λ_0 has to cancel the vacuum energy to a precision of at least 55 orders of magnitude.

In this paragraph, our purpose is to show in what sense Bohm's quantum potential and its geometrodynamic features allow us to throw new light as regards the cosmological constant. In this regard, we base our discussion on the recent fascinating paper "Cosmology from quantum potential" by Farag Ali and Das [241]. Ali and Das showed that replacing classical geodesics with Bohmian trajectories gives rise to a quantum corrected Raychaudhuri equation, which leads to second order Friedmann equations, where a couple of quantum correction terms appear, the first of which can be interpreted as cosmological constant (and gives a correct estimate of its observed value), while the second can be seen as a radiation term in the early universe, which gets rid of the big-bang singularity and predicts an infinite age of our universe.

By assuming that the universe is filled by a fluid or condensate which is described by the wave function $\Psi = Re^{iS/\hbar}$, the quantum corrected Raychaudhuri equation assumes the form

$$\frac{d\theta}{d\lambda} = -\frac{1}{3}\theta^2 - R_{cd}u^c u^d + \frac{\hbar^2}{m^2} q^{ab} \frac{\left(\nabla^2 - \frac{1}{c^2}\frac{\partial^2}{\partial t^2}\right) R}{R} + \frac{\varepsilon_1 \hbar^2}{m^2} q^{ab} R_{;ab} \tag{1.371}$$

where $\theta = q^{a,b}u_{a;b}$, $q_{ab} = g_{ab} - u_a u_b$, $u_a = \frac{\hbar}{m}\partial_a S$, ε_1 is a constant. From equation (1.371) the second order Friedmann equation satisfied by the scale factor $a(t)$ can be derived with the substitutions $\theta \to \frac{3\dot{a}}{a}$, $R_{cd}u^c u^d \to \frac{4\pi G}{3(\rho+3p)} - \frac{\Lambda c^2}{3}$:

$$\frac{\ddot{a}}{a} = -\frac{4\pi G}{3}(\rho + 3p) + \frac{\Lambda c^2}{3} + \frac{\hbar^2}{3m^2} q^{ab} \frac{\left(\nabla^2 - \frac{1}{c^2}\frac{\partial^2}{\partial t^2}\right) R}{R} + \frac{\varepsilon_1 \hbar^2}{m^2} q^{ab} R_{;ab} \tag{1.372}$$

where density ρ includes visible and dark matter, and may also include additional densities that arise in massive non-linear theories of gravity [see also [242–244]]. The two last terms in equations (1.371) and (1.372), depending on $\hbar^2$, can be interpreted as quantum corrections of the state of the universe. In particular, the first of these may be recognized as the quantum potential

$$Q = \frac{\hbar^2}{3m^2} q^{ab} \frac{\left(\nabla^2 - \frac{1}{c^2}\frac{\partial^2}{\partial t^2}\right) R}{R} \tag{1.373}$$

which vanish in the $\hbar \to 0$ limit, giving back the classical Raychaudhuri equation and the Friedmann equations. These additional terms are not ad-hoc or hypothetical, but rather emerge as an unavoidable consequence of a quantum description of the contents of the universe. From the quantum potential one can derive directly a cosmological constant that, by assuming a Gaussian form $\Psi \approx \exp(-r^2/L_0^2)$ or, in the context of a scalar field theory with an interaction of strength

g, $\Psi = \Psi_0 \tanh(r/L_0\sqrt{2})$ (for $g > 0$) and $\Psi = \sqrt{2}\Psi_0 \sec h(r/L_0)$ (for $g < 0$) where $L_0 = \hbar/mc$ is the characteristic length scale in the problem which is of the order of the Compton wavelength, is expressed by relation

$$\Lambda_Q = \frac{1}{L_0^2} = (mc/\hbar)^2 \tag{1.374}$$

which has the correct sign as the observed cosmological constant. As regards the estimate of its magnitude, one may identify L_0 with the current linear dimension of our observable universe, since anything outside it would not influence an accessible wavefunction. In this way, m can be interpreted as the small mass of gravitons (or axions), with gravity (or Coulomb field) following a Yukawa type of force law

$$F = -G\frac{m_1 m_2}{r^2} \exp(-r/L_0). \tag{1.375}$$

Since gravity has not been tested beyond this length scale, this interpretation seems to be natural [245]. If one substitutes $L_0 = 1,4 \cdot 10^{26} m$ in $L_0 = \hbar/mc$, one obtains $m \approx 10^{-68} Kg$, which turn out to be coherent with the estimated bounds on graviton masses from various experiments [246–249], and also from theoretical considerations [250–254]. In other words, in Farag Ali's and Das' model the quantum condensate filling the universe can be described as made up of gravitons, and described by a macroscopic wavefunction. Finally, by substituting the above value of $L_0 = 1,4 \cdot 10^{26} m$ in equation (1.374), one obtains

$$\Lambda_Q = 10^{-52} m^{-2} \tag{1.376}$$

which indeed is in good agreement with the observed value.

1.8 About the Quantum Potential in the Treatment of Entangled Qubits

As is known, a two-state quantum system can be considered as a quantum bit (qubit). While a classical bit takes a definite value, quantum bits can be prepared in a superposition of several values generating entangled states. In a system composed by two qubits, when one qubit interplays with the other qubit, the interplay can also be characterized by the exchange-type interactions of spins.

Inside the framework of standard quantum mechanics, spin properties of an entangled electron pair cannot be imagined in terms of classical variables: something mysterious surrounds the concept of quantum entanglement since it was first introduced by Erwin Schrödinger back in 1935. Despite this, today we are assisting at a significant explosion in research activity related to entanglement. This is owed to a growing awereness of its importance as a vital resource in quantum information through quantum teleportation [255], quantum cryptographic key distribution [256] and quantum computation [257, 258] as well as its potential for enhanced quantum sensing through the engineering of highly entangled quantum states, beating the usual quantum limit [259]. Although the relevant increase of the research activity as regards quantum entanglement, however, discussions about the meaning of entanglement and how it might be visualized beyond the standard quantum mechanics have been made [260, 261]. In particular, Valentini in [261] claims that quantum theory can be seen merely a special case of a much wider physics — a physics in which non-local signalling is possible, and in which the uncertainty principle can be violated. Valentini suggests a view of de Broglie's theory which provides a very novel perspective, according to which our local and indeterministic quantum physics can be drawn via relaxation processes out of a fundamentally non-local and deterministic physics — a physics the details of which are now screened off by the all-pervading statistical noise. In this picture, the non-local effect regarding the results of measurements on entangled particles averages to zero in equilibrium, but this cancellation is merely a feature of the equilibrium state. As equilibrium was approached, the possibility of superluminal signalling faded away and statistical uncertainty took over. Valentini claims that key features of what we regard as the laws of physics — locality, uncertainty and the principles of relativity theory — are merely features of our current state and not fundamental features of the world.

In this paragraph, we focus our attention on the geometry of systems of two entangled particles in the quantum potential approach, remembering that, in this regard, qubits are not restricted to real electron spins, but may be realized by any two-state quantum system such as for example, entangled photon [262], flux qubit in a superconducting ring [263], charge pseudo-spin of electron pairs in a double quantum dot [264], flying qubits in quantum point contacts [265] or qubits in a composite system [266]. We will analyse the geometry subtended by the quantum potential in quantum entanglement by considering the most fundamental and simple example of qubit pairs which is represented by a qubit pair of spin 1/2 particles in a pure state of the form

$$|\psi\rangle = \cos\frac{\vartheta}{2}\,|\uparrow\downarrow\rangle + e^{i\phi}\sin\frac{\vartheta}{2}\,|\downarrow\uparrow\rangle \tag{1.377}$$

where $|\uparrow\downarrow\rangle$ corresponds to the state of the system when the first particle (qubit) is in the "up" state, i.e., in the direction of the z-axis, and the second qubit is in the "down" state, while $|\downarrow\uparrow\rangle$ corresponds to the state of the system when the first qubit is in the down state and the second qubit is in the up state.

In the papers [32, 267] Holland showed that quantum entanglement can be described ontologically in a Bohmian framework by starting from the mapping between a quantum spherically symmetric rigid rotor and a classical spinning top in the presence of a quantum potential. Defining a differential operator $\hat{\vec{M}}$ representing the angular momentum, whose components are the infinitesimal generators of the rotation group $SO(3)$, the Hamiltonian of a quantum spherically symmetric rigid rotor is given by

$$\hat{H} = \frac{\hat{\vec{M}}^2}{2I} \tag{1.378}$$

where I is the moment of inertia. Eigenstates of the three mutually commuting operators $\hat{\vec{M}}^2$, M_z and $\vec{e}\cdot\hat{\vec{M}}$ are functions of Euler angles $\xi = (\alpha, \beta, \gamma)$, which specify the orientation of a rigid body with the principal axis defined by a normalized vector $\vec{e}$. By following a usual procedure inside a Bohmian approach, the wavefunction can be expressed in polar form as $\psi = Re^{iS}$, where $R(\xi)$ and $S(\xi)$ are real functions. By applying this decomposition of the wavefunction, the angular momentum in the Bohmian space turns out to be given by a real three dimensional vector

$$\vec{M} = i\hat{\vec{M}}S. \tag{1.379}$$

Relation (1.379) is an analogue of a more familiar de Broglie's guidance equation for the velocity of a point-like particle with mass m treated in the Bohmian approach, $m\vec{v} = \nabla S$ (see also [30] for a review of these concepts).

The dynamics is determined by the Hamilton-Jacobi-type equations for the classical Hamiltonian with an additional quantum potential Q,

$$\hat{H} = \frac{\hat{\vec{M}}^2}{2I} + Q, \tag{1.380}$$

where

$$Q = \frac{\hat{\vec{M}}^2 R}{2IR} \tag{1.381}$$

is the quantum potential. The quantum potential (1.381) generates a quantum torque

$$\vec{T} = -i\hat{\vec{M}}Q \tag{1.382}$$

which rotates the angular momentum vector via the equation of motion

$$\frac{d\vec{M}}{dt} = \vec{T} \tag{1.383}$$

along the trajectory $\xi(t)$. This is a counterpart of the Newton equation for the case of a free particle in the Bohm formulation given by

$$m\frac{d\vec{v}}{dt} = \nabla\left(\frac{\nabla^2 R}{2mR}\right). \tag{1.384}$$

The equations of motion regarding the angular momentum simplify to the following set of first order non-linear differential equations for the trajectories in the configuration space

$$I\dot{\alpha} = \frac{\partial S}{\partial \alpha} \tag{1.385}$$

$$I\dot{\beta} = 7\left(\frac{\partial S}{\partial \beta} - \cos\alpha \frac{\partial S}{\partial \gamma}\right) / \sin^2\alpha \tag{1.386}$$

$$I\dot{\gamma} = \left(\frac{\partial S}{\partial \gamma} - \cos\alpha \frac{\partial S}{\partial \beta}\right) / \sin^2\alpha \tag{1.387}$$

where the solutions $\xi(t)$ represent orbits in the configuration space, uniquely determined by the initial positions $\xi(0) = (\alpha_0, \beta_0, \gamma_0)$, and the angular momentum emerges as $\vec{M}[\xi(t)]$.

Let us analyse now the dynamics and the quantum effects of a general two qubits pure state, with vanishing total angular momentum projections, given by equation (1.377). In a Bohmian framework the guiding wave function

$$\psi(\xi) = \cos\frac{\vartheta}{2} u_{\uparrow}(\xi_1) u_{\downarrow}(\xi_2) + e^{i\phi} \sin\frac{\vartheta}{2} u_{\downarrow}(\xi_1) u_{\uparrow}(\xi_2) \tag{1.388}$$

is given in a six-dimensional space spanned by $\xi = \{\xi_1, \xi_2\}$, where ξ_1, ξ_2 are the coordinates of the first and the second rotor, respectively. The corresponding Hamiltonian is given by relation

$$H = \frac{\vec{M}_1^2 + \vec{M}_2^2}{2I} + Q \tag{1.389}$$

where

$$Q = \frac{(\hat{M}_1^2 + \hat{M}_2^2)R}{2IR} \tag{1.390}$$

is the quantum potential. It is interesting to observe that, even for two non-interacting, but entangled qubits, the quantum potential (1.390) represents a direct interaction between the rotors as a consequence of the quantum, and thus non-local, nature of the problem. The solutions for each of the angular momentum vectors $\vec{M}_1$ and $\vec{M}_2$ are functions of six common coordinates forming the trajectory $\xi(t)$ determined by six initial values $\xi(0)$.

In the papers "Geometrical view of quantum entanglement" [268] and "Spin-spin correlations of entangled qubit pairs in the Bohm interpretation of quantum mechanics" [269], Ramsak showed how quantum entanglement of a pair of qubits may be visualized in geometrical terms: by analysing the dynamics of the qubits in the framework of the de Broglie-Bohm interpretation of quantum mechanics, he found that the angular momenta of two qubits can be viewed geometrically in the Bohmian space of hidden variables and characterized by their relative angles. According to Ramsak's results, for perfectly entangled pairs, the qubits exhibit a unison precession making a constant angle between their angular momenta. In particular, in the paper [268] Ramsak computed trajectories $\vec{M}_{1,2}[\xi(t)]$ covering the full configuration space with $\approx 10^6$ initial values $\xi(0)$ per $|\Psi\rangle$, namely for a particular choice of ϑ and ϕ. Although these trajectories exhibit extremely rich variety, the following common properties can be outlined:

(i) The quasi-periodic motion appears chaotic and, except in special cases, the projections of the total momentum $\vec{M}$ onto the xy-plane winds around the origin an infinite number of times in a spirographic manner, forming a dense annulus limited by fixed outer and inner radii;
(ii) The curve corresponding to the relative momentum $\vec{M}_2 - \vec{M}_1$ is closed and periodic if plotted in the reference frame rotating synchronously with $\vec{M}$ around the z-axis.

These results suggest that the entanglement properties of a qubit pair in the state (1.377) are linked with the dynamics of the azimuthal angles $\phi_1[\xi(t)]$ and $\phi_2[\xi(t)]$ of the angular momenta. In the Bohmian interpretation, the angular momentum vectors of the two particles (qubits) precess in a well-defined way with some initial probability distribution. In particular, on the basis of Ramsak's results, the probability distribution

$$\frac{dP(\phi)}{d\phi} = \int \delta[\phi - \phi(\xi)]R^2(\xi)d\xi \tag{1.391}$$

of the ensemble average difference of azimuthal angles $\phi[\xi(t)] = \phi_2 - \phi_1$ is constant for unentangled qubits and becomes progressively peaked at ϕ for increasing entanglement, culminating in precession of angular momenta at equal relative angle $\phi[\xi(t)] = \phi$ for all ξ

consistent with perfect entanglement. The shape of the distribution is independent of ϕ.

Moreover, by classifying the ensemble representatives of the system of two qubits on the basis of the relative direction of angular momenta precession of the two qubits, Ramsak found that in the low entanglement regime the xy-plane projections of momenta $\vec{M}_1$ and $\vec{M}_2$ precess mainly in opposite directions, and that, in general, momentum pairs move part time in the same and part time in the opposite direction. Ramsak defined representatives that always precess in the same direction as "concurrent" movers whereas those which always precess in the opposite direction as "anticoncurrent" and measured the concurrency of the trajectories $\xi(t)$ by introducing the quantity

$$C_\xi = \frac{1}{\tau}\int_0^\tau sign\frac{d\phi_1[\xi(t)]}{dt}\frac{d\phi_2[\xi(t)]}{dt}dt. \tag{1.392}$$

At each moment the angular momenta for a given trajectory $\xi(t)$ precess either in the same or in the opposite direction. The concurrency of a trajectory $\xi(t)$ can be visualized as a measuring parameter of the share of the time that both angular momenta move in the same direction. For example, one has $C_\xi = \pm 1$ for perfectly concurrent and anti-concurrent movers, respectively, and $C_\xi > 0$ for trajectories where angular momenta move concurrently more than half of the time for some members of the ensemble. In the low-entanglement regime anticoncurrent movers dominate whereas the distribution of concurrent movers progressively dominates as entanglement increases. The concurrency (1.392) is characterized by the probability distribution

$$\frac{dP(C_\xi)}{dC_\xi} = P_+\delta(C_\xi - 1) + P_-\delta(C_\xi + 1) + \rho(C_\xi) \tag{1.393}$$

where $P_\pm$ is the probability that the concurrency is exactly ± 1, respectively, and $\rho(C_\xi)$ is a continuous function for which motion is sometimes concurrent and sometimes anti-concurrent as t changes.

Moreover, in the paper [269] Ramsak considered in particular the angle made by the angular momenta and the azimuthal angle made by the xy-plane projections of the momenta. The angular momenta

$\vec{M}_1$ e $\vec{M}_2$ of the two qubits make an angle Φ in the Bohmian space of hidden variables ξ satisfying the following relations

$$\langle \cos \Phi \rangle = \left\langle \frac{\vec{M}_1 \cdot \vec{M}_2}{|\vec{M}_1||\vec{M}_2|} \right\rangle, \tag{1.394}$$

$$(\Delta \cos \Phi)^2 = \langle \cos^2 \Phi \rangle - \langle \cos \Phi \rangle^2. \tag{1.395}$$

For $\vartheta = \pi/2$ one obtains the exact relation

$$\langle \cos \Phi \rangle = \frac{1}{3}(2 \cos \phi - 1) \tag{1.396}$$

identical to the expression of standard quantum mechanics. Equation (1.396) implies that a perfect antiparallel alignment occurs for the singlet state, while momenta for the triplet state, $\phi = 0$, are only partially aligned, $\langle \cos \Phi \rangle = \frac{1}{3}$. As regards qubit pairs with vanishing total z-axis projection of spin, the azimuthal angle made by the xy-plane projections of the momenta is

$$\langle \cos \phi \rangle = \left\langle \frac{M_{1x}M_{2x} + M_{1y}M_{2y}}{\sqrt{(M_{1x}^2 + M_{1y}^2)(M_{2x}^2 + M_{2y}^2)}} \right\rangle \tag{1.397}$$

and the corresponding variance is

$$(\Delta \cos \phi)^2 = \langle \cos^2 \phi \rangle - \langle \cos \phi \rangle^2. \tag{1.398}$$

Taking into account the guiding wave function (1.388) of the entangled system, Ramsak found that a finite ϕ represents only a shift of one of azimuthal angles of each member of the ensemble $\beta_2 \to \beta_2 + \phi$, which results in the identity $\langle \phi \rangle = \phi$ and the decoupling $\langle \cos \phi \rangle = C_B \cos \phi$, where C_B is a function of ϑ only. An analogous result was obtained for $\langle \sin \phi \rangle = C_B \sin \phi$ and thus $\langle \cos(\phi - \phi) \rangle = C_B$ and $\langle \sin(\phi - \phi) \rangle = 0$. A higher degree of entanglement can be visualized as a highly correlated distribution of angular momenta making azimuthal angles difference close to ϕ, with suppressed fluctuations for progressively increasing entanglement.

Ramsak's treatment in [269] of the entanglement properties of a qubit pair in the framework of Bohm's interpretation shows furthermore that the Bohmian analogue of Bell's inequalities, expressed in terms of Bohmian spin-spin correlators, is for fully entangled states identical to the standard quantum mechanics counterpart. In the picture proposed by Ramsak, the source of non-locality lies in the quantum potential which generates a direct, instantaneous coupling between the angular momenta of the entangled qubit pairs. In fully entangled states the angular momenta precess in a particular manner forming a constant relative azimuthal angle ϕ which is the origin of a specific form of both, the standard quantum mechanical and the Bohmian correlators,

$$\begin{aligned} B(\vec{a},\vec{b}) &= 3\left\langle \frac{(\vec{a}\cdot\vec{M}_1)(\vec{b}\cdot\vec{M}_2)}{|\vec{M}_1||\vec{M}_2|} \right\rangle \\ &= (a_x b_x + a_y b_y) B_x \cos\phi + (a_y b_x + a_x b_y) B_x \sin\phi - a_z b_z B_z, \end{aligned} \tag{1.399}$$

leading to the violation of the Bell inequalities.

Chapter 2

Quantum Entropy and Quantum Potential

De Broglie-Bohm theory can receive a new interesting and suggestive reading which is based on the idea that all the features of the quantum potential derive from a fundamental physical quantity that can be appropriately called "quantum entropy".

As one can draw from the literature on this topic, there are several definitions of quantum entropy. Let us remember, for example, the well-known von Neumann entropy, which corresponds to the average information the experimenter obtains in repeated observations of many copies of an identically unknown state at any point in the measurement process. Since the von Neumann entropy meets problems for the treatment of pure states, a more recent definition of quantum entropy has been proposed by Kak in which, by considering an interplay between unitary and non-unitary evolution, the quantum entropy indicates the average uncertainty that the receiver has in relation to the quantum state for each measurement. Kak's definition of entropy turns out to be in a complete agreement with the fact that, by analysing the information transfer problem from the point of view of the preparer of the state and the experimenter, both mixed and pure states provide information to the experimenter [155]. In this chapter, however, we consider another suggestive definition of quantum entropy, which is associated with the background of the processes (and thus can be considered more "fundamental" than other definitions of quantum entropy), originally introduced by Sbitnev in the papers *Bohmian split of the Schrödinger equation onto two equations describing evolution of real functions* [270] and *Bohmian trajectories and the path integral paradigm. Complexified lagrangian mechanics* [271]. We can call this new way of reading de Broglie-Bohm theory as the "entropic version" of de Broglie-Bohm theory or, more briefly, "entropic de Broglie-Bohm theory". As we will show in the following paragraphs, the approach based on Sbitnev's quantum entropy can be applied to the ideas and models illustrated in chapter 1, namely in the non-relativistic domain, in the relativistic quantum mechanics, in the relativistic de Broglie-Bohm theory in curved space-time, in relativistic quantum field theory, in Bohmian quantum gravity and in Bohmian quantum cosmology. Moreover, in the last paragraph of this chapter we will consider a second definition of quantum entropy recently suggested by the author together with Licata and Resconi, which can provide a new re-reading of quantum information. In summary, in this chapter our aim is to review the most significant results obtained as regards Sbitnev's quantum entropy and Resconi's approach, showing how the quantum entropy can be considered as the fundamental source and origin of the geometrodynamic features of the quantum potential. In this regards we base our discussion on the references [66, 171, 272–274].

2.1 The Quantum Entropy as the Ultimate Source of the Geometry of Space in the Non-Relativistic Domain

By following the treatment made by the author of this book in the papers [66, 171, 272, 273], in the non-relativistic domain, the entropic approach starts by considering the logarithmic function originally introduced by Sbitnev

$$S_Q = -\frac{1}{2}\ln\rho \tag{2.1}$$

where ρ is the probability density (describing the space-temporal distribution of an ensemble of particles, namely the density of particles in the element of volume d^3x around a point $\vec{x}$ at time t)

associated with the wave function $\psi(\vec{x},t)$ of an individual physical system. In the entropic version of Bohmian quantum mechanics, the space-temporal distribution of the ensemble of particles describing the individual physical system under consideration is assumed to generate a modification, a sort of deformation of the background space characterized by the quantity given by equation (2.1). On the basis of equation (2.1), it is plausible to make a parallelism with the standard definition of entropy given by the Boltzmann law, in other words equation (2.1) may be considered indeed as the quantum counterpart of a Boltzmann-type law. In the light of its relation with the wave function, the quantity given by equation (2.1) can be appropriately defined as "quantum entropy". The quantum entropy (2.1) can be interpreted as the physical parameter that, in the quantum domain, measures the degree of order and chaos of the vacuum — a storage of virtual trajectories supplying optimal ones for particle movement — which supports the density ρ describing the space-temporal distribution of the ensemble of particles associated with the wave function under consideration.

In the recent article [270] Sbitnev has shown that by introducing the quantity (2.1), the quantum potential can be expressed in the following convenient way

$$Q = -\frac{\hbar^2}{2m}(\nabla S_Q)^2 + \frac{\hbar^2}{2m}(\nabla^2 S_Q). \tag{2.2}$$

Substitution of equation (2.2) into equation (1.3) leads directly to the following equation of motion for the corpuscle associated with the wave function $\psi(\vec{x},t)$ in the picture of Bohm's non-relativistic quantum mechanics:

$$\frac{|\nabla S|^2}{2m} - \frac{\hbar^2}{2m}(\nabla S_Q)^2 + V + \frac{\hbar^2}{2m}(\nabla^2 S_Q) = -\frac{\partial S}{\partial t}. \tag{2.3}$$

Equation (2.3) represents an energy conservation law in which the term $-\frac{\hbar^2}{2m}(\nabla S_Q)^2$ can be interpreted as the quantum corrector to the kinetic energy $\frac{|\nabla S|^2}{2m}$ of the particle while the term $\frac{\hbar^2}{2m}(\nabla^2 S_Q)$ can be interpreted as the quantum corrector to the potential energy V. Sbitnev's mathematical formalism of the Bohmian approach based on equations (2.1)–(2.3) allows us thus to develop the following reading of the quantum potential and of the energy conservation law in quantum mechanics. The quantum potential and its active geometrodynamic information on the behaviour of quantum particles is associated with a more fundamental and profound geometry represented by the quantum entropy describing the degree of order and chaos of the background space (namely the modification in the background space) produced by the density of the ensemble of particles associated with the wave function under consideration. In other words, one can say that the density of the ensemble of particles associated with the wave function under consideration defines a ultimate vacuum information which generates the geometrodynamics action of the quantum potential and its fundamental existence in the quantum domain. In fact, in the light of equation (2.3), the information of the vacuum defined by the quantum entropy associated with the wavefunction can be considered the real origin of quantum processes in the sense that the quantum entropy determines two quantum correctors in the energy of the physical system under consideration (of the kinetic energy and of the potential energy respectively) and without these two quantum correctors (linked just with the quantum entropy) the total energy of the system would not be conserved [66, 272].

Moreover, by substituting the quantum entropy given by equation (2.1) in the continuity equation (1.6) one obtains the entropy balance equation

$$\frac{\partial S_Q}{\partial t} = -(\vec{v} \cdot \nabla S_Q) + \frac{1}{2} \nabla \cdot \vec{v} \tag{2.4}$$

where $\vec{v} = \frac{\nabla S}{m}$ is the particle's speed. In equation (2.4), the second term at the right hand describes the rate of the entropy flow due to spatial divergence of the speed and is nonzero in regions where the particle changes direction of movement. Since — on the basis of Brillouin's results obtained in [275] — a negative value of S_Q is related to information, equation (2.4) can be interpreted as a law which describes balance of the information flows. In virtue of the feature of the continuity equation (2.4) as describing balance of the information flows, it emerges naturally that the quantum potential expressed by equation (2.2), namely as a sum of two quantum correctors linked with the quantum entropy, constitutes indeed an information channel into the behaviour of the particle into consideration. And, since the quantum potential (2.2) is ultimately generated by the quantum entropy measuring the information of the vacuum which provides a storage of virtual trajectories supplying optimal ones for particle movement, one can say that it is just the quantum entropy that can be considered the fundamental element that determines the fact that the quantum potential acts as an information channel into the behaviour of the physical system under consideration. The quantum entropy is the fundamental structure which makes the action of the quantum potential an information channel into the behaviour of quantum particles. In other words, one can see that by introducing the quantum entropy given by equation (2.1), it is just this quantity describing the degree of order and chaos of the vacuum — a storage of virtual trajectories supplying optimal ones for particle movement — supporting the density ρ (of the particles associated with the wave function under consideration) the ultimate element which, at a fundamental level, produces an active information in the behaviour of the particles. The geometry of space determined by the quantum potential in the non-relativistic domain can be thus seen as a secondary manifold which emerges from the quantum entropy (2.1). It is the quantum entropy (2.1) the fundamental element which, by indicating the modification of the geometrical properties of the background space, of the fundamental vacuum, produced by the density ρ of the ensemble of particles associated with the wave function under consideration, generates the geometric properties of space from which the quantum force, and thus the behaviour of quantum particles, are derived. In particular, the non-local features of the geometry of quantum processes (such as in EPR-type experiments) and which are associated with the quantum potential can be seen as a consequence of the quantum entropy representing the information of the vacuum, in virtue of the presence of the Laplace operator of the quantum entropy.

As regards many-body systems, the development of the mathematical formalism of the entropic non-relativistic Bohmian quantum mechanics is direct. By considering a wave function

$\psi = R(\vec{x}_1,\ldots,\vec{x}_N,t)e^{iS(\vec{x}_1,\ldots,\vec{x}_N,t)/\hbar}$, which is defined on the configuration space R^{3N} of a system of N particles, one can define a quantum entropy of the form (2.1) where $\rho = |\psi(\vec{x}_1,\vec{x}_2,\ldots,\vec{x}_N,t)|^2$ is the probability density associated with the wave function $\psi(\vec{x}_1,\vec{x}_2,\ldots,\vec{x}_N,t)$ of the many-body system under consideration. This quantum entropy associated with the action of the wave ψ of a many-body system determines a many-body quantum potential given by the following expression

$$Q = \sum_{i=1}^{N}\left[-\frac{\hbar^2}{2m_i}\left(\nabla_i S_Q\right)^2 + \frac{\hbar^2}{2m_i}\left(\nabla_i^2 S_Q\right)\right]. \tag{2.5}$$

Here, the quantum Hamilton-Jacobi equation of motion of the system determined by this quantum entropy is the following

$$\sum_{i=1}^{N}\frac{|\nabla_i S|^2}{2m_i} - \sum_{i=1}^{N}\frac{\hbar^2}{2m_i}(\nabla_i S_Q)^2 + V + \sum_{i=1}^{N}\frac{\hbar^2}{2m_i}(\nabla_i^2 S_Q) = -\frac{\partial S}{\partial t} \tag{2.6}$$

and the continuity equation for the probability density is:

$$\frac{\partial S_Q}{\partial t} = \sum_{i=1}^{N}\left[-\left(\vec{v}_i \cdot \nabla_i S_Q\right) + \frac{1}{2}\nabla_i \cdot \vec{v}_i\right]. \tag{2.7}$$

In summary, in the approach based on the quantum entropy (2.1), both for one-body systems and for many-body systems, the ultimate source of the geometry of the quantum world in the non-relativistic domain is the quantum entropy describing the degree of order and chaos of the vacuum supporting the probability density: the quantum entropy, by providing a measure of the deformation of the geometry of the processes, can be indeed interpreted as a sort of an intermediary entity between the background space and the behaviour of quantum particles, and thus between the action of the quantum potential and the behaviour of quantum particles [66, 171, 272, 273].

In the entropic version of non-relativistic de Broglie-Bohm theory, on the basis of equations (2.3) and (2.6), one can say that the probability density associated with the wave function under consideration determines a quantum entropy (which describes the degree of order and chaos of the vacuum supporting this probability density) and this quantum entropy determines a deformation, a change of the geometry of the physical space which is linked with the two quantum corrector terms in the energy of the system. These two quantum corrector terms can thus be interpreted as a sort of modification of the background space, as a sort of degree of chaos of the background space determined by the ensemble of particles associated with the wave function under consideration. In the non-relativistic entropic Bohmian mechanics, the geometrodynamic features of the quantum potential — namely the fact that it has a geometric nature, it contains a global information on the environment in which the experiment is performed, and at the same time it is a dynamical entity, namely its information about the process and the environment is active — are just determined by the deformation of the geometry determined by the quantum entropy (2.1). The quantum entropy emerges just as informational line of the quantum potential describing the change of the fundamental geometry of the physical space in the presence of quantum effects.

Moreover, in this entropic approach to the Bohm theory, one can say that the classical limit of the geometry of space can be expressed by the conditions

$$(\nabla S_Q)^2 \rightarrow (\nabla^2 S_Q) \tag{2.8}$$

for one-body systems and

$$(\nabla_i S_Q)^2 \rightarrow (\nabla_i^2 S_Q) \tag{2.9}$$

for many-body systems. The quantum dynamics will approach the classical dynamics when the quantum entropy satisfies conditions (2.8) (for one-body systems) or (2.9) (for many-body systems) which can be considered as the expression of a correspondence principle in quantum mechanics.

Moreover, in the paper [171] the author of this book has introduced opportune quantum-entropic lengths to characterize the geometric properties of the background space associated with the quantum entropy, given by

$$L_{quantum} = \frac{1}{\sqrt{(\nabla S_Q)^2 - \nabla^2 S_Q}} \tag{2.10}$$

for one-body systems and by

$$L_{quantum} = \frac{1}{\sqrt{\sum_{i=1}^{N}\left((\nabla_i S_Q)^2 - \nabla_i^2 S_Q\right)}} \tag{2.11}$$

for many-body systems. The quantum-entropic lengths (2.10) and (2.11) are the fundamental parameters that can be used to evaluate the strength of quantum effects and, therefore, the modification of the geometry with respect to the Euclidean geometry characteristic of classical physics. They define the fundamental geometrical properties of the vacuum supporting the probability density of the process into consideration. Once the quantum-entropic lengths ((2.10) or (2.11)) become non-negligible the system goes into a quantum regime. In this picture, Heisenberg's uncertainty principle derives from the fact that we are unable to perform a classical measurement to distances smaller than the quantum-entropic lengths. In other words, the size of a measurement has to be bigger than the quantum-entropic lengths

$$\Delta L \geq L_{quantum} = \frac{1}{\sqrt{(\nabla S_Q)^2 - \nabla^2 S_Q}} \tag{2.12}$$

(for one-body systems)

$$\Delta L \geq L_{quantum} = \frac{1}{\sqrt{\sum_{i=1}^{N}\left((\nabla_i S_Q)^2 - \nabla_i^2 S_Q\right)}} \tag{2.13}$$

(for many-body systems). The quantum regime is entered when the quantum-entropic lengths must be taken under consideration [66, 171].

Here we can also make a parallelism with the model proposed by Novello, Salim and Falciano in [93]. In analogy with the geometrical approach of Novello, Salim and Falciano [93] in the Weyl integrable space (analysed in chapter 1.3.3) which suggests that the presence of quantum effects are linked with the Weyl length and thus with the curvature scalar, in the picture proposed by the author of this book, the quantum effects are owed to the quantum entropy (2.1). If in Novello's, Salim's and Falciano's approach quantum effects reveal themselves as manifestations of the deformation of the geometrical structure of the physical space from Euclidean to a non-Euclidean Weyl integrable space, inside the entropic approach to Bohm's quantum mechanics here analysed, the presence of the quantum effects corresponds with a change in the geometry of the physical space determined by the quantum entropy and therefore here the strength of quantum effects are measured by a quantum-entropic length, namely the parameter describing the geometry of the background is ultimately determined by the quantum entropy.

It is also interesting to observe that in equations (2.10) and (2.11) the presence of the two quantum correctors of the energy seem to suggest that (2.10) and (2.11) are a indicators of non-local correlation. One may easely observe that the maximum value of (2.10) and (2.11) is obtained for $L_{quantum}^{\max} = 1$, which corresponds to the maximum de-localization of a quantum system: for this reason the author and Licata, as regards the quantities (2.10) and (2.11), coined the term *Bell length*, in honour of John S. Bell (1928–1990) [75, 276].

In the picture of the background space, of the vacuum associated with the quantum entropy, a new reading to the active information of the quantum potential can be introduced and thus new fundamental light into the interpretation of quantum information can be thrown. Inside this approach, the quantum potential is an information channel which describes the deformation of the geometry of the physical space in the presence of quantum effects and its geometrodynamic information is ultimately determined by the quantum entropy. As a consequence, the real ultimate grid of quantum information, the real fundamental structure which determines quantum information can be considered the deformation of the background space determined by the quantum entropy and described by the quantum-entropic lengths (2.10) and (2.11). The geometry of the vacuum produced by the quantum entropy (2.1) emerges thus as the most fundamental source of quantum information.

In order to explain in major detail the concept of quantum information as a measure of the deformation of the background space determined by the fundamental geometry of the vacuum described by the quantum entropy, let us take under consideration the classic two-slit experiment. Here, the wavefunction of the physical system involved in the experimental apparatus is characterized by n probability densities $h_1, h_2, \ldots, h_n$:

$$|\Psi\rangle = |h_1\rangle + |h_2\rangle + \cdots + |h_n\rangle \tag{2.14}$$

where $h_1 = \alpha_1 + i\beta_1, h_2 = \alpha_2 + i\beta_2, \ldots, h_n = \alpha_n + i\beta_n$. The probability for the interference is

$$P(x) = \langle\psi|\psi\rangle = \sum_{i,j} g_{ij}\xi^i(x)\xi^j(x) \tag{2.15}$$

where

$$\xi^i = \sqrt{I^i}, \xi^j = \sqrt{I^j} \tag{2.16}$$

where $I_i = \langle h_i|h_i\rangle$ and

$$g = \begin{bmatrix} 1 & \cos(\alpha_1 - \alpha_2) & \cdots & \cos(\alpha_1 - \alpha_2) \\ \cos(\alpha_2 - \alpha_1) & 1 & \cos(\alpha_2 - \alpha_1) & \cos(\alpha_2 - \alpha_1) \\ \cdots & \cdots & \cdots & \cdots \\ \cos(\alpha_n - \alpha_1) & \cos(\alpha_n - \alpha_2) & \cdots & 1 \end{bmatrix}. \tag{2.17}$$

Taking account of the metric (2.17), the square-distance between the end-points of two vectors ξ^i and η^i is

$$s^2(\theta_k) = \sum_{i,j} g_{i,j}\left(\xi^i(\theta_k) - \eta^i(\theta_k)\right)\left(\xi^j(\theta_k) - \eta^j(\theta_k)\right). \tag{2.18}$$

Moreover, by considering $\eta^i = \xi^i + \frac{\partial\xi^i}{\partial\theta_k}$ the following relation regarding the square-distance characterizing the process of double-slit interference may be obtained

$$ds^2 = \sum_{i,j} g_{i,j}\left(\frac{\partial\xi^i}{\partial\theta_k}\partial\theta_k\right)\left(\frac{\partial\xi^i}{\partial\theta_h}\partial\theta_h\right) = G_{h,k}\partial\theta^h\partial\theta^k \tag{2.19}$$

where

$$G = A^T g A \tag{2.20}$$

and

$$A = \begin{bmatrix} \dfrac{\partial \xi^1}{\partial \theta_1} & \dfrac{\partial \xi^1}{\partial \theta_2} & \cdots & \dfrac{\partial \xi^1}{\partial \theta_q} \\ \dfrac{\partial \xi^2}{\partial \theta_1} & \dfrac{\partial \xi^2}{\partial \theta_2} & \cdots & \dfrac{\partial \xi^2}{\partial \theta_q} \\ \cdots & \cdots & \cdots & \cdots \\ \dfrac{\partial \xi^n}{\partial \theta_1} & \dfrac{\partial \xi^n}{\partial \theta_2} & \cdots & \dfrac{\partial \xi^n}{\partial \theta_q} \end{bmatrix}. \tag{2.21}$$

Equations (2.19)–(2.21) physically means that the process of interference of n probabilities densities is described by n quantum states and n parameters of the states [277].

Now, if we want to describe the double-slit interference in the geometrodynamic entropic approach to Bohm's quantum potential where the fundamental origin of the processes is the geometry of the vacuum associated with the quantum entropy, we must put in correspondence relation (2.19) with the quantum entropic length. In this way, by introducing the quantum-entropic length given by equation (2.11) inside equation (2.19) we obtain:

$$G_{h,k}\partial\theta^h\partial\theta^k = \frac{1}{\sum_{i=1}^{N}\left((\nabla_i S_Q)^2 - \nabla_i^2 S_Q\right)}. \tag{2.22}$$

Equation (2.22) physically means that in the entropic view of Bohm's theory, the quantum entropy — and thus the geometry of the vacuum supporting the space-temporal distribution of the ensemble of particles associated with the wavefunction — is the ultimate element which generates the parameter states characterizing the quantum states in the process of the double-slit interference. It becomes so permissible the following re-reading of the formalism of the double-slit interference (in the case of a wave function characterized by n probabilities densities) inside the geometrodynamic approach to quantum potential based on the quantum entropy (2.1): the probability density associated with the wave function under consideration determines a quantum entropy characterizing the vacuum associated with the beams of electrons during the process, and this quantum entropy generates a deformation of the geometry of space described by a quantum-entropic length given by equation (2.22). The interference process is associated with the quantum entropic length (2.22) and thus is determined by the quantum entropy in the sense that it is the quantum entropic length — just associated with the quantum entropy — that produces the n parameters of the n quantum states associated with the n probabilities densities characterizing the processes. The quantum entropy can be thus considered as the real ultimate origin of the processes in the experiment of the double-slit interference. To sum up, the quantum-entropic distance given by equation (2.22) and determined by the quantum entropy can be considered as the ultimate source of quantum information as regards the double-slit interference [66].

Moreover, the formulations (2.2) and (2.5) of the quantum potential in terms of the quantum entropy lead us directly to introduce interesting perspectives about the relation between the geometrodynamic quantum information associated with the quantum entropy (2.1) and Feynman's path integrals approach. In the geometrodynamic entropic approach to Bohm's theory, the two fundamental equations of motion — which emerge from the introduction of the quantum entropy given by equation (2.1) as the fundamental entity that determines the behaviour of quantum particles — namely the energy conservation law (equation (2.3) for one-body systems and (2.6) for many body-systems) and the entropy balance equation (which is(2.4) for one-body systems and (2.7) for many-body systems) introduce a new suggestive way to read the de Broglie-Bohm path integrals and their link with Feynman's path integrals. In this regard, the fundamental paper on which we base our discussion is [273]. In this paper, it has been shown that, by taking into account the entropic definitions of the quantum potential ((2.2) for one-body systems and (2.5) for many-body systems), the Bohmian path-integrals proposed by Abolhasani and Golshani in [64] can be formulated in the following convenient forms:

$$\psi(\vec{x},t) = \exp\left\{\frac{i}{\hbar}\int_{\vec{x}_0,t_0}^{\vec{x},t}\left[\frac{(\nabla S)^2}{2m} + \frac{\hbar^2}{2m}(\nabla S_Q)^2\right]dt - \int_{\vec{x}_0,t_0}^{\vec{x},t}\left(\frac{\nabla^2 S}{2m} + \frac{\hbar^2}{2m}\nabla^2 S_Q + V\right)dt\right\}\psi(\vec{x}_0,t_0) \quad (2.23)$$

for one-body systems, and

$$\psi(\vec{x}_1,\ldots,\vec{x}_N,t) = \exp\left\{\frac{i}{\hbar}\int_{\vec{x}_0,t_0}^{\vec{x},t}\sum_{i=1}^{N}\left[\frac{(\nabla_i S)^2}{2m_i} + \frac{\hbar^2}{2m_i}(\nabla_i S_Q)^2\right]dt - \int_{\vec{x}_0,t_0}^{\vec{x},t}\sum_{i=1}^{N}\left[\frac{\nabla_i^2 S}{2m_i} + \frac{\hbar^2}{2m_i}\nabla_i^2 S_Q\right]dt\right\}\psi(\vec{x}_{01},\ldots,\vec{x}_{0N},t_0) \quad (2.24)$$

for many-body systems. The physical significance of equations (2.23) and (2.24) is that the quantum entropy given by equation (2.1) is the crucial entity associated with the geometry subtending and generating a Bohmian path in the sense that its role is to determine appropriate corrective terms into the kinetic energy and the potential energy (and therefore also into the Lagrangian) of the particle under consideration. In other words, on the basis of equations (2.23) and (2.24), one can say that Bohmian trajectories associated with Bohmian path integrals are indeed determined by the geometry of the background space created by the quantum entropy. One can say that the quantum entropy — describing the degree of order and chaos of the vacuum supporting the probability density associated with the wave function of the system under consideration — produces a modification of the geometrical properties of the physical space which reveal themselves in the form of appropriate corrective terms in the energy of the system and the modified geometry retro-acts on the system redesigning its trajectory and building thus a peculiar Bohmian trajectory corresponding to a peculiar Bohmian path integral.

Following the treatment of [273] (as well as Sbitnev's treatment in [271]), another relevant result about the geometry of space which derives from the quantum entropy in the picture of non-relativistic

Bohmian theory lies in the emergence of a complexified state space as a fundamental background of processes. As a consequence of the dependence of the quantum potential on the quantum entropy, an opportune unification of the quantum Hamilton-Jacobi equation ((2.3) for one-body systems and (2.6) for many-body systems) and the entropy balance equation (which is (2.4) for one-body systems and (2.7) for many-body systems) leads directly to a complexified Hamilton-Jacobi equation containing complex kinetic and potential terms. And in this picture the two quantum corrector terms of kinetic energy and potential energy both depending on the quantum entropy turn out to be the fundamental terms that modify the classical Feynman's path integral by expanding coordinates and momenta to a imaginary background.

Let us consider, before all, for simplicity, a one-body system. In this regard, by multiplying equation (2.4) for $-i\hbar$ and introducing this result into equation (2.3), this latest equation becomes

$$\frac{|\nabla S|^2}{2m} + i\hbar\frac{1}{m}\left(\nabla S \cdot \nabla S_Q\right) - \frac{\hbar^2}{2m}\left(\nabla S_Q\right)^2 + V - i\hbar\frac{1}{2}\left(\nabla \vec{v}\right) + \frac{\hbar^2}{2m}\left(\nabla^2 S_Q\right) = -\frac{\partial S}{\partial t}. \quad (2.25)$$

If one defines $J = S + i\hbar S_Q$ as a complexified action, the first three terms in (2.25) can be expressed as gradient of the complexified action squared

$$\frac{|\nabla S|^2}{2m} + i\hbar\frac{1}{m}\left(\nabla S \cdot \nabla S_Q\right) - \frac{\hbar^2}{2m}\left(\nabla S_Q\right)^2 = \frac{1}{2m}\left(\nabla J\right)^2. \quad (2.26)$$

As regards the other three terms of (2.25), they can be expressed as a Taylor's series of the potential energy extended in the complex space (as regards the complex extension, the reader may find some details for example in the reference [278]):

$$V\left(\vec{x} + i\vec{\varepsilon}\right) \approx V\left(\vec{x}\right) + i\hbar\left(\vec{n}\cdot\left(\frac{s}{2m}\nabla V\left(\vec{x}\right)\right)\right) - \frac{\hbar^2}{2m}\left(\frac{s^2}{2m}\nabla^2 V\left(\vec{x}\right)\right) + \cdots \quad (2.27)$$

where $\vec{\varepsilon} = \frac{\hbar}{2m} s\vec{n}$ is a small vector having the dimension of a length and s is the universal constant, the reverse velocity [279], $s = 4\pi\varepsilon_0 \frac{\hbar}{e^2} = 4,57{\cdot}10^{-7}$ $[s/m]$, e is the elementary charge carried by a single electron, ε_0 is the vacuum permittivity. In order to characterize the second term of left hand-side of equation (2.27), one can observe that we have a force $-\nabla V(\vec{x})$ multiplied by a vector $l\vec{n}$ providing thus an elementary work performed by this force at shifting on a length l along $\vec{n}$. As a consequence, the force multiplied by the factor $l\vec{n}$ and divided into mass m physically measures a rate of velocity's variation per unit length, namely it represents divergence of the velocity. This implies that this second term of left hand-side of equation (2.27) can be rewritten in the following form

$$\frac{s}{2m}(\vec{n} \cdot \nabla V(\vec{x})) = -\frac{1}{2}(\nabla \cdot \vec{v}). \tag{2.28}$$

The term $\left(\frac{s^2}{2m}\nabla^2 V(\vec{x})\right)$ is comparable with S_Q, so it can be made the position

$$-\left(\frac{s^2}{2m}\nabla^2 V(\vec{x})\right) = \nabla^2 S_Q. \tag{2.29}$$

Now, if one defines the complexified momentum $\vec{p}' = \nabla J = \nabla S + i\hbar\nabla S_Q$ and the complexified coordinates $\vec{x}' = \vec{x} + i\vec{\varepsilon}$, equation (2.25) can be formulated as a complexified Hamilton-Jacobi equation

$$-\frac{\partial J}{\partial t} = \frac{1}{2m}(\nabla J)^2 + V(\vec{x}') = H(\vec{x}', \vec{p}', t) \tag{2.30}$$

where $H(\vec{x}', \vec{p}', t)$ on the right side is the complexified Hamiltonian. By making the total derivative of the complex action one obtains the following equation

$$\frac{dJ}{dt} = -H(\vec{x}', \vec{p}', t) + \sum_{i=1}^{N} p_i{}' \dot{x}_i' = L(\vec{x}', \dot{\vec{x}}', t). \tag{2.31}$$

If we integrate equations (2.30) and (2.31) we obtain the solutions

$$J = -\int_{t_0}^{t} H(\vec{x}', \vec{p}', \tau)\, d\tau + C_1 \tag{2.32}$$

$$J = -\int_{t_0}^{t} L\left(\vec{x}', \dot{\vec{x}}', \tau\right) d\tau + C_2 \tag{2.33}$$

where C_1 and C_2 are two integration constants that satisfy the following condition:

$$C_1 - C_2 = \int_{t_0}^{t} \sum_{i=1}^{N} p'_i \dot{x}'_i dt = \int_L \sum_{i=1}^{N} p'_i dx'_i. \tag{2.34}$$

In equation (2.34) L is a curve beginning at t_0 and terminating at t.

As it has been shown clearly by Sbitnev in his recent work *Bohmian trajectories and the Path Integral Paradigm. Complexified Lagrangian Mechanics*, the complexified state space described by equations (2.30)–(2.34) can be considered as the fundamental stage which determines the features of Bohmian trajectories: Bohmian trajectories are trajectories submitted to the principle of least action that expands on the action integral (2.33) containing the complexified Lagrangian function derived from the quantum entropy [271]. Bohmian trajectories turn out to be geodesic trajectories of an incompressible fluid loaded by the complexified lagrangian that in turn is determined by the quantum potential expressed by equation (2.2), in other words by the two quantum correctors determined by the fundamental geometry associated with the quantum entropy. On the basis of equations (2.30)–(2.34), it becomes thus permissible the following re-reading of the geometry of space associated with the quantum potential: the real origin of the features of the background space determined by the quantum potential is the quantum entropy and the fundamental background of processes is a complexified space defined by a complexified Lagrangian function determined by the quantum entropy [66].

In the complexified state space defined by equations (2.30)–(2.34), a solution of the Schrödinger equation can be written as

$$\psi\left(\vec{x}', \vec{p}', t\right) = \exp\left(\frac{i}{\hbar} J\right) = \exp\left(\frac{i}{\hbar} S - S_Q\right). \tag{2.35}$$

If we insert the action integral (2.32) into equation (2.35) we obtain

$$\psi\left(\vec{x}', \vec{p}', t\right) = \frac{1}{Z_1} \exp\left\{-\frac{i}{\hbar} \int_{t_0}^{t} H\left(\vec{x}', \vec{p}', \tau\right) d\tau\right\} \tag{2.36}$$

where $Z_1 = \exp\left(-\frac{i}{\hbar}C_1\right)$. The probability density becomes

$$\rho = \exp\left(-2S_Q\right). \tag{2.37}$$

In this complexified state space, Hamilton's principle $\delta J = 0$ states that the motion of an arbitrary mechanical system occurs in such a way that the definite integral (2.33) becomes stationary for arbitrary possible variations of the configuration of the system, provided the initial and final configurations of the system are prescribed. Moreover, the wave function expressed in terms of the complexified action (2.33) which is

$$\psi\left(\vec{x}',\vec{p}',t\right) = \frac{1}{Z_2}\exp\left\{-\frac{i}{\hbar}\int_{t_0}^{t} L\left(\vec{x}',\dot{\vec{x}}',\tau\right)d\tau\right\} \tag{2.38}$$

where $Z_2 = \exp\left(-\frac{i}{\hbar}C_2\right)$, allows us to reformulate in a alternative way Hamilton's principle. In this case the principle states: this exponent becomes stationary for arbitrary possible variations of the configuration of the system, provided the initial and final configurations of the system are prescribed. Obviously, it results from stationarity of the integral (2.33) stated above.

On the other hand, the two Bohmian quantum correctors determined by the fundamental geometry of the vacuum associated with the quantum entropy emerge as indispensable terms that modify Feynman's path integral by expanding coordinates and momenta to the imaginary sector. As shown by Grosche [280], Feynman's path integral can be written mathematically in the following way inside the complexified state space

$$K\left(\vec{x}',t;\vec{x}_0',t_0\right) = \iint\ldots\int D\left[\vec{x}'(\tau)\right]\exp\left\{\frac{i}{\hbar}\int_{t_0}^{t} L\left(\vec{x}',\dot{\vec{x}}',\tau\right)d\tau\right\} \tag{2.39}$$

where the path-integral symbol indicates the multiple integral

$$\iint\ldots\int D\left[\vec{x}'(\tau)\right] \Leftrightarrow \left(\frac{2\pi i\hbar\delta t}{m}\right)^{-M/2}\int_{\vec{x}_o'}^{\vec{x}'} d\vec{x}_1'\int_{\vec{x}_o'}^{\vec{x}'} d\vec{x}_2'\cdots\int_{\vec{x}_o'}^{\vec{x}'} d\vec{x}_M'. \tag{2.40}$$

The fundamental principle of quantum mechanics, namely the principle of superposition, underlies the path integral (2.39). Whereas evolution of a classical object is described by a unique trajectory satisfying the principle of least action, the path integral tests all possible virtual classical trajectories, among which there is a unique trajectory satisfying the least action principle. Other trajectories cancel each other by their interference.

As Sbitnev showed in the reference [271], the complexified state space characterized by the complexified momenta $\vec{p}' = \nabla J = \nabla S + i\hbar\nabla S_Q$ and by the complexified coordinates $\vec{x}' = \vec{x} + i\vec{\varepsilon}$ permits to provide a simple and natural interpretation of Feynman's path integral approach based on equations (2.39) and (2.40): the key of reading is the fundamental geometry of the background corresponding with the two Bohmian quantum correctors linked with the quantum entropy. The path integral computation arises directly from decomposition of the Schrödinger equation to the modified quantum Hamilton-Jacobi equation plus the entropy balance equation. The two Bohmian quantum correctors linked with the quantum entropy (2.1) and resulted from this decomposition allow the expanding of the geometry of state space to the imaginary sector. The imaginary terms emergent in this computations suppress the wilder contributions to the path integral. Thus, a non-trivial N-dimensional manifold embedded in the 2N-dimensional complex state space is originated where its real part is the conventional coordinate state space.

As shown by the author of this book in [273], the approach, based on equations (2.25)–(2.30), of the complexified state space as fundamental background determined by the quantum entropy can be extended in a simple way also to many-body systems. In fact, equation (2.25) can be generalized to many-body systems as

$$\sum_{i=1}^{N} \frac{|\nabla_i S|^2}{2m_i} + i\hbar \sum_{i=1}^{N} \frac{1}{m_i} (\nabla_i S \cdot \nabla_i S_Q) - \sum_{i=1}^{N} \frac{\hbar^2}{2m_i} (\nabla_i S_Q)^2 + V$$
$$-\frac{1}{2} i\hbar \sum_{i=1}^{N} \nabla_i \cdot \vec{v}_i + \sum_{i=1}^{N} \frac{\hbar^2}{2m_i} \left(\nabla_i^2 S_Q\right) = -\frac{\partial S}{\partial t} \tag{2.41}$$

where, by defining $J = S + i\hbar S_Q$ as a complexified action, the first three terms in (2.41) can be rewritten as

$$\sum_{i=1}^{N} \frac{|\nabla_i S|^2}{2m_i} + i\hbar \sum_{i=1}^{N} \frac{1}{m_i} (\nabla_i S \cdot \nabla_i S_Q) - \sum_{i=1}^{N} \frac{\hbar^2}{2m_i} (\nabla_i S_Q)^2$$
$$= \sum_{i=1}^{N} \frac{1}{2m_i} (\nabla_i J)^2. \tag{2.42}$$

As regards the other three terms of (2.41), one can use an expansion into Taylor's series of the potential energy extended in the complex space:

$$\begin{aligned} V(\vec{x}_1 + i\vec{\varepsilon}_1, \ldots, \vec{x}_i + i\vec{\varepsilon}_i, \ldots) \\ \approx \sum_{i=1}^{N} V(\vec{x}_i) + i\hbar \sum_{i=1}^{N} \left(\vec{n}_i \cdot \left(\frac{s}{2m_i} \nabla_i V(\vec{x}_i) \right) \right) \\ - \sum_{i=1}^{N} \frac{\hbar^2}{2m_i} \left(\frac{s^2}{2m_i} \nabla_i^2 V(\vec{x}_i) \right) + \cdots \end{aligned} \tag{2.43}$$

where $\vec{\varepsilon}_i = \frac{\hbar}{2m_i} s\vec{n}_i$, $s = 4\pi\varepsilon_0 \frac{\hbar}{e^2} = 4,57 \cdot 10^{-7}\,[s/m]$. The second term of equation (2.43) can be rewritten in the following form

$$\sum_{i=1}^{N} \frac{s}{2m_i} (\vec{n}_i \cdot \nabla_i V(\vec{x}_i)) = -\frac{1}{2} \sum_{i=1}^{N} (\nabla_i \cdot \vec{v}_i). \tag{2.44}$$

As regards the other term $\sum_{i=1}^{N} \frac{\hbar^2}{2m_i} \left(\frac{s^2}{2m_i} \nabla_i^2 V(\vec{x}_i) \right)$ it can be made the position

$$\sum_{i=1}^{N} \frac{\hbar^2}{2m_i} \left(\frac{s^2}{2m_i} \nabla_i^2 V(\vec{x}_i) \right) = \sum_{i=1}^{N} \frac{\hbar^2}{2m_i} \nabla_i^2 S_Q. \tag{2.45}$$

Here, if one defines the complexified total momentum of the system $\vec{p}' = \sum_{i=1}^{N} \nabla_i J = \sum_{i=1}^{N} \nabla_i S + i\hbar \nabla_i S_Q$ and the complexified coordinates for each particle $\vec{x}'_i = \vec{x}_i + i\vec{\varepsilon}_i$, equation (2.41) can be formulated as a complexified Hamilton-Jacobi equation for many-body systems

$$-\frac{\partial J}{\partial t} = \sum_{i=1}^{N}\left[\frac{1}{2m_i}(\nabla_i J)^2 + V\left(\vec{x}_i'\right)\right] = \sum_{i=1}^{N} H_i(\vec{x}_i', \vec{p}_i', t) \tag{2.46}$$

where $\sum_{i=1}^{N} H_i\left(\vec{x}_i', \vec{p}_i', t\right)$ on the right side is the total complexified Hamiltonian for the many-body system under consideration. The total derivative of the complex action gives the following equation

$$\frac{dJ}{dt} = \sum_{i=1}^{N} L_i(\vec{x}', \dot{\vec{x}}', t), \tag{2.47}$$

where

$$\sum_{i=1}^{N} L_i(\vec{x}', \dot{\vec{x}}', t) = -\sum_{i=1}^{N} H_i(\vec{x_i}', \vec{p_i}', t) + \sum_{i=1}^{N} p_i'\dot{x}_i'. \tag{2.48}$$

By integrating equations (2.46) and (2.47) we obtain the solutions

$$J = -\int_{t_0}^{t}\sum_{i=1}^{N} H_i(\vec{x}_i', \vec{p_i}', \tau)d\tau + C_1 \tag{2.49}$$

$$J = -\int_{t_0}^{t}\sum_{i=1}^{N} L_i(\vec{x}_i', \dot{\vec{x}}_i', \tau)d\tau + C_2 \tag{2.50}$$

where C_1 and C_2 are two integration constants that satisfy the following condition:

$$C_1 - C_2 = \sum_{i=1}^{N}\int_{L_i}\sum_{k=1}^{N} p_k' dx_k'. \tag{2.51}$$

In the complexified state space of many-body systems defined by equations (2.46)–(2.50), a solution of the Schrödinger equation can be written as

$$\psi\left(\vec{x}_i', \vec{p_i}', t\right) = \frac{1}{Z_1}\exp\left\{-\frac{i}{\hbar}\int_{t_0}^{t}\sum_{i=1}^{N} H_i\left(\vec{x}_i', \vec{p_i}', \tau\right)d\tau\right\} \tag{2.52}$$

where $Z_1 = \exp\left(-\frac{i}{\hbar}C_1\right)$.

In analogy to the treatment of one-body systems originally proposed by Sbitnev in [271], in this complexified state space for many-body systems, Hamilton's principle $\delta J = 0$ states that the motion of an arbitrary mechanical system takes place in such a way that the definite integral (2.49) becomes stationary for arbitrary possible variations of the configuration of the system, provided the initial and final configurations of each particle of the system are prescribed. Moreover, if we express the wave function in terms of the complexified action (2.46) as

$$\psi\left(\vec{x}_i', \vec{p}_i, t\right) = \frac{1}{Z_2} \exp\left\{-\frac{i}{\hbar}\int_{t_0}^{t}\sum_{i=1}^{N} L_i\left(\vec{x}_i', \dot{\vec{x}}_i', \tau\right) d\tau\right\} \tag{2.53}$$

where $Z_2 = \exp\left(-\frac{i}{\hbar}C_2\right)$, this principle states: this exponent becomes stationary for arbitrary possible variations of the configuration of the system, provided the initial and final configurations of each particle of the system are prescribed. And, of course, it results from stationarity of the integral (2.49).

In summary, in the light of the results obtained by Sbitnev in [271] as well as by the author of this book in [273], the link between Feynman's path integrals with a complexified state space determined by the quantum entropy generating Bohm's quantum potential provides another significant element which demonstrates the fundamental role of the ultimate geometry of the vacuum associated with the quantum entropy as origin of the quantum processes inside a unifying picture.

2.2 The Quantum Entropy in Bohm's Approach to the Klein-Gordon Relativistic Quantum Mechanics

In order to analyse the perspectives introduced by the entropy in the Klein-Gordon relativistic quantum mechanics, let us start by considering the interesting Bohmian approach to the Klein-Gordon equation developed by F. Shojai and A. Shojai in the references [97, 280] and reviewed in chapter 1.4.1. In the entropic version of the Bohmian Klein-Gordon equation developed by the author in [274], the space-temporal distribution of the ensemble of particles describing the spinless physical system under consideration is assumed to generate a modification of the geometry of the background described by the quantum entropy

$$S_Q = -\frac{1}{2}\ln\rho. \tag{2.54}$$

The change of the geometry associated with the quantum entropy (2.54) determines a quantum potential given by relation

$$Q = -\frac{\hbar^2}{m^2c^2}(\partial_\mu S_Q)^2 + \frac{\hbar^2}{m^2c^2}\left(\left(\nabla^2 - \frac{1}{c^2}\frac{\partial^2}{\partial t^2}\right)S_Q\right). \tag{2.55}$$

So, the quantum Hamilton-Jacobi equation (1.130) emerging from the Klein-Gordon equation may be written as

$$\partial_\mu S\partial^\mu S = m^2c^2\exp\left[-\frac{\hbar^2}{m^2c^2}(\partial_\mu S_Q)^2 + \frac{\hbar^2}{m^2c^2}\left(\left(\nabla^2 - \frac{1}{c^2}\frac{\partial^2}{\partial t^2}\right)S_Q\right)\right], \tag{2.56}$$

while the continuity equation (1.131) becomes

$$\frac{1}{c}\frac{\partial S_Q}{\partial t} = -(p^\mu\partial_\mu S_Q) + \frac{1}{2}\partial_\mu p^\mu. \tag{2.57}$$

Equations (2.56) and (2.57) can be considered as the fundamental equations of motion in the entropic view of F. Shojai's and A. Shojai's Bohmian model of the Klein-Gordon relativistic quantum mechanics. Furthermore, the connection between the mass of the particle and the quantum potential, expressed by relation (1.139), reads

$$M = m\exp\left[-\frac{\hbar^2}{m^2c^2}(\partial_\mu S_Q)^2 + \frac{\hbar^2}{m^2c^2}\left(\left(\nabla^2 - \frac{1}{c^2}\frac{\partial^2}{\partial t^2}\right)S_Q\right)\right]. \tag{2.58}$$

Equation (2.58), which is derived directly from the relativistic quantum Hamilton-Jacobi equation (1.130) inside the entropic approach based on the quantum potential (2.55), implies that the quantum mass is generated by the quantum entropy. In other words, one can say that, in the presence of a relativistic spinless particle, the quantum entropy describing the degree of order and chaos of space of the background of processes corresponding with the probability density associated with the wave function under consideration introduces a deformation of the geometry of space producing thus quantum mass (2.58).

As regards a Bohmian approach to the Klein-Gordon relativistic quantum mechanics, as we have seen in chapter 1.4.1, interesting results are obtained also in the context of the Bertoldi-Faraggi-Matone theory. In the light of the quantum entropy (2.54), the results of Bertoldi-Faraggi-Matone theory regarding the Klein-Gordon equation can receive a new suggestive reading which provides a more direct physical interpretation. In the entropic view, the implementing of the equivalence principle, which states that all physical systems can be connected by a coordinate transformation to the free situation with vanishing energy and leads to the quantum Hamilton-Jacobi equation (1.143), is physically associated with a modification of the geometry of the configuration space of the processes described the quantum entropy (2.54). In other words, one can say that the equivalence principle invoked by the Bertoldi-Faraggi-Matone theory constitutes a consequence, which emerges and reveals itself at the level of quantum mechanics, of a more fundamental deformation of the geometry of the vacuum defined by the quantum entropy. The quantum potential (1.142) (which derives from the implementation of the equivalence principle inside the relativistic classical Hamilton-Jacobi equation) is determined by the quantum entropy (2.54) describing the change of the geometrical properties of the background on the basis of the fundamental equation

$$Q_{rel}= \frac{\hbar^2}{2m}\left(\partial_\mu S_Q\right)^2 - \frac{\hbar^2}{2m}\left(\left(\nabla^2 - \frac{1}{c^2}\frac{\partial^2}{\partial t^2}\right) S_Q\right). \tag{2.59}$$

And thus also the generation of mass can be seen as an effect of the modification of the fundamental geometry of the vacuum associated with the quantum entropy on the basis of the following relation

$$\frac{1}{2mc^2} = \left(p\,|p^0\right) Q^0\left(q^0\right) - \left[\frac{\hbar^2}{m^2c^2}\left(\partial_\mu S_Q\right)^2 - \frac{\hbar^2}{m^2c^2}\left(\left(\nabla^2 - \frac{1}{c^2}\frac{\partial^2}{\partial t^2}\right) S_Q\right)\right]. \tag{2.60}$$

Finally, in this entropic approach of the Bertoldi-Faraggi-Matone theory, the geometrical properties of the vacuum of processes associated to the quantum entropy can be characterized by introducing a quantum-entropic length associated with the implementing of the equivalence principle

$$L_{quantum} = \frac{1}{\sqrt{-(\nabla_\mu S_Q)^2 + \left(\nabla^2 - \frac{1}{c^2}\frac{\partial^2}{\partial t^2}\right) S_Q}}. \tag{2.61}$$

In the Bertoldi-Faraggi-Matone theory, the quantum-entropic length (2.61) can be used to evaluate the strength of quantum effects and, therefore, the non-local correlation degree and the modification of the geometry in a quantum relativistic regime — corresponding with the implementing of the equivalence principle — with respect to the Euclidean geometry characteristic of classical physics. It defines the fundamental geometrical properties of the vacuum supporting the space-temporal distribution of the ensemble of particles describing the physical system in the Klein-Gordon equation regime. Once the quantum-entropic length (2.61) becomes non-negligible thus influencing the processes, the spinless particle into consideration goes into a quantum regime. And Heisenberg's uncertainty principle can be seen as reflex of the fact that we are unable to perform a classical measurement to distances smaller than this quantum length (2.61) corresponding with the implementing of the equivalence principle.

With the introduction of the quantum entropy (2.54), it is possible to provide an entropic reading also of Nikolic's model of the Klein-Gordon relativistic quantum mechanics. The quantum entropy can be seen as the ultimate entity which generates the particle trajectories (1.149) and the quantum potential (1.152). In fact, equation (1.153) may be written conveniently as

$$\frac{dx^\mu}{ds} = \frac{j^\mu}{2m\exp(-2S_Q)} \tag{2.62}$$

and the quantum potential (1.152) becomes

$$Q = \frac{1}{2m}(\partial_\mu S_Q)^2 - \frac{1}{2m}\left(\left(\nabla^2 - \frac{\partial^2}{\partial t^2}\right) S_Q\right). \tag{2.63}$$

The equation of motion (1.151) of Nikolic's approach can also be formulated as

$$\frac{\partial_\mu S \partial^\mu S}{2m} - \frac{m}{2} + \frac{1}{2m}(\partial_\mu S_Q)^2 - \frac{1}{2m}\left(\left(\nabla^2 - \frac{\partial^2}{\partial t^2}\right) S_Q\right) = 0 \tag{2.64}$$

which together with the identity $\frac{d}{ds} = \frac{dx^{\mu}}{ds}\partial_{\mu}$ leads to the following equivalent expression for the equation of motion

$$\frac{d^2x^{\mu}}{ds^2} = \partial^{\mu}\left[\frac{1}{2m}\left(\partial_{\mu}S_Q\right)^2 - \frac{1}{2m}\left(\left(\nabla^2 - \frac{\partial^2}{\partial t^2}\right)S_Q\right)\right]. \quad (2.65)$$

On the basis of the entropic approach to Nikolic's model based on the mathematical formalism constituted by equations (2.62)–(2.65), it becomes so permissible the following reading of the Klein-Gordon quantum mechanics in the Bohmian approach developed by Nikolic. The space-temporal distribution of the ensemble of particles describing the physical system under consideration produces a deformation of the geometry of the vacuum described by a quantum entropy given by equation (2.54), the quantum entropy together with the particle current determines particle trajectories having the form (2.62), the quantum potential (2.63) and equations of motion in the geometry of the background space (modified by the quantum entropy) given by equations (2.64) and (2.65). And, if in Nikolic's approach one obtains a Lorentz covariance of the equations of the particle trajectories, therefore it follows that also this Lorentz covariance of the equations of particle trajectories emerge as a consequence of the quantum entropy defining vacuum of the processes, in the sense that the trajectories do not depend on the choice of the affine parameter s in the geometry of the background space corresponding with the quantum entropy. Moreover, also in Nikolic's approach the geometrical properties of the background space of processes can be described by introducing a quantum length associated with the quantum entropy (and the particle current $j_{\mu} = i_{\psi^*}\partial_{\mu}\psi$), given by the following relation

$$L_{quantum} = \frac{1}{\sqrt{-\left(\nabla_{\mu}S_Q\right)^2 + \left(\nabla^2 - \frac{\partial^2}{\partial t^2}\right)S_Q}}. \quad (2.66)$$

The quantum-entropic length (2.66) defining the geometrical properties of the vacuum in Nikolic's approach can be used to evaluate the strength of quantum effects and, therefore, the non-local correlation degree and the modification of the geometry in a quantum relativistic regime — and corresponding with the quantum entropy together with the particle current — with respect to the Euclidean geometry characteristic of classical physics. It shows how the vacuum supporting the space-temporal distribution of the ensemble of particles describing the physical system under consideration deforms the geometry of the processes inside Nikolic's approach of the Klein-Gordon relativistic quantum mechanics. Once the quantum-entropic length (2.66) becomes non-negligible the spinless particle into consideration goes into a quantum regime. And, even here, Heisenberg's uncertainty principle can be seen as a result which derives from the fact that we are unable to perform a classical measurement to distances smaller than this quantum-entropic length (2.66).

In summary, in the approach based on the introduction of the quantum entropy (2.54) as fundamental entity measuring the geometry of the vacuum supporting the evolution of the system into consideration, one indeed obtains a direct simple physical interpretation of the results of three different and very important approaches to the Klein-Gordon relativistic quantum mechanics in a Bohmian picture: F. Shojai's and A. Shojai's model, the Bertoldi-Faraggi-Matone theory and Nikolic's approach. Moreover, the fundamental geometry described by the quantum entropy can in some way provide a unifying reading of these three different approaches. In particular, in all these three approaches the particle trajectories turn out to be determined by the change of the geometrical properties of the background space associated with the quantum entropy and an appropriate quantum-entropic length can be defined which describes the strength of the modification of the geometry. Moreover, the quantum entropy allows us to obtain the following other important unifying result: both in F. Shojai's and A. Shojai's model and in the Bertold-Faraggi-Matone theory the quantum mass turns out to be generated by the quantum potential just as a consequence of the fact that the density of particles associated with the wave function of the system under consideration produces a deformation of the geometry of the background.

At the end of this chapter it is interesting to observe that, in the light of the model, proposed by Sbitnev in [70–73], of the physical vacuum of quantum processes in terms of a special superfluid medium composed by virtual particle-antiparticle pairs, another interesting characterization of the geometry of the vacuum described by the quantum entropy can be provided. In fact, by comparing equation (2.55) with the fundamental equation (1.160) of the approach developed by Sbitnev in [70–73], where the quantum potential is just an emergence of the zero-point fluctuations of a superfluid physical vacuum consisting of an enormous amount of virtual pairs of particles-antiparticles, one can obtain a direct link between the pressures that arise between ensembles of virtual particles populating the vacuum (which generate Bohm's quantum potential in Sbitnev's approach) and the two quantum correctors to the energy of the system expressing the modification of the geometry of space:

$$-\frac{\hbar^2}{4m\rho}\left[\nabla^2\rho-\frac{1}{c^2}\frac{\partial^2}{\partial t^2}\rho\right]+\frac{\hbar^2}{8m\rho^2}\left[(\nabla\rho)^2-\frac{1}{c^2}\left(\frac{\partial}{\partial t}\rho\right)^2\right]$$

$$=-\frac{\hbar^2}{m^2c^2}(\partial_\mu S_Q)^2+\frac{\hbar^2}{m^2c^2}\left(\left(\nabla^2-\frac{1}{c^2}\frac{\partial^2}{\partial t^2}\right)S_Q\right), \quad (2.67)$$

ρ being the density distribution of the virtual particles in the vacuum. Equation (2.67) leads directly to the following identifications

$$-\frac{\hbar^2}{4m\rho}\left[\nabla^2\rho-\frac{1}{c^2}\frac{\partial^2}{\partial t^2}\rho\right]=\frac{\hbar^2}{m^2c^2}\left(\left(\nabla^2-\frac{1}{c^2}\frac{\partial^2}{\partial t^2}\right)S_Q\right) \quad (2.68)$$

and

$$\frac{\hbar^2}{8m\rho^2}\left[(\nabla\rho)^2-\frac{1}{c^2}\left(\frac{\partial}{\partial t}\rho\right)^2\right]=-\frac{\hbar^2}{m^2c^2}(\partial_\mu S_Q)^2, \quad (2.69)$$

which can be conveniently rewritten respectively as

$$-\frac{1}{4\rho}\left[\nabla^2\rho-\frac{1}{c^2}\frac{\partial^2}{\partial t^2}\rho\right]=\frac{1}{mc^2}\left(\left(\nabla^2-\frac{1}{c^2}\frac{\partial^2}{\partial t^2}\right)S_Q\right) \quad (2.70)$$

and

$$\frac{1}{8\rho^2}\left[(\nabla\rho)^2-\frac{1}{c^2}\left(\frac{\partial}{\partial t}\rho\right)^2\right]=-\frac{1}{mc^2}(\partial_\mu S_Q)^2. \quad (2.71)$$

Equations (2.70) and (2.71), by expressing a direct link between the quantum entropy and the pressures that arise between ensembles of virtual particles populating the vacuum and thus originate the quantum potential, let us realize immediately that, at a fundamental level, the zero-point fluctuations of the vacuum have the effect to determine a modification of the geometry of the background of processes, in other words the quantum entropy may be considered just as the result of these zero-point fluctuations.

2.3 About the Quantum Entropy in a Bohmian Approach to the Dirac Relativistic Quantum Mechanics

In this chapter, in order to develop an entropic approach to the Dirac equation and analyse the perspectives introduced by the quantum entropy into the treatment of a Dirac particle, we take into consideration the de Broglie-Bohm like model for the Dirac equation developed by Chavoya-Aceves in the paper [131] and the Bohmian models of Dirac's particles developed by Nikolic [132–134] and Hernadez-Zapata [135], which we have reviewed in chapter 1.4.2. In this regard, we base our discussion on the paper [274].

The entropic view of the Dirac relativistic quantum mechanics starts by expressing the components of the two sets of 4-spinors $\psi^{(P)}(x)$ and $\psi^{(A)}(x)$ representing respectively the wave function of particles of spin 1/2 and the wave function associated with the corresponding antiparticles (where we have denoted $x = (t, \vec{x})$), and which appear in the general solution (1.175) of the Dirac equation (1.170), in polar form as follows:

$$\psi_1^{(P)} = \sqrt{\rho_1^{(P)}}e^{\frac{i}{\hbar}S_1^{(P)}} \tag{2.72}$$

$$\psi_2^{(P)} = \sqrt{\rho_2^{(P)}}e^{\frac{i}{\hbar}S_2^{(P)}} \tag{2.73}$$

$$\psi_3^{(P)} = \sqrt{\rho_3^{(P)}}e^{-\frac{i}{\hbar}S_3^{(P)}} \tag{2.74}$$

$$\psi_4^{(P)} = \sqrt{\rho_4^{(P)}}e^{-\frac{i}{\hbar}S_4^{(P)}} \tag{2.75}$$

and

$$\psi_1^{(A)} = \sqrt{\rho_1^{(A)}} e^{\frac{i}{\hbar} S_1^{(A)}} \tag{2.76}$$

$$\psi_2^{(A)} = \sqrt{\rho_2^{(A)}} e^{\frac{i}{\hbar} S_2^{(A)}} \tag{2.77}$$

$$\psi_3^{(A)} = \sqrt{\rho_3^{(A)}} e^{-\frac{i}{\hbar} S_3^{(A)}} \tag{2.78}$$

$$\psi_4^{(A)} = \sqrt{\rho_4^{(A)}} e^{-\frac{i}{\hbar} S_4^{(A)}} . \tag{2.79}$$

In the entropic view of the Dirac equation we assume that each component of the 4-spinor $\psi^{(P)}(x)$ given by equations (2.72)–(2.75) and each component of the 4-spinor $\psi^{(A)}(x)$ given by equations (2.76)–(2.79) corresponds to a peculiar quantum entropy, on the basis of equations:

$$S_{Q1}^{(P)} = -\frac{1}{2} \ln \rho_1^{(P)} \tag{2.80}$$

$$S_{Q2}^{(P)} = -\frac{1}{2} \ln \rho_2^{(P)} \tag{2.81}$$

$$S_{Q3}^{(P)} = -\frac{1}{2} \ln \rho_3^{(P)} \tag{2.82}$$

$$S_{Q4}^{(P)} = -\frac{1}{2} \ln \rho_4^{(P)} \tag{2.83}$$

and

$$S_{Q1}^{(A)} = -\frac{1}{2} \ln \rho_1^{(A)} \tag{2.84}$$

$$S_{Q2}^{(A)} = -\frac{1}{2} \ln \rho_2^{(A)} \tag{2.85}$$

$$S_{Q3}^{(A)} = -\frac{1}{2} \ln \rho_3^{(A)} \tag{2.86}$$

$$S_{Q4}^{(A)} = -\frac{1}{2} \ln \rho_4^{(A)} \tag{2.87}$$

respectively.

The quantities (2.80)–(2.83) express the degree of order and chaos of the vacuum supporting the densities associated with the components of the spinor $\psi^{(P)}(x)$ corresponding with the appearance of particles of spin 1/2; the quantities (2.84)–(2.87) express the degree of order and chaos of the vacuum supporting the densities associated with each component of the spinor $\psi^{(A)}(x)$ corresponding with the appearance of the corresponding antiparticles. For example, taking into consideration the density of the ensemble of particles associated with the wave function $\psi^{(P)}(x)$, the quantities (2.80)–(2.83) may be interpreted in the following way. The quantity (2.80) expresses the degree of order and chaos of the vacuum supporting the density $\rho_1^{(P)}$ associated with the first component of the spinor $\psi^{(P)}(x)$; the quantity (2.81) expresses the degree of order and chaos of the vacuum supporting the density $\rho_2^{(P)}$ associated with the second component of the spinor $\psi^{(P)}(x)$; the quantity (2.82) expresses the degree of order and chaos of the vacuum supporting the density $\rho_3^{(P)}$ associated with the third component of the spinor $\psi^{(P)}(x)$; finally, the quantity (2.83) expresses the degree of order and chaos of the vacuum supporting the density $\rho_4^{(P)}$ associated with the fourth component of the spinor $\psi^{(P)}(x)$. The quantities, given by equations (2.84)–(2.87), associated with the four components of the spinor $\psi^{(A)}(x)$, have an analogous interpretation. In summary, on the basis of equations (2.80)–(2.87), the following characterization of the fundamental geometry of the background in the Dirac relativistic quantum mechanics can be provided. Each component of the spinor — both of the spinor of the particles of spin 1/2, and of the spinor of the corresponding antiparticles — generates a deformation of the geometry of the configuration space and therefore, in the Dirac relativistic quantum mechanics, the vacuum responsible of the processes is described by eight components of the quantum entropy: there are four components of the quantum entropy associated with the wave function of the particles and four components of the quantum entropy associated with the wave function of the corresponding antiparticles. On the basis of the quantum entropies (2.80)–(2.87), the components of the two sets of 4-spinors $\psi^{(P)}(x)$ and $\psi^{(A)}(x)$ may be formulated respectively as:

$$\psi_1^{(P)} = e^{-S_{Q1}^{(P)} + \frac{i}{\hbar} S_1^{(P)}} \tag{2.88}$$

$$\psi_2^{(P)} = e^{-S_{Q2}^{(P)} + \frac{i}{\hbar} S_2^{(P)}} \tag{2.89}$$

$$\psi_3^{(P)} = e^{-S_{Q3}^{(P)} - \frac{i}{\hbar} S_3^{(P)}} \tag{2.90}$$

$$\psi_4^{(P)} = e^{-S_{Q4}^{(P)} - \frac{i}{\hbar} S_4^{(P)}} \tag{2.91}$$

and

$$\psi_1^{(A)} = e^{-S_{Q1}^{(A)} + \frac{i}{\hbar} S_1^{(A)}} \tag{2.92}$$

$$\psi_2^{(A)} = e^{-S_{Q2}^{(A)} + \frac{i}{\hbar} S_2^{(A)}} \tag{2.93}$$

$$\psi_3^{(A)} = e^{-S_{Q3}^{(A)} - \frac{i}{\hbar} S_3^{(A)}} \tag{2.94}$$

$$\psi_4^{(A)} = e^{-S_{Q4}^{(A)} - \frac{i}{\hbar} S_4^{(A)}}. \tag{2.95}$$

Taking account of relations (2.88)–(2.95), the wave equation (1.173) and the continuity equation (1.174) read respectively as:

$$i\hbar \frac{\partial}{\partial T} \begin{pmatrix} e^{-S_{Q1}^{(P)} + \frac{i}{\hbar} S_1^{(P)}} + e^{-S_{Q1}^{(A)} + \frac{i}{\hbar} S_1^{(A)}} \\ e^{-S_{Q2}^{(P)} + \frac{i}{\hbar} S_2^{(P)}} + e^{-S_{Q2}^{(A)} + \frac{i}{\hbar} S_{21}^{(A)}} \\ e^{-S_{Q3}^{(P)} + \frac{i}{\hbar} S_3^{(P)}} + e^{-S_{Q3}^{(A)} + \frac{i}{\hbar} S_3^{(A)}} \\ e^{-S_{Q4}^{(P)} + \frac{i}{\hbar} S_4^{(P)}} + e^{-S_{Q4}^{(A)} + \frac{i}{\hbar} S_4^{(A)}} \end{pmatrix}$$

$$= \left(c\vec{\alpha} \cdot \vec{p} + mc^2\beta\right) \begin{pmatrix} e^{-S_{Q1}^{(P)} + \frac{i}{\hbar} S_1^{(P)}} + e^{-S_{Q1}^{(A)} + \frac{i}{\hbar} S_1^{(A)}} \\ e^{-S_{Q2}^{(P)} + \frac{i}{\hbar} S_2^{(P)}} + e^{-S_{Q2}^{(A)} + \frac{i}{\hbar} S_{21}^{(A)}} \\ e^{-S_{Q3}^{(P)} + \frac{i}{\hbar} S_3^{(P)}} + e^{-S_{Q3}^{(A)} + \frac{i}{\hbar} S_3^{(A)}} \\ e^{-S_{Q4}^{(P)} + \frac{i}{\hbar} S_4^{(P)}} + e^{-S_{Q4}^{(A)} + \frac{i}{\hbar} S_4^{(A)}} \end{pmatrix} \tag{2.96}$$

and

$$\frac{\partial}{\partial x_i}\left[c\begin{pmatrix} e^{-S_{Q1}^{(P)}+\frac{i}{\hbar}S_1^{(P)}}+e^{-S_{Q1}^{(A)}+\frac{i}{\hbar}S_1^{(A)}} \\ e^{-S_{Q2}^{(P)}+\frac{i}{\hbar}S_2^{(P)}}+e^{-S_{Q2}^{(A)}+\frac{i}{\hbar}S_2^{(A)}} \\ e^{-S_{Q3}^{(P)}+\frac{i}{\hbar}S_3^{(P)}}+e^{-S_{Q3}^{(A)}+\frac{i}{\hbar}S_3^{(A)}} \\ e^{-S_{Q4}^{(P)}+\frac{i}{\hbar}S_4^{(P)}}+e^{-S_{Q4}^{(A)}+\frac{i}{\hbar}S_4^{(A)}} \end{pmatrix}^{+}\right.$$

$$\left.\times\,\alpha_i\begin{pmatrix} e^{-S_{Q1}^{(P)}+\frac{i}{\hbar}S_1^{(P)}}+e^{-S_{Q1}^{(A)}+\frac{i}{\hbar}S_1^{(A)}} \\ e^{-S_{Q2}^{(P)}+\frac{i}{\hbar}S_2^{(P)}}+e^{-S_{Q2}^{(A)}+\frac{i}{\hbar}S_2^{(A)}} \\ e^{-S_{Q3}^{(P)}+\frac{i}{\hbar}S_3^{(P)}}+e^{-S_{Q3}^{(A)}+\frac{i}{\hbar}S_3^{(A)}} \\ e^{-S_{Q4}^{(P)}+\frac{i}{\hbar}S_4^{(P)}}+e^{-S_{Q4}^{(A)}+\frac{i}{\hbar}S_4^{(A)}} \end{pmatrix}\right]$$

$$+\frac{\partial}{\partial T}\left[\begin{pmatrix} e^{-S_{Q1}^{(P)}+\frac{i}{\hbar}S_1^{(P)}}+e^{-S_{Q1}^{(A)}+\frac{i}{\hbar}S_1^{(A)}} \\ e^{-S_{Q2}^{(P)}+\frac{i}{\hbar}S_2^{(P)}}+e^{-S_{Q2}^{(A)}+\frac{i}{\hbar}S_2^{(A)}} \\ e^{-S_{Q3}^{(P)}+\frac{i}{\hbar}S_3^{(P)}}+e^{-S_{Q3}^{(A)}+\frac{i}{\hbar}S_3^{(A)}} \\ e^{-S_{Q4}^{(P)}+\frac{i}{\hbar}S_4^{(P)}}+e^{-S_{Q4}^{(A)}+\frac{i}{\hbar}S_4^{(A)}} \end{pmatrix}^{+}\right.$$

$$\left.\times\begin{pmatrix} e^{-S_{Q1}^{(P)}+\frac{i}{\hbar}S_1^{(P)}}+e^{-S_{Q1}^{(A)}+\frac{i}{\hbar}S_1^{(A)}} \\ e^{-S_{Q2}^{(P)}+\frac{i}{\hbar}S_2^{(P)}}+e^{-S_{Q2}^{(A)}+\frac{i}{\hbar}S_2^{(A)}} \\ e^{-S_{Q3}^{(P)}+\frac{i}{\hbar}S_3^{(P)}}+e^{-S_{Q3}^{(A)}+\frac{i}{\hbar}S_3^{(A)}} \\ e^{-S_{Q4}^{(P)}+\frac{i}{\hbar}S_4^{(P)}}+e^{-S_{Q4}^{(A)}+\frac{i}{\hbar}S_4^{(A)}} \end{pmatrix}\right]=0. \qquad (2.97)$$

According to equations (2.96) and (2.97), the behaviour of a Dirac relativistic particle in a Bohmian picture turns out to be linked with the quantum entropies and the phases of the different components of the 4-spinors $\psi^{(P)}(x)$ and $\psi^{(A)}(x)$.

Now, as shown in [274], in the entropic approach we assume that the four-velocities of the Dirac particle are given by equations

$$mv_1^\mu = -\frac{\partial}{\partial X_\mu}(S_1^{(P)} + \hbar S_{Q1}^{(P)} + S_1^{(A)} + \hbar S_{Q1}^{(A)}) \tag{2.98}$$

$$mv_2^\mu = -\frac{\partial}{\partial X_\mu}(S_2^{(P)} + \hbar S_{Q2}^{(P)} + S_2^{(A)} + \hbar S_{Q2}^{(A)}) \tag{2.99}$$

$$mv_3^\mu = -\frac{\partial}{\partial X_\mu}(S_3^{(P)} + \hbar S_{Q3}^{(P)} + S_3^{(A)} + \hbar S_{Q3}^{(A)}) \tag{2.100}$$

$$mv_4^\mu = -\frac{\partial}{\partial X_\mu}(S_4^{(P)} + \hbar S_{Q4}^{(P)} + S_4^{(A)} + \hbar S_{Q4}^{(A)}). \tag{2.101}$$

where the quantum entropies have opportune values in such a way that

$$v_i^\mu v_{i,\mu} = c^2. \tag{2.102}$$

The physical significance of equations (2.98)–(2.101) is that, in the entropic Bohmian approach, the velocity of the Dirac relativistic particle satisfying the equation of motion (1.173) is associated with the phases of the different components of the 4-spinors $\psi^{(P)}(x)$ and $\psi^{(A)}(x)$ as well as the quantum entropies indicating the deformation of the geometry. The four-velocities (2.98)–(2.101) lead directly to formulate the following equation

$$\begin{aligned}\partial^\mu S_i \partial_\mu S_i &= (mv_i^\mu + \hbar\partial^\mu S_{Qi}^{(P)} + \hbar\partial^\mu S_{Qi}^{(A)})(mv_{i,\mu} + \hbar\partial_\mu S_{Qi}^{(P)} + \hbar\partial_\mu S_{Qi}^{(A)}) \\ &= m^2c^2 + 2\hbar m v_i^\mu \partial^\mu S_{Qi}^{(P)} + 2\hbar m v_i^\mu \partial_\mu S_{Qi}^{(A)} \\ &\quad + \hbar^2(\partial^\mu S_{Qi}^{(P)} + \partial^\mu S_{Qi}^{(A)})^2 \end{aligned} \tag{2.103}$$

and hence the Dirac equation (1.170) allows us to obtain the two following equations

$$\sum_{i=1,2}(e^{-2S_{Qi}^{(P)}} + e^{-2S_{Qi}^{(A)}})\left(\hbar v_i \frac{\partial}{\partial X_\mu}(S_{Qi}^{(P)} + S_{Qi}^{(A)}) \right.$$
$$\left. + \frac{\hbar^2}{2m}\frac{\partial}{\partial X_\mu}\left(S_{Qi}^{(P)} + S_{Qi}^{(A)}\right)\frac{\partial}{\partial X^\mu}(S_{Qi}^{(P)} + S_{Qi}^{(A)})\right)$$

$$= \sum_{i=1,2} \frac{\hbar^2}{2m} \left(\left(\frac{\partial}{\partial X_\mu} (S_{Qi}^{(P)} + S_{Qi}^{(A)}) \right)^2 - \left(\nabla^2 - \frac{1}{c^2} \frac{\partial^2}{\partial t^2} \right) \right.$$

$$\left. \times (S_{Qi}^{(P)} + S_{Qi}^{(A)}) \right) + \frac{\hbar q}{2mc} \begin{pmatrix} \psi_1^{(P)} + \psi_1^{(A)} \\ \psi_2^{(P)} + \psi_2^{(A)} \end{pmatrix}^+ \vec{\sigma} \cdot \begin{pmatrix} \psi_1^{(P)} + \psi_1^{(A)} \\ \psi_2^{(P)} + \psi_2^{(A)} \end{pmatrix} \tag{2.104}$$

and

$$\sum_{i=3,4} (e^{-2S_{Qi}^{(P)}} + e^{-2S_{Qi}^{(A)}}) \left(\hbar v_i \frac{\partial}{\partial X_\mu} (S_{Qi}^{(P)} + S_{Qi}^{(A)}) \right.$$

$$\left. + \frac{\hbar^2}{2m} \frac{\partial}{\partial X_\mu} (S_{Qi}^{(P)} + S_{Qi}^{(A)}) \frac{\partial}{\partial X^\mu} (S_{Qi}^{(P)} + S_{Qi}^{(A)}) \right)$$

$$= \sum_{i=3,4} \frac{\hbar^2}{2m} \left(\left(\frac{\partial}{\partial X_\mu} (S_{Qi}^{(P)} + S_{Qi}^{(A)}) \right)^2 \right.$$

$$\left. - \left(\nabla^2 - \frac{1}{c^2} \frac{\partial^2}{\partial t^2} \right) (S_{Qi}^{(P)} + S_{Qi}^{(A)}) \right)$$

$$+ \frac{\hbar q}{2mc} \begin{pmatrix} \psi_1^{(P)} + \psi_1^{(A)} \\ \psi_2^{(P)} + \psi_2^{(A)} \end{pmatrix}^+ \vec{\sigma} \cdot \begin{pmatrix} \psi_1^{(P)} + \psi_1^{(A)} \\ \psi_2^{(P)} + \psi_2^{(A)} \end{pmatrix}. \tag{2.105}$$

Equations (2.104) and (2.105) suggest that the behaviour of a relativistic Dirac particle in a Bohmian approach is directly determined by the geometry of the vacuum associated with the quantum entropies given by equations (2.80)–(2.87). Moreover, the first term on the right side of equations (2.104) and (2.105) can be identified with the quantum potential associated with the first two components and the second two components of the spinors respectively:

$$Q_{1,2} = \sum_{i=1,2} \frac{\hbar^2}{2m} \left(\left(\frac{\partial}{\partial X_\mu} \left(S_{Qi}^{(P)} + S_{Qi}^{(A)} \right) \right)^2 \right.$$

$$\left. - \left(\nabla^2 - \frac{1}{c^2} \frac{\partial^2}{\partial t^2} \right) \left(S_{Qi}^{(P)} + S_{Qi}^{(A)} \right) \right) \tag{2.106}$$

$$Q_{3,4} = \sum_{i=3,4} \frac{\hbar^2}{2m} \left(\left(\frac{\partial}{\partial X_\mu} (S_{Qi}^{(P)} + S_{Qi}^{(A)}) \right)^2 - \left(\nabla^2 - \frac{1}{c^2} \frac{\partial^2}{\partial t^2} \right) (S_{Qi}^{(P)} + S_{Qi}^{(A)}) \right). \quad (2.107)$$

According to equations (2.106) and (2.107), in the geometrodynamic entropic approach to Dirac's equation in a Bohmian picture, the quantum potential emerges as a geometric information channel into the behaviour of the particle which is determined by the vacuum expressed by the quantum entropies corresponding to the different components of the 4-spinors. Starting from equations (2.106) and (2.107) one can also introduce two appropriate quantum lengths expressing the geometrical properties corresponding to the quantum entropies of the first two components and the second two components of the spinor respectively:

$$L_{quantum(1,2)} = \frac{1}{\sqrt{\sum_{i=1,2} \left(- \left(\frac{\partial}{\partial X_\mu} (S_{Qi}^{(P)} + S_{Qi}^{(A)}) \right)^2 + \left(\nabla^2 - \frac{1}{c^2} \frac{\partial^2}{\partial t^2} \right) (S_{Qi}^{(P)} + S_{Qi}^{(A)}) \right)}} \quad (2.108)$$

$$L_{quantum(3,4)} = \frac{1}{\sqrt{\sum_{i=3,4} \left(- \left(\frac{\partial}{\partial X_\mu} (S_{Qi}^{(P)} + S_{Qi}^{(A)}) \right)^2 + \left(\nabla^2 - \frac{1}{c^2} \frac{\partial^2}{\partial t^2} \right) (S_{Qi}^{(P)} + S_{Qi}^{(A)}) \right)}}. \quad (2.109)$$

The quantum-entropic lengths (2.108) and (2.109) can be used to evaluate the modification of the geometry characterizing a relativistic Dirac particle with respect to the Euclidean geometry of classical physics. They show how the vacuum supporting the space-temporal distribution of the ensemble of particles and antiparticles describing the physical system under consideration deforms the geometry of the processes inside a Bohmian approach of the Dirac relativistic quantum mechanics. Once the quantum lengths (2.108) and (2.109) become non-negligible the Dirac particle into consideration goes into a quantum relativistic regime.

2.4 Perspectives of the Quantum Entropy in Bohm's Relativistic Quantum Field Theory

In the context of the first Bohmian models of relativistic bosonic quantum field theory developed by Bohm, Hiley, Bell and Kaloyerou [69, 145, 146], the entropic interpretation — developed by the author of this book in the paper [274] — starts by defining the quantum entropy associated with the Bose-Einstein fields:

$$S_{Q,f} = -\frac{1}{2}\ln\rho \tag{2.110}$$

where $\rho = |\Psi(\{\varphi(\vec{x})\},t)|^2$ is the density of the fields in the element of volume d^3x around a point $\vec{x}$ at time t associated with the wave functional under consideration (it physically represents the probability density for fields to have the configuration $\{\varphi(\vec{x})\}$ at time t). The quantum entropy given by (2.110) can be defined as "quantum entropy of fields". The introduction of the quantum entropy (2.110) leads to formulate the quantum potential (1.238) of relativistic bosonic quantum fields theory in the models of Bohm, Hiley, Bell and Kaloyerou as:

$$Q(\{\varphi(\vec{x})\},t) = -\frac{1}{2}\sum_k \int d^3x \frac{1}{R}\left[\left(\frac{\delta S_{Q,f}}{\delta\varphi_k(\vec{x})}\right)^2 - \frac{\delta^2 S_{Q,f}}{\delta\varphi_k^2(\vec{x})}\right]. \tag{2.111}$$

The quantum Hamilton-Jacobi equation (1.236) characterizing these models becomes therefore:

$$\frac{1}{2}\sum_k \int d^3x \left[-\frac{1}{2}\frac{\delta^2}{\delta\varphi_k^2(\vec{x})} + \frac{1}{2}|\nabla\varphi_k(\vec{x})|^2\right] + V$$
$$-\frac{1}{2}\sum_k \int d^3x \frac{1}{R}\left[\left(\frac{\delta S_{Q,f}}{\delta\varphi_k(\vec{x})}\right)^2 - \frac{\delta^2 S_{Q,f}}{\delta\varphi_k^2(\vec{x})}\right] = -\frac{\partial S}{\partial t} \tag{2.112}$$

which can be considered as the fundamental equation of motion, for Bose-Einstein fields, of the entropic view of Bohm's relativistic quantum field theory.

With the introduction of the quantum entropy of fields (2.110), the following re-reading of Bohm's picture of relativistic quantum field theory for Bose-Einstein fields emerges naturally. The density of the fields is assumed to generate a modification of the geometrical properties of the configuration space expressed by the quantum entropy for fields (2.110) and the quantum entropy for fields (2.110) determines a quantum potential (2.111). Here, the real origin of the quantum potential (1.138) is the ultimate geometry of the vacuum associated with the quantum entropy of fields (2.110). The geometrodynamic information contained in the quantum potential (1.138) of Bohm's relativistic bosonic quantum field theory can be therefore seen as a consequence of the quantum entropy for fields (2.110) namely of the modification of the geometrical properties caused by the presence of these fields. Moreover, the strength of quantum effects in this regime can be evaluated by introducing the quantum-entropic length given by relation

$$L_{quantum,Bose-Einstein} = \frac{1}{\sqrt{\sum_k \int d^3x \frac{1}{R}\left(\left(\frac{\delta S_{Q,f}}{\delta \varphi_k(\vec{x})}\right)^2 - \frac{\delta^2 S_{Q,f}}{\delta \varphi_k^2(\vec{x})}\right)}}. \tag{2.113}$$

The quantum-entropic length (2.113) is the ultimate parameter which can be used to describe the modification of the geometry — introduced by Bose-Einstein fields in a quantum relativistic regime — with respect to the Euclidean geometry characteristic of classical fields. It shows how the vacuum supporting the space-temporal distribution of the fields under consideration deforms the geometry of the processes inside a Bohmian approach to relativistic bosonic quantum field theory. Once the value of the quantum-entropic length (2.113) becomes non-negligible and assumes a significant role for a system we deal with Bose-Einstein fields in a quantum relativistic regime. And here Heisenberg's uncertainty principle can be seen as a consequence of the fact that we are unable to perform a classical measurement to distances smaller than this quantum length corresponding with Bose-Einstein fields in a quantum-relativistic regime.

As regards Nikolic's treatment of Bohmian particle trajectories in relativistic bosonic quantum field theory provided in the papers [148–151], in the entropic interpretation developed by the author in [274], one assumes that the density of the fields to have the configuration $\phi(\vec{x})$ at time t generates a modification of the geometrical properties of the vacuum expressed by the quantum entropy for fields

$$S_{Q,f} = -\frac{1}{2}\ln\rho \tag{2.114}$$

where $\rho = |\Psi[\phi(\vec{x}), t]|^2$ is the density of the fields. The introduction of the quantum entropy for fields (2.114) leads to express the quantum potential (1.145) of Nikolic's model of relativistic bosonic quantum field theory as

$$Q = -\frac{1}{2|\Psi|}\int d^3x \left[\left(\frac{\delta S_{Q,f}}{\delta\phi\left(\vec{x}\right)}\right)^2 - \frac{\delta^2 S_{Q,f}}{\delta\phi^2\left(\vec{x}\right)}\right] \tag{2.115}$$

and thus in this approach the law describing the causal evolution of the field $\phi(\vec{x})$ becomes the following

$$\begin{aligned}(\partial_0^2 - \nabla^2 + m^2c^2)\phi(x) &= J(\phi(x)) \\ &+ \frac{\delta}{\delta\phi(\vec{x})}\left(\frac{1}{2|\Psi|}\int d^3x\left[\left(\frac{\delta S_{Q,f}}{\delta\phi(\vec{x})}\right)^2 - \frac{\delta^2 S_{Q,f}}{\delta\phi^2(\vec{x})}\right]\right)_{\phi(\vec{x})=\phi(x)}\end{aligned} \tag{2.116}$$

which can be considered the fundamental law of motion of Nikolic's model of bosonic relativistic quantum field theory in the entropic interpretation.

In the light of the entropic interpretation, the following suggestive re-reading of Nikolic's model of Bohm's bosonic relativistic quantum field theory may be provided. The density of the fields is assumed to provoke a deformation of the geometrical properties of the ultimate background space expressed by the quantum entropy for fields (2.114) and the quantum entropy for fields (2.114) generates a quantum potential (2.115). The real origin of the quantum potential of relativistic bosonic quantum field of Nikolic's approach is the fundamental vacuum associated with the quantum entropy for fields (2.114). The geometrodynamic information contained in the quantum potential (1.245) of Nikolic's model of a Bohmian approach to relativistic quantum field theory for bosonic fields can be therefore seen as a consequence of the quantum entropy for fields namely of the modification of the geometry of the vacuum caused by the presence of these fields, of the geometrical properties of the vacuum supporting the distribution of these fields. Moreover, if in Nikolic's approach the effectivity parameter (1.247) measuring the probability that there are n particles in the system at time t if the field is equal (but not measured) to be $\phi(\vec{x})$ at that time is linked with the evolution of the field, hence it derives that, since the evolution of the field is expressed by equation (2.116), the evolution of the effectivity parameter is ultimately determined by the modification of the geometrical properties of the background space expressed by the quantum entropy for fields. According to the fundamental equations (2.115) and (2.116), the quantum entropy for fields (2.110) emerges as the ultimate parameter characterizing the dynamics, as the ultimate grid of Nikolic's model of bosonic relativistic quantum field theory in a Bohmian picture. Moreover, also here the geometrical properties of the background space can be characterized by introducing a quantum-entropic length given by relation

$$L_{quantum,Bose-Einstein} = \frac{1}{\sqrt{\int d^3x \frac{1}{|\Psi|}\left(\left(\frac{\delta S_{Q,f}}{\delta\phi(\vec{x})}\right)^2 - \frac{\delta^2 S_{Q,f}}{\delta\phi^2(\vec{x})}\right)}}. \tag{2.117}$$

The quantum-entropic length (2.117) can be considered as the fundamental geometrical quantity of the vacuum which can be used to measure the modification of the geometrical properties — introduced by bosonic fields in a quantum relativistic regime — with respect to the Euclidean geometry characteristic of classical fields. It shows how the vacuum supporting the space-temporal distribution of the ensemble of Bose-Einstein fields characterizing the physical system under consideration deforms the geometry of the processes in the relativistic bosonic quantum field theory. Once the quantum length (2.117) becomes non-negligible we have bosonic fields in a quantum relativistic regime.

Finally, as regards Nikolic's model of a Bohmian covariant interpretation for the many-fingered-time Tomonaga-Schwinger equation for relativistic quantum field theory (developed in the paper [152]), always following the treatment in the paper [274], an entropic interpretation can be provided by assuming that the density of the fields to have the configuration $\phi(\vec{s})$ on the 3-dimensional manifold Σ at time $\tau(\vec{s})$ generates a change of the geometrical properties of the background space expressed by the quantum entropy for fields

$$S_{Q,f} = -\frac{1}{2}\ln\rho \tag{2.118}$$

where

$$\rho[\phi,\tau] = |\Psi[\phi,\tau]|^2 \tag{2.119}$$

represents the probability density for the field to have a value ϕ on Σ at time $\tau(\vec{s})$. The introduction of the quantum entropy for fields (2.118) leads directly to formulate the quantum potential (1.263) of the many-fingered-time Tomonaga-Schwinger equation for relativistic quantum field theory as

$$Q = -\frac{1}{2R|q(\vec{s})|^{1/2}}\int d^3x\left[\left(\frac{\delta S_{Q,f}}{\delta\phi(\vec{s})}\right)^2 - \frac{\delta^2 S_{Q,f}}{\delta\phi^2(\vec{s})}\right]. \tag{2.120}$$

In this picture, the covariant quantum many-fingered-time Klein-Gordon equation (1.262) becomes:

$$\left(\left(\frac{\partial}{\partial\tau(\vec{s})}\right)^2 - \nabla^i\nabla_i + m^2\right)\Phi[\vec{s},X]$$
$$= \left[\frac{1}{|q(\vec{s})|^{1/2}}\frac{\partial\left(\frac{1}{2R|q(\vec{s})|^{1/2}}\int d^3x\left[\left(\frac{\delta S_{Q,f}}{\delta\phi(\vec{s})}\right)^2 - \frac{\delta^2 S_{Q,f}}{\delta\phi^2(\vec{s})}\right]\right)}{\partial\phi(\vec{s})}\right]_{\phi=\Phi} \tag{2.121}$$

which can be considered the fundamental law regarding Nikolic's model of a Bohmian covariant interpretation for the many-fingered-time Tomonaga-Schwinger equation for relativistic quantum field theory inside the entropic interpretation.

In summary, in the light of the results obtained by the author in [274], inside the entropic interpretation the following re-reading of Nikolic's Bohmian covariant model of the many-fingered-time Tomonaga-Schwinger equation for relativistic quantum field theory emerges naturally. The density of the fields (2.119) is assumed to generate a modification of the geometrical properties of the 3-dimensional manifold Σ in space-time expressed by the quantum entropy for fields (2.118) and the quantum entropy for fields (2.118) determines a quantum potential (2.120). In this picture, the real origin of the quantum potential of the many-fingered-time Tomonaga-Schwinger equation for relativistic quantum field theory is the geometry of the vacuum associated with the quantum entropy. The geometrodynamic information contained in the quantum potential (2.120) of Nikolic's covariant Bohmian approach to many-fingered-time Tomonaga-Schwinger equation for relativistic quantum field theory represents the reflex of a more fundamental geometry of the background supporting the density of the fields into consideration, namely of the quantum entropy for fields (2.118), namely of the modification of the geometrical properties of the background space caused by the presence of these fields. The many-fingered-time field $\Phi[\vec{s}, X]$ satisfies the fundamental entropic covariant quantum many-fingered-time Klein-Gordon equation (2.121). According to equation (2.121), the behaviour of the many-fingered-time field $\Phi[\vec{s}, X]$ is linked with the change of the geometrical properties of the 3-dimensional manifold Σ expressed by the quantum entropy for fields (2.118). To conclude, on the basis of the fundamental equations (2.120) and (2.121), the quantum entropy for fields can be considered therefore as the ultimate parameter describing the dynamics of a Bohmian approach to relativistic quantum field theory also in Nikolic's covariant version of the many-fingered-time Tomonaga-Schwinger equation.

Finally, in analogy to the other relativistic quantum field theory models analysed in this chapter, also for Nikolic's covariant Bohmian model to many-fingered-time Tomonaga-Schwinger equation for relativistic quantum field theory the geometry of the vacuum associated with the quantum entropy of the fields can be characterized by introducing a quantum-entropic length given by relation:

$$L_{quantum,MFT-TS} = \frac{1}{\sqrt{\frac{1}{2R|q(\vec{s})|^{1/2}} \int d^3x \left[\left(\frac{\delta S_{Q,f}}{\delta \phi(\vec{s})} \right)^2 - \frac{\delta^2 S_{Q,f}}{\delta \phi^2(\vec{s})} \right]}}. \tag{2.122}$$

The quantum-entropic length (2.122) is the fundamental property of the vacuum which allows us to evaluate the modification of the geometrical properties — introduced by bosonic fields in a three-dimensional many-fingered-time background in a quantum relativistic regime — with respect to the Euclidean geometry characteristic of classical fields. It indicates the action of the vacuum supporting the space-temporal distribution of the ensemble of fields under consideration into the geometry of the processes inside Nikolic's covariant version of the many-fingered-time Tomonaga-Schwinger equation. Once the quantum length (2.122) becomes non-negligible we have fields in a quantum relativistic regime of Tomonaga-Schwinger many-fingered-time background.

2.5 The Quantum Entropy as the Ultimate Source of the Geometry of Space in the Relativistic de Broglie-Bohm Theory in Curved Space-Time

The next important step is to analyse the perspectives introduced by the quantum entropy in the context of A. Shojai's and F. Shojai's Bohmian model regarding the motion of a spinless particle in a relativistic curved space-time. In order to provide an entropic view of relativistic de Broglie-Bohm theory in a curved space-time based on the Klein-Gordon equation, just like in the non-relativistic problem, we start by considering the quantum entropy

$$S_Q = -\frac{1}{2} \ln \rho \tag{2.123}$$

where ρ is the density of distribution of the ensemble of particles describing the physical system. In the entropic version of Bohmian version of Klein-Gordon equation, the space-temporal distribution of the ensemble of particles describing the physical system is assumed to generate a modification, and thus a degree of order and chaos, of the background space, of a vacuum of processes, characterized by the quantity given by equation (2.123).

If, as we have seen in chapter 1.6.1, F. Shojai and A. Shojai showed that, as regards Bohm's version of Klein-Gordon equation, the de Broglie–Bohm quantum theory of motion and gravity can be unified and the quantum potential can be interpreted as the conformal degree of freedom of the space–time metric [97], in the reference [273] the author demonstrated that, inside F. Shojai's and A. Shojai's model, the effects of gravity on geometry and the quantum effects on the geometry of space-time are highly coupled, the quantum potential emerges as the conformal degree of freedom of the space-time metric, as a consequence of a physical vacuum described by the quantum entropy. In other words, in the approach developed by the author in [273], the ensemble of particles associated with the wave function of the physical system determines a modification of the fundamental geometry of the vacuum of processes in a relativistic curved space-time. In the entropic approach, the equations of motion for a spinless particle in a curved background assume the following form:

$$\tilde{g}^{\mu\nu}\frac{1}{c}\frac{\partial S_Q}{\partial t} = \tilde{g}^{\mu\nu}\left[-(p_\mu\tilde{\nabla}^\mu S_Q) + \frac{1}{2}\nabla_\mu p^\mu\right] \tag{2.124}$$

and

$$\tilde{g}^{\mu\nu}\tilde{\nabla}_\mu S\tilde{\nabla}_\nu S = m^2c^2 \tag{2.125}$$

where

$$\tilde{g}_{\mu\nu} = \frac{M^2}{m^2}g_{\mu\nu} \tag{2.126}$$

is a conformal transformation, $\tilde{\nabla}_\mu$ represents the covariant differentiation with respect to the metric $\tilde{g}_{\mu\nu}$. The conformal transformation (2.126) is itself generated by the fundamental geometry of the vacuum associated with the quantum entropy on the basis of equation

$$M^2 = m^2 \exp\left[-\frac{\hbar^2}{m^2c^2}(\nabla_\mu S_Q)^2 + \frac{\hbar^2}{m^2c^2}\left(\left(\nabla^2 - \frac{1}{c^2}\frac{\partial^2}{\partial t^2}\right)_g S_Q\right)\right] \tag{2.127}$$

which can be interpreted as a quantum mass which therefore is ultimately generated by the quantum entropy. The quantum potential describing the dynamics is determined by the geometrical properties of the vacuum on the basis of equation:

$$Q = -\frac{\hbar^2}{m^2c^2}(\nabla_\mu S_Q)^2 + \frac{\hbar^2}{m^2c^2}\left(\left(\nabla^2 - \frac{1}{c^2}\frac{\partial^2}{\partial t^2}\right)_g S_Q\right). \tag{2.128}$$

In the entropic approach, in the light of equations (2.123)–(2.128), the presence of the quantum potential is equivalent to a curved space-time and the metric characterizing the curvature is given by (2.126) where the mass square is determined by the quantum entropy, and thus by the geometry of the vacuum, on the basis of equation (2.127). According to equations (2.126)–(2.128), the ultimate origin of the the conformal metric of the space-time, of the quantum mass and, consequently of the geometrodynamic action of the quantum potential inside F. Shojai's and A. Shojai's model of relativistic de Broglie-Bohm theory of spinless particles in curved space-time is represented by the vacuum defined by the quantum entropy (2.123). The fundamental geometry of the background associated with the quantum entropy emerges as the ultimate parameter which describes, at a fundamental level, the dynamics regarding a spinless particle in a relativistic curved space-time.

In this way, the entropic approach suggests a geometrization of the quantum aspects of matter in a picture based on the idea that the density of particles associated with a given wave function determines a modification of the geometry of the background of the processes. The effects of gravity on geometry and the quantum effects on the geometry of space-time are highly coupled because they are both determined by the fundamental background space associated to the quantum entropy, they are both produced by the degree of order and chaos of the vacuum supporting the density of particles associated with the wave function under consideration. In other words, one can say that the geometric properties which determine the behaviour of a spinless particle are linked with the curvature of the background and are expressed by the quantum entropy. In summary, one can say that the particles determine the curvature of space-time and at the same time the space-time metric is linked with the vacuum defined by the quantum entropy which influences the behaviour of the particles. The quantum entropy generates itself a curvature which may have a large influence on the classical contribution to the curvature of the space-time. The quantum entropy appears indeed as a real fundamental intermediary between gravitational and quantum effects of matter.

Moreover, as the author has shown in the paper [274], inside the entropic approach to F. Shojai's and A. Shojai's model of relativistic spinless particles in a curved space-time, the geometrical properties can be characterized by introducing an appropriate quantum length associated with the conformal metric (2.126) determined by the quantum entropy:

$$L_{quantum} = \frac{1}{\sqrt{(\nabla_\mu S_Q)^2 - \left(\nabla^2 - \frac{1}{c^2}\frac{\partial^2}{\partial t^2}\right)_g S_Q}}. \quad (2.129)$$

The quantum-entropic length (2.129) is the fundamental geometrical quantity of the vacuum which allows us to evaluate the strength of quantum effects and, therefore, the non-local correlation degree and the modification of the geometry — introduced in a relativistic curved space-time regime — with respect to the Euclidean geometry characteristic of classical physics. It shows how the vacuum supporting the space-temporal distribution of the ensemble of particles describing the physical system under consideration deforms the fundamental geometry responsible of the motion of spinless particles in a relativistic curved space-time. Once the quantum-entropic length (2.129) becomes non-negligible the spinless particle into consideration goes into a quantum regime where the quantum and gravitational effects are highly related. In this picture, Heisenberg's uncertainty principle derives from the fact that we are unable to perform a classical measurement to distances smaller than this quantum-entropic length. A quantum regime where there is a coupling between gravitational and quantum effects is entered when the quantum length (2.129) must be taken under consideration.

On the basis of equation (2.129), it becomes thus permissible the following reading of the geometry of the background space regarding the motion of a spinless particle in the context of F. Shojai's and A. Shojai's model. The quantum entropy can be considered as the fundamental entity which provides a measure of the geometrical properties of a vacuum regime where the quantum behaviour of matter becomes highly related with its gravitational effects. In virtue of the link between the quantum potential and the quantum entropy, the geometric properties of the background space which correspond with a regime where the quantum behaviour of matter is highly coupled with gravitational effects are determined by the vacuum defined by the quantum entropy and the condition (2.129) about the quantum length indicates when this coupling between quantum and gravitational effects must be taken into consideration. The mathematical formalism constituted by equations (2.123)–(2.129) of the entropic approach of F. Shojai's and A. Shojai's model and its physical interpretation can be considered a relevant development of relativistic de Broglie-Bohm theory in curved space-time.

2.6 The Quantum Entropy as the Ultimate Source of the Geometry of Space in Bohmian Quantum Gravity

As we have shown in chapter 1.6.2, F. Shojai and A. Shojai developed a toy model of quantum gravity (providing a scalar-tensor picture of the ideas mentioned in section 1.6.1) in which the form of the quantum potential and its relation to the conformal degree of freedom of the space-time metric can be derived using the equations of motion. As we have seen, by showing that it is just the quantum gravity equations of motion which make the quantum potential the entity expressing the geometrical properties which influence the behaviour of the particles and which is related to the space-time metric, F. Shojai's and A. Shojai's model suggests a sort of unification of the gravitational and quantum aspects of matter at a fundamental level of physical reality.

Here we want to review an entropic version of F. Shojai's and A. Shojai's toy model of quantum gravity recently proposed by the author of this paper in [272] and to analyse the most significant results suggested by this entropic view about the geometry of space. This approach can be considered a relevant and interesting development in the interpretation of the quantum entropy as the ultimate parameter which describes the geometry of space, also in the quantum gravity domain.

In the entropic picture, the conformal factor $\Omega^2 = \exp Q$ appearing in F. Shojai's and A. Shojai's equations of motion (1.309)–(1.313) which describe a general relativistic system with quantum effects, expresses the deformation of the background in the sense that it is determined by the quantum entropy on the basis of the relation

$$\Omega^2 = \exp\left[-\frac{\hbar^2}{m^2c^2}(\nabla_\mu S_Q)^2 + \frac{\hbar^2}{m^2c^2}\left(\left(\nabla^2 - \frac{1}{c^2}\frac{\partial^2}{\partial t^2}\right)_g S_Q\right)\right], \tag{2.130}$$

and thus is produced by the degree of order and chaos of the vacuum supporting the density of particles associated with the wave function under consideration. By taking into account equation (2.130), the constraint equation (1.313) can be also expressed in the equivalent form

$$\exp\left[-\frac{\hbar^2}{m^2c^2}(\nabla_\mu S_Q)^2 + \frac{\hbar^2}{m^2c^2}\left(\left(\nabla^2 - \frac{1}{c^2}\frac{\partial^2}{\partial t^2}\right)_g S_Q\right)\right] = \frac{\hbar^2}{m^2}\frac{\left(\bar{\nabla}^2 - \frac{\partial^2}{\partial t^2}\right)\sqrt{\rho}}{\sqrt{\rho}} \tag{2.131}$$

which shows the direct link between the density of the particles and the quantum entropy in the quantum gravity domain. The equations of motion (1.309)–(1.313), in virtue of the link between the quantum potential with the quantum entropy, tell us that there are back-reaction effects of the quantum factor on the background which are due to the geometry of the vacuum described by the quantum entropy. On the basis of these high-coupled five equations, one can say that in the quantum gravity domain the quantum entropy is the fundamental entity which introduces the links (and thus the back-reaction terms) between the quantum effects and the background.

Moreover, in F. Shojai's and A. Shojai's model, the other more specific equations of quantum gravity (1.316)–(1.318), which are obtained by combining equation (1.309)–(1.311) in the context of a perturbative way, can be themselves seen as laws which originate from the vacuum defined by the quantum entropy. In virtue of the link between the conformal factor and the quantum entropy, namely the fact that the conformal factor emerges from a more fundamental deformation of the geometry of the background space, the quantum entropy turns out to be the ultimate dynamics parameter also as regards the contents of these three equations. Finally, also the next step of F. Shojai's and A. Shojai's toy model represented by the quantum gravity equations which make dynamical the conformal factor and the quantum potential, namely the four equations (1.324)–(1.327), can be associated with the vacuum supporting the density of the particles, with the geometry of background space whose ultimate entity is the quantum entropy. More precisely, one can draw the following important conclusions:

— On the basis of equation (1.326) the causal structure of the space-time $g^{\mu\nu}$ is determined by the gravitational effects of matter and thus by the fundamental geometry of the vacuum described by quantum entropy, which must be considered as the ultimate entity which shows that the quantum effects and the gravitational effects are coupled (also in the quantum gravity domain).

— On the basis of equation (1.324) quantum effects, and thus the quantum entropy, determine directly the scale factor of space-time.

In summary, according to the entropic interpretation provided in chapters 1.6.1 and 1.6.2, the geometrization of the quantum effects and the coupling between, on one hand, gravitational effects (and thus the causal structure of the space-time) and, on the other hand, quantum effects (and thus the conformal structure of space-time) are determined by the vacuum of the processes defined by the quantum entropy. The conformal mapping of the metric which is equivalent to the presence of the quantum potential is produced by the quantum entropy. The existence of different conformally related frames where one measures different quantum masses and different curvatures is determined by the geometrical properties of the background expressed by the quantum entropy. One can also say that the geometry subtended by the quantum entropy makes not distinguishable the conformally related frames. In this picture, it is the geometry of the background space represented by the quantum entropy the fundamental level of reality which determines the fact that, at any point (or even globally), the quantum effects of matter can be removed by a suitable conformal transformation. Therefore, the quantum equivalence principle, according to which gravitational effects can be removed by going to a freely falling frame while quantum effects can be eliminated by choosing an appropriate scale, in this picture has a secondary ontological status in the sense that is originated from a more fundamental geometry of the vacuum supporting the density of the particles, namely from the geometry associated with the quantum entropy. To conclude, in the entropic interpretation, all the aspects of the geometry of physical space regarding frames characterized by quantum and conformal covariance can be seen as the consequence of the geometrical properties of the more fundamental background associated with the quantum entropy.

2.7 The Quantum Entropy as the Ultimate Source of the Geometry of Space in Bohmian Quantum Cosmology

In order to develop an entropic view of the models of WDW equation in the Bohmian picture provided in chapter 1.7, we start by defining a "quantum entropy for the gravitational field"

$$S_Q = q \ln \rho. \tag{2.132}$$

The quantum entropy (2.132) indicates that a degree of order and chaos can be associated with the gravitational field characterized by the 3-metric q_{ab}. More precisely, we can say that the density of particles $\rho = R^2$ associated with the wave-functional Ψ of the universe determines a modification of the gravitational space described the quantum entropy for the gravitational field (2.132). With the introduction of the quantum entropy (2.132), the quantum potential for the gravitational field (1.349) can be expressed in the entropic way:

$$Q_G = \hbar^2 N q G_{abcd} \frac{1}{R} \left(\frac{\delta^2}{\delta q_{ab} \delta q_{cd}} \right)_q S_Q. \tag{2.133}$$

On the basis of equation (2.133), the quantum entropy for the gravitational field can be considered as the fundamental entity, as the ultimate parameter which determines the action of the quantum potential for the gravitational field. The following interesting perspective is therefore opened: just like in the non-relativistic de Broglie-Bohm theory the quantum entropy represents the fundamental entity from which the behaviour of subatomic particles derive, in analogous way as regards the Bohmian approach to WDW equation the quantum entropy for the gravitational field can be considered as the ultimate parameter which produces the behaviour of the universe in the presence of a gravitational field.

On the basis of equation (2.133), one can say that, in F. Shojai's and A. Shojai's model, in the Bohm-Einstein equations (1.360) which describe the geometry of the physical space derived from the WDW equation, the real origin of the curvature of space is the vacuum defined by the quantum entropy for the gravitational field. More precisely, in the Bohm-Einstein equations (1.360), the quantum potential for gravity is determined by a quantum entropy for the gravitational field indicating the degree of order and chaos of the gravitational field on the basis of equation (2.133), while the quantum potential for matter is associated with a quantum entropy for matter (a degree of order and chaos of the vacuum supporting the probability density of the particles) in terms of relation

$$Q_m = \hbar^2 \frac{N\sqrt{H}}{2} \left[\left(\frac{\delta S_{Q,m}}{\delta \varphi} \right)^2 - \frac{\delta^2 S_{Q,m}}{\delta \varphi^2} \right] \tag{2.134}$$

where $S_{Q,m}$ is the quantum entropy determining the quantum potential for matter.

If the quantum potential for the gravitational field is determined by the quantum entropy for the gravitational field (2.132) on the basis of equation (2.133), and the quantum entropy for matter is determined by a quantum entropy for matter on the basis of equation (2.134), hence it derives that the quantum corrector tensor $S^{\mu\nu}$ which describes the modification, the deformation of the geometry of the physical space produced by matter and gravity in WDW equation's regime, depends both on the quantum entropy for gravity and on the quantum entropy for matter, which can be indeed considered as the most fundamental parameters characterizing the geometry of processes in WDW equation regime. It is just the quantum entropy for the gravitational field (2.132), together with the quantum entropy for matter, the fundamental entity which makes the Bohm-Einstein equations (1.360) invariant under temporal $\otimes$ spatial diffeomorphisms of the general coordinate transformations.

Moreover, if in Pinto-Neto's model the specific behaviour of the quantum potential leads to a non-degenerate for-geometry or to a degenerate four-geometry, in the entropic approach one can speculate that it is just the vacuum defined by the quantum entropy for the gravitational field, as well as by the quantum entropy for matter, the fundamental manifold which, with its peculiar features, makes the geometry of the physical space a non-degenerate four-geometry (which can be Euclidean or hyperbolic), or a degenerate four-geometry indicating the presence of special vector fields and the breaking of the space-time structure as a single entity (in a wider class of possibilities). In Pinto-Neto's approach, the quantum potential reads

$$Q = -\frac{\hbar^2}{R} k G_{abcd} \left(\frac{\delta^2}{\delta q_{ab} \delta q_{cd}} \right)_q S_q$$

$$- \frac{\hbar^2}{2R\sqrt{q}} \left[\left(\frac{\delta S_{Q,m}}{\delta \phi} \right)^2 - \frac{\delta^2 S_{Q,m}}{\delta \phi^2} \right]. \quad (2.135)$$

Here, a non degenerate four-geometry can be attained if the quantum potential (2.135) assumes the specific form

$$Q = -\sqrt{q}\left[(\varepsilon+1)\left(-\frac{1}{k}{}^{(3)}R + \frac{1}{2}q^{ab}\partial_a\phi\partial_b\phi\right)\right.$$
$$\left. + \frac{2}{k}(\varepsilon\bar{\Lambda}+\Lambda) + \varepsilon\bar{U}(\phi) + U(\phi)\right] \qquad (2.136)$$

As a consequence, the Euclidean or hyperbolic 3-geometries emerge in this picture from specific behaviours of the quantum entropy, describing the peculiar features of the fundamental background of the processes.

Finally, also in Farag Ali's and Das' model, developed in [241], regarding the link between the quantum potential and the cosmological constant in the context of a quantum corrected Raychaudhuri equation, one can speculate that the cosmological constant is ultimately associated with the geometrical properties of the fundamental background of processes expressed by the quantum entropy (2.132). In this entropic approach, the quantum corrected Raychaudhuri equation (1.371) reads

$$\frac{d\theta}{d\lambda} = -\frac{1}{3}\theta^2 - R_{cd}u^c u^d$$
$$+ \frac{\hbar^2}{m^2}q^{ab}\left[\left[(\partial_\mu S_Q)^2 - \left(\nabla^2 - \frac{1}{c^2}\frac{\partial^2}{\partial t^2}\right)S_Q\right]\right] + \frac{\varepsilon_1\hbar^2}{m^2}q^{ab}R_{;ab} \qquad (2.137)$$

where the quantum potential is:

$$Q = \frac{\hbar^2}{m^2}q^{ab}\left[(\partial_\mu S_Q)^2 - \left(\nabla^2 - \frac{1}{c^2}\frac{\partial^2}{\partial t^2}\right)S_Q\right]. \qquad (2.138)$$

Therefore, if one considers a Gaussian form $\Psi \approx \exp\left(-r^2/L_0^2\right)$ or, in the context of a scalar field theory with an interaction of strength g, the wave function $\Psi = \Psi_0 \tanh\left(r/L_0\sqrt{2}\right)$ (for g > 0) and $\Psi = \sqrt{2}\Psi_0 \sec h\,(r/L_0)$ (for g < 0) where L_0 is the characteristic length scale in the problem which is of the order of the Compton wavelength, the cosmological constant emerges directly from the quantum entropy on the basis of relation

$$\frac{\hbar^2}{m^2} q^{ab} \left[(\partial_\mu S_Q)^2 - \left(\nabla^2 - \frac{1}{c^2} \frac{\partial^2}{\partial t^2} \right) S_Q \right] = \frac{1}{L_0^2} = (mc/\hbar)^2 \quad (2.139)$$

where m can be regarded as the small mass of gravitons (or axions), with gravity (or Coulomb field) following the Yukawa type of force law (1.375). This implies also that, inside this entropic approach, gravitons or axions cannot be considered as fundamental realities but as secondary emergent structures which are originated from the vacuum defined by the quantum entropy for the gravitational field.

2.8 About the Geometrodynamics of Bohm's Quantum Potential in an Entropic Approach in the Condition of Fisher Information

In this last paragraph of chapter 2, we want to make some considerations as regards an entropic approach to Bohm's quantum potential in a Fisher geometry proposed recently by the author, Licata and Resconi in the papers [282–284]. According to this model, by applying a minimum condition of Fisher information in a picture based on a superposition of Boltzmann entropies, Bohm's quantum potential emerges as a geometrodynamic entity indicating the deformation of the geometry of space. Here, Fisher information plays the role of a natural medium to build a metric able to connect the system's statistical outcomes and its global geometry.

By following the references [282–284], on the basis of an extension of the tensor calculus to operators represented by non-quadratic matrices, the quantum potential emerges as an information channel determined by the vector of the superposition of entropies (one for each observer)

$$\begin{cases} S_1 = k \log W_1(\theta_1, \theta_2, \ldots, \theta_p) \\ S_2 = k \log W_2(\theta_1, \theta_2, \ldots, \theta_p) \\ \qquad \ldots \\ S_n = k \log W_n(\theta_1, \theta_2, \ldots, \theta_p) \end{cases} \quad (2.140)$$

where k is the Boltzmann constant, W define the number of the microstates of the physical system into consideration, which depend on the parameters θ as of the distribution probability (and thus, for example, on the space-temporal distribution of an ensemble of particles, namely the density of particles in the element of volume d^3x around a point $\vec{x}$ at time t). In this picture, quantum effects are equivalent to a background which is described by the following equation

$$\frac{\partial}{\partial x^k}+\frac{\partial^2 S_j}{\partial x^k \partial x^p}\frac{\partial x^i}{\partial S_j}=\frac{\partial}{\partial x^k}+\frac{\partial \log W_j}{\partial x_h}=\frac{\partial}{\partial x^k}+B_{j,h} \quad (2.141)$$

where $B_{j,h}=\frac{\partial S_j}{\partial x_h}=\frac{\partial \log W_j}{\partial x_h}$ is a Weyl-like gauge potential. This Weyl-like gauge potential produces a deformation of the moments for the change of the geometry stated by the following expression of the action:

$$A=\int \rho\left[\frac{\partial A}{\partial t}+\frac{1}{2m}\left(p_i+B_{k,i}\right)\left(p_j+B_{k,j}\right)+V\right]dtd^n x \quad (2.142)$$

which gives

$$A=\int \rho\left[\frac{\partial A}{\partial t}+\frac{1}{2m}\left(p_i p_j+B_{k,i}B_{k,j}\right)+V\right]dtd^n x \quad (2.143)$$

namely

$$A=\int \rho\left[\frac{\partial A}{\partial t}+\frac{1}{2m}\left(p_i p_j+\frac{\partial \log W_k}{\partial x_i}\frac{\partial \log W_k}{\partial x_j}\right)+V\right]dtd^n x \quad (2.144)$$

namely

$$A=\int \rho\left[\frac{\partial A}{\partial t}+\frac{1}{2m}p_i p_j+V+\frac{1}{2m}\left(\frac{\partial \log W_k}{\partial x_i}\frac{\partial \log W_k}{\partial x_j}\right)\right]dtd^n x. \quad (2.145)$$

The quantum action assumes the minimum value when

$$\delta A=0 \quad (2.146)$$

namely

$$\delta\int \rho\left[\frac{\partial A}{\partial t}+\frac{1}{2m}p_i p_j+V\right]dtd^n x + \delta\int \frac{\rho}{2m}\frac{\partial \log W_k}{\partial x_i}\frac{\partial \log W_k}{\partial x_j}dtd^n x=0 \quad (2.147)$$

and thus

$$\frac{\partial A}{\partial t} + \frac{1}{2m} p_i p_j + V + \frac{1}{2m}\left(\frac{1}{W_k^2}\frac{\partial W_k}{\partial x_i}\frac{\partial W_k}{\partial x_j} - \frac{2}{W_k}\frac{\partial^2 W_k}{\partial x_i \partial x_j}\right)$$
$$= \frac{\partial S_k}{\partial t} + \frac{1}{2m} p_i p_j + V + Q \quad (2.148)$$

where Q is Bohm's quantum potential. Equation (2.148) states clearly that the quantum potential — and thus its geometrodynamic features — can be derived as a consequence for the extreme condition of Fisher information [283]. In the light of equation (2.148), Bohm's quantum potential can be interpreted as an information channel determined by the functions W_k which define the number of microstates of the physical system under consideration and generate the vector of the superposition of Boltzmann entropies (2.140). In other words, the distribution probability of the wave function determines the functions W_k defining the number of microstates of the physical system under consideration, a vector of superposition of Boltzmann entropies emerges from these functions W_k given by equations (2.140), and these functions W_k, and therefore the vector of the superpose entropies, can be considered as the fundamental physical entities which determine the geometrodynamic action of the quantum potential (in the extreme condition of the Fisher information) on the basis of equation[1]

$$Q = \frac{1}{2m}\left(\frac{1}{W^2}\frac{\partial W}{\partial x_i}\frac{\partial W}{\partial x_j} - \frac{2}{W}\frac{\partial^2 W}{\partial x_i \partial x_j}\right). \quad (2.149)$$

In this picture, in the extreme condition of Fisher information, the Boltzmann entropies defined by equations (2.140) (and thus the functions W_k) emerge as "informational lines" of the quantum potential. In other words, each of the entropies appearing in the superposition vector (2.140) can be considered as a specific information channel of the quantum potential.

[1]In the next equations of this paragraph, for simplicity we are going to denote the generic function W_k (defined by equations (2.140)) with W.

Moreover, on the basis of equations (2.141) and (2.148), one can say that the quantum phenomena correspond with a change of the geometry, which is expressed by a Weyl-like gauge potential and is determined by the functions W and thus by the vector of the superposition of the entropies. According to this model, it becomes so permissible the following reading of the mathematical formalism in non-relativistic Bohmian quantum mechanics: the distribution probability of the wave function determines the functions W defining the number of microstates of the physical system under consideration, a vector of superposition of entropies emerges from these functions W given by equations (2.140), and these functions W (and thus also the vector of the superpose entropies given by equations (2.140)) determine a change of the geometry expressed by a Weyl-like gauge potential and characterized by a deformation of the moments given by equation (2.145). The quantum potential, in the extreme condition of Fisher information, can be therefore considered as an information channel describing the change of the geometry of the physical space in the presence of quantum effects.

On the other hand, if one inserts the definition (2.149) of the quantum potential inside the quantum Hamilton-Jacobi equation (1.7), one obtains

$$\frac{|\nabla S|^2}{2m} + V + \frac{1}{2m}\left(\frac{1}{W^2}\frac{\partial W}{\partial x_i}\frac{\partial W}{\partial x_j} - \frac{2}{W}\frac{\partial^2 W}{\partial x_i \partial x_j}\right) = -\frac{\partial S}{\partial t} \qquad (2.150)$$

which provides a new alternative way (in the regime of Fisher information) to interpret the energy conservation law in quantum mechanics. Equation (2.150) states that two quantum corrector terms appear in the energy of the system, which are owed to the functions W linked with the vector of the superpose entropies, and which thus characterize the deformation of the geometry in the presence of quantum effects. On the basis of equation (2.150), we can say that the distribution probability of the wave function determines the functions W defining the number of microstates of the physical system under consideration, a vector of superpose entropies emerges from these functions W given by equations (2.140), these functions W (and thus also the vector of the superpose entropies given by equations (2.140)) determine a change of the geometry of the physical space, and this change of the geometry produces two quantum corrector terms in the energy of the system. These two quantum corrector terms can thus be interpreted as a sort of modification of the physical features of the background space determined by the ensemble of particles associated with the wave function under consideration. According to this approach, it is just the functions W and thus the vector of superposition of the different Boltzmann entropies (2.140) the fundamental entities which, by expressing the deformation of the geometry of the background space determined by the ensemble of particles associated with the wave function under consideration, express the geometric properties of space from which the quantum force, and thus the behaviour of quantum particles, are derived. It is also interesting to observe that also in this approach the geometrical properties of the vacuum defined by the vector of superposition of different Boltzmann entropies can be characterized by considering an appropriate quantum-entropic length given by relation

$$L_{quantum} = \frac{1}{\sqrt{\frac{1}{\hbar^2}\left(\frac{2}{W}\frac{\partial^2 W}{\partial x_i \partial x_j} - \frac{1}{W^2}\frac{\partial W}{\partial x_i}\frac{\partial W}{\partial x_j}\right)}} \tag{2.151}$$

that can be used to evaluate the strength of quantum effects and, therefore, the modification of the geometry with respect to the Euclidean geometry characteristic of classical physics inside this picture. Once the quantum-entropic length (2.151) becomes non-negligible the system enters a quantum regime.

It is also interesting to make a comparison of this geometric approach to quantum entropy and information analysed in this paragraph with the entropic approach, originally suggested by Sbitnev in [270, 271], and furtherly developed by the author of this book in [66, 171, 272–274]. If in the approach based on Sbitnev's entropy (2.1), the quantum potential emerges as an information channel into the behaviour of the particles given by relation (2.2) and thus associated with two quantum corrector terms of the energy of the system depending on the quantum entropy (2.1), in a similar way in the approach of Fisher-Bohm geometry based on a vector of superposition of Boltzmann entropies the quantum potential emerges as an information medium associated with two quantum corrector terms of the energy of the system deriving from the functions W which define the number of microstates of the system (and thus also from the superposition of entropies). The correspondence between the two approaches can be obtained by equating equations (2.150) and (2.3):

$$\frac{1}{2m}\left(\frac{1}{W^2}\frac{\partial W}{\partial \theta_i}\frac{\partial W}{\partial \theta_j}-\frac{2}{W}\frac{\partial^2 W}{\partial \theta_i \partial \theta_j}\right)=-\frac{\hbar^2}{2m}\left(\nabla S_Q\right)^2+\frac{\hbar^2}{2m}\left(\nabla^2 S_Q\right). \tag{2.152}$$

Equation (2.152) allows us to define the link between the two quantum corrector terms of the energy of the particle $-\frac{\hbar^2}{2m}\left(\nabla S_Q\right)^2$ and $\frac{\hbar^2}{2m}\left(\nabla^2 S_Q\right)$ of Sbitnev's approach to quantum entropy with the two quantum corrector terms of the approach here analysed depending on the functions W which define the number of the microstates of the system, namely $\frac{1}{2m}\left(\frac{1}{W^2}\frac{\partial W}{\partial \theta_i}\frac{\partial W}{\partial \theta_j}\right)$ and $\left(-\frac{1}{mW}\frac{\partial^2 W}{\partial \theta_i \partial \theta_j}\right)$ respectively:

$$\frac{1}{2m}\left(\frac{1}{W^2}\frac{\partial W}{\partial \theta_i}\frac{\partial W}{\partial \theta_j}\right)=-\frac{\hbar^2}{2m}\left(\nabla S_Q\right)^2, \tag{2.153}$$

$$\left(-\frac{1}{mW}\frac{\partial^2 W}{\partial \theta_i \partial \theta_j}\right)=\frac{\hbar^2}{2m}\left(\nabla^2 S_Q\right). \tag{2.154}$$

Thus, in the light of equations (2.153) and (2.154), the functions W appearing in the vector of superposition of different Boltzmann entropies (2.140) are connected with the quantity (2.1) and thus with the density of the particles associated with the wavefunction under consideration. Moreover, the introduction of the two quantum correctors of the energy given by equations (2.153) and (2.154) leads directly to formulate the quantum Hamilton-Jacobi equation of non-relativistic Bohmian quantum mechanics as

$$\frac{|\nabla S|^2}{2m}+V+\frac{1}{2m}\left(\frac{1}{W^2}\frac{\partial W}{\partial \theta_i}\frac{\partial W}{\partial \theta_j}-\frac{2}{W}\frac{\partial^2 W}{\partial \theta_i \partial \theta_j}\right)=-\frac{\partial S}{\partial t}. \tag{2.155}$$

Equation (2.155) provides an energy conservation law in non-relativistic quantum mechanics in the space of parameters where the two quantum corrector terms given by relations (2.153) and (2.154) can be seen as a sort of modification of the geometry of the parameter space determined by the density of the set of particles associated with the wavefunction under consideration.

However, as the author of this book and Licata showed clearly in [155], a fundamental difference can be found between the model of Fisher-Bohm geometry based on the vector of superposition of the entropies (2.140) and the entropic approach based on Sbitnev's entropy (2.1). In fact, while in Sbitnev's approach the quantum entropy and information associated with the quantum potential emerge directly from the density matrix, instead here the quantum potential emerges from a minimum condition of Fisher information in a picture where the conventional Schrödinger equation is not taken as a fundamental concept. Here, all observers must agree that there is a non-local "difference" from the classical case expressed by a quantum potential which thus emerges as a gauge condition on the Boltzmann entropies of different observers. While in Sbitnev's approach the non-local nature of quantum information emerges directly from the matrix density, instead here it emerges in a geometric picture where a specific Boltzmann entropy may be defined for each observer and therefore the primary, fundamental source of quantum information is the deformation of the geometrical properties of the parameter space determined by the vector of superposition of different Boltzmann entropies. Moreover, although, as regards the feature of the quantum potential as an information channel into the behaviour of quantum particles determined by two quantum corrector terms in the energy associated with a quantum entropy, the results of the two approaches are similar, the derivation of the two quantum corrector terms of the energy of the system is different in the two approaches [155].

Finally, at the end of this chapter let us see how the approach based on equations (2.140)–(2.151) allow us to provide a geometrodynamic entropic approach of a qubit pair of spin 1/2 particles in the pure state (1.377). In this regard, before all, one assumes that the quantum effects of a symmetric rigid rotor can be described by a change of the geometrical properties of the configuration space associated with the quantum superposition of entropies (2.140) which is associated with a deformation of the angular momenta stated by the action

$$A = \int \rho \left[\frac{\partial A}{\partial t} + \frac{1}{2I} M_i M_j + V + \frac{1}{2I} \left(\frac{\partial \log W_k}{\partial x_i} \frac{\partial \log W_k}{\partial x_j} \right) \right] dt d^n x. \tag{2.156}$$

The quantum action assumes the minimum value when

$$\delta \int \rho \left[\frac{\partial A}{\partial t} + \frac{1}{2I} M_i M_j + V \right] dt d^n x$$
$$+ \delta \int \frac{\rho}{2I} \frac{\partial \log W_k}{\partial x_i} \frac{\partial \log W_k}{\partial x_j} dt d^n x = 0 \tag{2.157}$$

namely

$$\frac{\partial A}{\partial t} + \frac{1}{2I} M_i M_j + V + \frac{1}{2I} \left(\frac{1}{W_k^2} \frac{\partial W_k}{\partial x_i} \frac{\partial W_k}{\partial x_j} - \frac{2}{W_k} \frac{\partial^2 W_k}{\partial x_i \partial x_j} \right)$$
$$= \frac{\partial S_k}{\partial t} + \frac{1}{2I} M_i M_j + V + Q. \tag{2.158}$$

In equation (2.158) Q is the quantum potential describing the dynamics of a spherically symmetric rigid rotor, that emerges as a consequence for the extreme condition of Fisher information, and can be expressed as follows

$$Q = \frac{\hat{\vec{M}}^2 R}{2IR} = \frac{1}{2I} \left(\frac{1}{W_k^2} \frac{\partial W_k}{\partial x_i} \frac{\partial W_k}{\partial x_j} - \frac{2}{W_k} \frac{\partial^2 W_k}{\partial x_i \partial x_j} \right) \tag{2.159}$$

and thus one obtains

$$\hat{\vec{M}}^2 R = \left(\frac{1}{W_k^2} \frac{\partial W_k}{\partial x_i} \frac{\partial W_k}{\partial x_j} - \frac{2}{W_k} \frac{\partial^2 W_k}{\partial x_i \partial x_j} \right) R. \tag{2.160}$$

According to the formalism based on equations (2.156)–(2.160), one can say that the quantum effects associated with the angular momentum are determined by the functions W_k defining the number of microstates of the rigid rotor, which depend on the parameters θ of the distribution probability (and thus, for example, on the space-temporal distribution of an ensemble of particles, namely the density of particles in the element $d^3\xi$ along the trajectory $\xi(t)$) and which correspond to the vector of superposition of the Boltzmann entropies (2.140). In other words, the functions W_k and therefore the Boltzmann entropies can be considered as informational lines of the quantum potential of the rigid rotor. Moreover, these functions W_k and therefore the Boltzmann entropies, determine a deformation of the background of the processes in the sense that they generate a quantum torque

$$\vec{T} = -\frac{i}{2I}\left(\frac{1}{W_k^2}\frac{\partial W_k}{\partial x_i}\frac{\partial W_k}{\partial x_j} - \frac{2}{W_k}\frac{\partial^2 W_k}{\partial x_i \partial x_j}\right)\hat{\vec{M}} \tag{2.161}$$

which rotates the angular momentum vector via the equation of motion

$$\frac{d\vec{M}}{dt} = -\frac{i}{2I}\left(\frac{1}{W_k^2}\frac{\partial W_k}{\partial x_i}\frac{\partial W_k}{\partial x_j} - \frac{2}{W_k}\frac{\partial^2 W_k}{\partial x_i \partial x_j}\right)\hat{\vec{M}} \tag{2.162}$$

along the trajectory $\xi(t)$. It is also interesting to observe that, in this approach, the modification of the geometry associated with the functions W_k and thus with the quantum entropy as regards a spherically symmetric rigid rotor can be characterized introducing the quantum-entropic length given by relation

$$L_{quantum} = \frac{1}{\sqrt{\frac{1}{2I\hbar^2}\left(\frac{2}{W}\frac{\partial^2 W}{\partial x_i \partial x_j} - \frac{1}{W^2}\frac{\partial W}{\partial x_i}\frac{\partial W}{\partial x_j}\right)}}. \tag{2.163}$$

Once the quantum-entropic length becomes non-negligible the rigid rotor goes into a quantum regime.

Let us analyse now the dynamics and the quantum effects of a general two qubits state, with vanishing total angular momentum projections, given by equation (1.377). In a Bohmian framework, in this case of two entangled qubits the extreme condition of Fisher information is

$$\frac{\partial A}{\partial t} + \frac{1}{2I}M_{i_1}M_{j_1} + \frac{1}{2I}M_{i_2}M_{j_2}$$
$$+ V + \frac{1}{2I}\left(\frac{1}{W_{k_1}^2}\frac{\partial W_{k_1}}{\partial x_{i_1}}\frac{\partial W_{k_1}}{\partial x_{j_1}} - \frac{2}{W_{k_1}}\frac{\partial^2 W_{k_1}}{\partial x_{i_1}\partial x_{j_1}}\right)$$

$$+\frac{1}{2I}\left(\frac{1}{W_{k_2}^2}\frac{\partial W_{k_2}}{\partial x_{i_2}}\frac{\partial W_{k_2}}{\partial x_{j_2}}-\frac{2}{W_{k_2}}\frac{\partial^2 W_{k_2}}{\partial x_{i_2}\partial x_{j_2}}\right)$$
$$=\frac{\partial S_k}{\partial t}+\frac{1}{2I}M_iM_j+V+Q \tag{2.164}$$

and the quantum potential may be expressed as

$$Q=\frac{1}{2I}\left(\frac{1}{W_{k_1}^2}\frac{\partial W_{k_1}}{\partial x_{i_1}}\frac{\partial W_{k_1}}{\partial x_{j_1}}-\frac{2}{W_{k_1}}\frac{\partial^2 W_{k_1}}{\partial x_{i_1}\partial x_{j_1}}\right.$$
$$\left.+\frac{1}{W_{k_2}^2}\frac{\partial W_{k_2}}{\partial x_{i_2}}\frac{\partial W_{k_2}}{\partial x_{j_2}}-\frac{2}{W_{k_2}}\frac{\partial^2 W_{k_2}}{\partial x_{i_2}\partial x_{j_2}}\right) \tag{2.165}$$

and thus one has

$$\left(\hat{M}_1^2+\hat{M}_2^2\right)R=\left(\frac{1}{W_{k_1}^2}\frac{\partial W_{k_1}}{\partial x_{i_1}}\frac{\partial W_{k_1}}{\partial x_{j_1}}-\frac{2}{W_{k_1}}\frac{\partial^2 W_{k_1}}{\partial x_{i_1}\partial x_{j_1}}\right.$$
$$\left.+\frac{1}{W_{k_2}^2}\frac{\partial W_{k_2}}{\partial x_{i_2}}\frac{\partial W_{k_2}}{\partial x_{j_2}}-\frac{2}{W_{k_2}}\frac{\partial^2 W_{k_2}}{\partial x_{i_2}\partial x_{j_2}}\right)R. \tag{2.166}$$

Moreover, the modification of the geometry associated with the entangled qubit pair can be described by introducing the quantum-entropic length given by relation

$$L_{quantum}=\frac{1}{\sqrt{\frac{1}{2I\hbar^2}\left(\begin{array}{c}\frac{2}{W_1}\frac{\partial^2 W_1}{\partial x_{i_1}\partial x_{j_1}}-\frac{1}{W_1^2}\frac{\partial W_1}{\partial x_{i_1}}\frac{\partial W_1}{\partial x_{j_1}}\\+\frac{2}{W_2}\frac{\partial^2 W_2}{\partial x_{i_2}\partial x_{j_2}}-\frac{1}{W_2^2}\frac{\partial W_2}{\partial x_{i_2}}\frac{\partial W_2}{\partial x_{j_2}}\end{array}\right)}}. \tag{2.167}$$

As regards a qubit pair in the state equation (1.377), the total angular momentum projection $M_{1z}+M_{2z}$ is zero while the angular momenta due to the action of the non-local quantum potential (2.165) and the corresponding quantum torques

$$\vec{T}_1 = -\frac{i}{2I}\left(\frac{1}{W_{k_1}^2}\frac{\partial W_{k_1}}{\partial x_i}\frac{\partial W_{k_1}}{\partial x_j} - \frac{2}{W_{k_1}}\frac{\partial^2 W_{k_1}}{\partial x_i \partial x_j} + \frac{1}{W_{k_2}^2}\frac{\partial W_{k_2}}{\partial x_i}\frac{\partial W_{k_2}}{\partial x_j} - \frac{2}{W_{k_2}}\frac{\partial^2 W_{k_2}}{\partial x_i \partial x_j}\right)\hat{\vec{M}}_1 \quad (2.168)$$

and

$$\vec{T}_2 = -\frac{i}{2I}\left(\frac{1}{W_{k_1}^2}\frac{\partial W_{k_1}}{\partial x_i}\frac{\partial W_{k_1}}{\partial x_j} - \frac{2}{W_{k_1}}\frac{\partial^2 W_{k_1}}{\partial x_i \partial x_j} + \frac{1}{W_{k_2}^2}\frac{\partial W_{k_2}}{\partial x_i}\frac{\partial W_{k_2}}{\partial x_j} - \frac{2}{W_{k_2}}\frac{\partial^2 W_{k_2}}{\partial x_i \partial x_j}\right)\hat{\vec{M}}_2 \quad (2.169)$$

exhibit a complex precessional motion.

Moreover, the entropic approach of Fisher-Bohm geometry based on the vector of superposition of Boltzmann entropies (2.140) allows us to re-read in a suggestive way Ramsak's results regarding the geometrization of the quantum entanglement of a pair of qubits, developed in the papers “Geometrical view of quantum entanglement” [268] and “Spin-spin correlations of entangled qubit pairs in the Bohm interpretation of quantum mechanics” [269]. In particular, the behaviour of the probability distribution (1.391) of the ensemble average difference of azimuthal angles $\phi[\xi(t)] = \phi_2 - \phi_1$ of the angular momenta is determined by the quantum torques (2.168) and (2.169) and thus by the functions W_{k_1} and W_{k_2}, defining the number of microstates of the particle 1 and particle 2 of the system respectively (and representing the informational lines of the quantum potential (2.165)). The entanglement properties of a pair of particles in the state (1.773) are therefore determined by the deformation of the geometrical properties of the information space produced by the quantum torques (2.168) and (2.169). The geometrical properties of the background of the two particles into consideration can also be characterized by defining the following quantum entropy for a two-qubit system

$$\begin{cases} S_1 = k \log \left(W_{1_1} W_{1_2} \right) \\ S_2 = k \log \left(W_{2_1} W_{2_2} \right) \\ \quad \cdots \\ \quad \cdots \\ S_n = k \log \left(W_{n_1} W_{n_2} \right) \end{cases} . \tag{2.170}$$

The peaking of the probability distribution (1.391) for increasing entanglement between the two qubits is associated with the precessional motion of the quantum torques (2.168) and (2.169) and thus with the informational lines of the quantum potential of the system of the two qubits. In the approach based on equations (2.164)–(2.170), it becomes therefore permissible the following re-reading of Ramsak's results about the behaviour of the probability distribution (1.391): as regards a system of two entangled qubits the extreme condition of the Fisher metric determines the quantum potential (2.165) which corresponds with the deformation of the geometry described by the quantum-entropic length (2.167) produced by the superposition of Boltzmann entropies (2.170), and the quantum information associated with the quantum potential (2.165) and thus the quantum-entropic length (2.167) generate the quantum torques (2.168) and (2.169) which, by exhibiting a precessional motion, imply the precession of angular momenta at equal relative angle $\phi[\xi(t)] = \varphi$ for all ξ consistent with perfect entanglement regarding the probability distribution (1.391). The dynamics of the azimuthal angles $\phi_1[\xi(t)]$ and $\phi_2[\xi(t)]$ of angular momenta — and, consequently, the chaotic features of the trajectories in which the projections of the total momentum $\vec{M}$ onto the xy-plane winds around the origin an infinite number of times in a spirographic manner and the closed and periodic curve corresponding to the relative momentum $\vec{M}_2 - \vec{M}_1$ — are therefore determined by the quantum torques (2.168) and (2.169), and thus by the quantum potential (2.165), namely by the deformation of the geometrical properties of the information space described by the vector of superposition of entropies (2.170).

Moreover, if in Ramsak's model the motion of the angular momenta can be described by a concurrency (1.392) characterized by the probability distribution (1.393), in the context of the entropic approach based on equations (2.164)–(2.170), the concurrency (1.392) of the trajectories of the angular momentum vectors and their distribution probability (1.393) can be seen as results of the behaviours of the quantum torques (2.168) and (2.169) and thus of the geometry of the information space associated with the vector of superposition of the entropies (2.170): in the extreme condition of Fisher metric, the quantum information associated with the quantum potential (2.165) and thus the quantum-entropic length (2.167) generate the quantum torques (2.168) and (2.169) which, by exhibiting a precessional motion, determine the fact that the angular momentum vectors of each representative of the ensemble of a two qubits system precess in unison or not and the characteristic probabilities of these concurrent or anti-concurrent motions. And, in the approach here analised, on the basis of equations (2.168) and (2.169), also the angle made by the angular momenta and the azimuthal angle made by the xy-plane projections of the momenta (as well as their consequences as regards the description of the entanglement) can be seen as a result of the quantum torques and thus of the deformation of the background associated with the informational lines of the quantum potential W_{k_1} and W_{k_2}, in other words of the quantum superposition of entropies for the two qubits (2.170).

Finally, if in Ramsak's treatment in [269] of the entanglement properties of a qubit pair in the framework of Bohm's interpretation, the source of non-locality lies in the quantum potential which generates an instant coupling between the angular momenta of entangled qubit pairs, in the entropic approach to quantum entanglement based on equations (2.164)–(2.170), since the quantum potential (2.165) corresponds with the deformation of the geometry described by the quantum-entropic length (2.167) and is produced by the quantum entropy, and the quantum information associated with the quantum potential (2.165) and thus the quantum-entropic length (2.167) generate the quantum torques (2.168) and (2.169) which determine the coupling of the angular momenta of entangled qubit pairs, the real, ultimate source of non-locality (and thus of the Bohmian counterpart of Bell's inequalities) is the deformation of the geometry of the background described by the quantum entropy and therefore the ultimate entities that determine the quantum information of the non-locality are the quantum torques (2.168) and (2.169). In summary, one can say that, on the basis of the geometrodynamic entropic approach to quantum entanglement based on equations (2.164)–(2.170), the ultimate source of quantum information is represented by the quantum torques (2.168) and (2.169) corresponding with the deformation of the geometry described by the vector of superposition of Boltzmann entropies.

Chapter 3

Immediate Quantum Information and Symmetryzed Quantum Potential

One of the most intriguing features of the geometry of space determined by quantum theory is certainly represented by the existence of non-local correlations between subatomic particles. In the quantum domain, the state of an object may be strongly correlated with the state of another object at a large distance from the first, such that no classical communication (involving signals which do not travel faster than light) between them is possible. As we have seen in chapter 1.1, in the de Broglie-Bohm approach, the non-local geometry of the quantum world is determined by the quantum potential. Bohm's theory manages to make manifest quantum non-locality by means of the geometric properties of space described and expressed by the quantum potential. On the basis of Bohm's and Hiley's research, the ordinary space-time manifold cannot explain and reproduce the non-local correlations characterizing the geometry of the quantum world. In particular, Bohm's and Hiley's research suggest that the non-local geometry determined by the quantum potential can be reproduced in the context of a fundamental geometrodynamic background (the Bohmian implicate order, or also the analogous Hiley's pre-space and notion of underlying process of quantum phenomena) which redesigns the behaviour of the particles.

In this chapter, our aim is to review an interesting approach proposed recently by the author and Sorli according to which the fundamental arena of quantum processes is a three-dimensional space which acts as a direct medium of information in the picture of an appropriate extension of Bohm's quantum potential called the symmetrized quantum potential. In the first paragraph we will analyse the symmetrized quantum potential in the non-relativistic quantum mechanics. Then we will extend this symmetrized approach to relativistic de Broglie-Bohm theory in curved space-time, relativistic quantum field theory and quantum cosmology. In this discussion we refer to the papers [285–288].

3.1 The Symmetrized Quantum Potential in the Non-relativistic Domain

The features of quantum potential imply that, at a fundamental level, the geometrodynamic properties of a three-dimensional space have a crucial role in determining the motion of a subatomic particle and that space functions as a direct, immediate medium of information between subatomic particles, providing thus an instantaneous connection between them. The quantum potential contains the idea of space as a direct fundamental medium for the transmission of quantum information between subatomic particles in an implicit way. In the references [285, 286] the author has developed in detail these concepts and has introduced, in this regard, a peculiar interpretation of quantum non-locality which can be defined as the "immediate interpretation" of quantum non-locality. According to this interpretation, one can say that in EPR-type experiments a three-dimensional physical space is an "immediate information medium", a direct information medium between elementary particles. In EPR-type experiments the behaviour of a subatomic particle is influenced instantaneously by the other particle thanks to the three-dimensional space which functions as an immediate information medium.

When one takes into consideration an atomic or subatomic process (such as for example the case of an EPR-type experiment, of two subatomic particles, before joined and then separated and carried away at big distances one from the other), one can say that, at a fundamental level, physical space assumes the special "state" represented by quantum potential, and this state is characterized by geometrodynamic properties which generate an instantaneous communication — and by consequence may be defined as an immediate information medium — between the particles into consideration. In other words, the properties of space derived the geometrodynamic action of the quantum potential indicate that space acts as an immediate information medium between quantum particles producing an instantaneous connection between them. The instantaneous communication between two particles A and B derives just from the special feature of space (determined by the quantum potential) to be an immediate information medium: by disturbing the particle A, the particle B may indeed be instantaneously influenced independently on the distance separating the two particles thanks to space which is characterized by a fundamental geometry which acts as an immediate information medium between them and puts them into an immediate contact [289, 290].

In order to illustrate in detail in what sense a three-dimensional space acts as an immediate medium of information transfer between subatomic particles by means of quantum entanglement, the best way is to consider the classic example of EPR-type experiments given by Bohm [49] in 1951. We have a physical system given by a molecule of total spin 0 composed by two spin 1/2 atoms in a singlet state:

$$\psi(\vec{x}_1, \vec{x}_2) = f_1(\vec{x}_1) f_2(\vec{x}_2) \frac{1}{\sqrt{2}} (u_+ v_- - u_- v_+). \tag{3.1}$$

where $f_1(\vec{x}_1)$, $f_2(\vec{x}_2)$ are non-overlapping packet functions, $u_\pm$ are the eigenfunctions of the spin operator $\hat{s}_{z_1}$ in the z-direction pertaining to particle 1 and $v_\pm$ are the eigenfunctions of the spin operator $\hat{s}_{z_2}$ in the z-direction pertaining to particle 2: $\hat{s}_{z_1} u_\pm = \pm\frac{\hbar}{2} u_\pm$, $\hat{s}_{z_2} v_\pm = \pm\frac{\hbar}{2} v_\pm$. If in a spin measurement on the particle 1 in the z-direction when the molecule is in such a state we obtain the result spin up for this particle 1, then according to the usual quantum theory, the wavefunction of the molecule (3.1) reduces to the first of its summands:

$$\psi \to f_1 f_2 u_+ v_-. \tag{3.2}$$

The result of the measurement carried out on the particle 1 leads us to have knowledge about the state of the unmeasured system 2: if the particle 1 is found in the state of spin up, we know immediately that the particle 2 is in the state v_- which indicates that the particle 2 has spin down. In other words, we find that, the state of the particle 2 is directly determined and fixed by a measurement performed on the particle 1 (independently from the distance between the two particles), namely that, as regards spin measurements, there are non-local correlations between the two particles. Although the two partial systems (the particle 1 and the particle 2) are clearly separated in space (in the conventional sense that the outcomes of position measurements on the two systems are widely separated), indeed they cannot be considered physically separated because the state of the particle 2 is indeed instantaneously influenced by the kind of measurements made on the particle 1. In the light of Bohm's treatment in [49], we can therefore conclude that that entanglement in spin space implies non-locality and non-separability in Euclidean three-dimensional space as a consequence of the fact that the spin measurements couple the spin and space variables.

On the other hand, the non-local correlations regarding the results of the spin measurements of Bohm's example have been formalized by the well-known Bell theorem, which implies that for a system of two entangled particles (such as in the state (3.1)), it is not possible to interpret all the correlations between the two particles regarding the spin measurements by assuming that the two particles are born with the relative instructions about how to behave [291]. But then in what way can we interpret the fact that the measurement of a particle defines in what state the other particle is found, independently from the distance? In 1935, after EPR's work, Bohr suggested that the two entangled particles, independently from their distance, continue to constitute an unity, a single system: the two particles have not an autonomous existence. The immediate interpretation of non-locality suggested by the author, in which space is considered as the medium of information transfers in quantum physics, can give Bohr's interpretation a natural basis.

According to the previsions of quantum theory, in EPR-type experiments the transmission of the information has not duration. What occurs in EPR-type experiments implies that time, at a fundamental level, is not a primary physical reality and thus the fundamental arena of processes is timeless. The state of the second particle changes instantaneously after the measurement on the first particle thanks to the medium of a timeless space. According to the approach developed by the author and Sorli, the information between the two particles is instantaneous thanks to the medium of space. A three-dimensional timeless space (where time is not a primary physical reality but exists only as a numerical order of material changes) can be considered the fundamental arena which can explain the non-local correlations determined by entanglement in Bohm's example. One can say that the state of the particle 2 is instantaneously influenced by the kind of measurements regarding the particle 1 because space acts as an immediate information medium between the two particles. It is the medium of the three-dimensional space which produces an instantaneous connection between the two particles as regards the spin measurements: by disturbing system 1, system 2 is instantaneously influenced despite the big distance separating the two systems thanks to space which acts as an immediate information medium and puts them in an immediate contact.

The interpretation of quantum entanglement and non-locality as immediate physical phenomena determined by a timeless three-dimensional space that acts as a direct medium of information implies that quantum entanglement and non-locality cannot be explained by invoking a mechanism of entities that are transmitters of information between the particles under consideration: there is no information signal in form of photon or some other particle traveling between particles 1 and 2 of Bohm's example. The time of information transfer between particle 1 and particle 2 is zero [292]. In this picture, a three-dimensional space (where time exists only as a numerical order of material changes) is the fundamental arena which allows us to explain the instantaneous communication of information between particles 1 and 2 in Bohm's example. It is the three-dimensional space the medium which determines an immediate information transfer and allows us to explain why and in what sense, in an EPR experiment, two particles coming from the same source and which go away, remain joined by a mysterious link, why and in what sense if we intervene on one of the two particles 1 and 2, also the other feels the effects instantaneously despite the relevant distances separating it [290].

In the approach proposed by the author and Sorli, the view of the three-dimensional space as a direct, immediate information medium between subatomic particles follows as a natural development from Bohm's quantum potential. As we have shown in chapter 1.1, the action of this potential is space-like, namely creates onto the particles a non-local, instantaneous action. The quantum potential defined by equations (1.4) and (1.9) turns out to be an entity which contains a spatial active information. On the basis of the fact that the quantum potential, as it is expressed by equations (1.4) and (1.9), has an instantaneous action and contains an active information about the environment, one can say that it is space the medium responsible of the behaviour of quantum particles. Indeed the definitions (1.4) and (1.9) of the quantum potential contain the idea of space as an immediate information medium in an implicit way.

In particular, if we consider a many-body quantum process (such as for example the case of an EPR-type experiment, of two subatomic particles, before joined and then separated and carried away at big distances one from the other), we can say that the three-dimensional physical space assumes the special "state" represented by quantum potential (1.9), and this allows an instantaneous communication between the particles into consideration [304]. If we take under examination the situation considered by Bohm in 1951 (illustrated before) we can say that it is the state of space in the form of the quantum potential (1.9) which produces an instantaneous connection between the two particles as regards the spin measurements.

In summary, one can say that in EPR-type experiments the quantum potential (1.9) makes the three-dimensional physical space an "immediate information medium" between elementary particles. In EPR-type experiments the behaviour of a subatomic particle is influenced instantaneously by the other particle thanks to the three-dimensional space which functions as an immediate information medium in virtue of the geometric properties expressed by the quantum potential (1.9).

Moreover, in the reference [287] the author of this book and Sorli showed that, in the immediate interpretation of quantum non-locality, there is a fundamental geometric property contained in the quantum potential which determines the action of the 3D space as an immediate information medium, namely the quantum entropy (defined, in the non-relativistic domain, by equation (2.1)). By following the treatment of the reference [287], let us review briefly this view of the three-dimensional space which acts as an immediate information medium, based on the quantum entropy. As we have shown in chapter 2.1, by introducing the quantum entropy, for one-body systems the quantum potential can be expressed in the following convenient way

$$Q = -\frac{\hbar^2}{2m}(\nabla S_Q)^2 + \frac{\hbar^2}{2m}(\nabla^2 S_Q). \tag{3.3}$$

and the equation of motion for the corpuscle associated with the wavefunction $\psi(\vec{x}, t)$ becomes:

$$\frac{|\nabla S|^2}{2m} - \frac{\hbar^2}{2m}(\nabla S_Q)^2 + V + \frac{\hbar^2}{2m}(\nabla^2 S_Q) = -\frac{\partial S}{\partial t} \tag{3.4}$$

which provides an energy conservation law where the term $-\frac{\hbar^2}{2m}(\nabla S_Q)^2$ can be interpreted as the quantum corrector of the kinetic energy $\frac{|\nabla S|^2}{2m}$ of the particle while the term $\frac{\hbar^2}{2m}(\nabla^2 S_Q)$ can be interpreted as the quantum corrector of the potential energy V.

In the case of many-body systems, the quantum potential is given by the following expression

$$Q = \sum_{i=1}^{N}\left[-\frac{\hbar^2}{2m_i}(\nabla_i S_Q)^2 + \frac{\hbar^2}{2m_i}(\nabla_i^2 S_Q)\right] \tag{3.5}$$

and the equation of motion is

$$\sum_{i=1}^{N}\frac{|\nabla_i S|^2}{2m_i}-\sum_{i=1}^{N}\frac{\hbar^2}{2m_i}(\nabla_i S_Q)^2+V+\sum_{i=1}^{N}\frac{\hbar^2}{2m_i}(\nabla_i^2 S_Q)=-\frac{\partial S}{\partial t} \tag{3.6}$$

which provides an energy conservation law where the term $-\sum_{i=1}^{N}\frac{\hbar^2}{2m_i}(\nabla_i S_Q)^2$ can be interpreted as the quantum corrector of the kinetic energy of the many-body system while the term $\sum_{i=1}^{N}\frac{\hbar^2}{2m_i}(\nabla_i^2 S_Q)$ can be interpreted as the quantum corrector of the potential energy.

With the introduction of the quantum entropy which leads to the energy conservation law (equation (3.4) for one-body systems and equation (3.6) for many-body systems), on the basis of the results of the paper [287], it is possible to throw new light on the interpretation of the action of the three-dimensional space as an immediate information medium in EPR-type correlations. In the view proposed by the author and Sorli, on the basis of equation (3.6), one can say that the action of the three-dimensional space as an immediate information medium derives just from the two quantum correctors to the energy of the system under consideration, namely from the quantum corrector to the potential energy $\sum_{i=1}^{N}\frac{\hbar^2}{2m_i}(\nabla_i^2 S_Q)$ and the quantum corrector to the kinetic energy $-\sum_{i=1}^{N}\frac{\hbar^2}{2m_i}(\nabla_i S_Q)^2$ (while the other two terms $\sum_{i=1}^{N}\frac{|\nabla_i S|^2}{2m_i}$ and V on the right-hand of equation (3.6) generate a local feature of space). The feature of the quantum potential to make the three-dimensional space an immediate information channel into the behaviour of quantum particles derives just from the quantum entropy. In other words, one can see that, by introducing the quantum entropy, it is just the two quantum correctors to the energy of the system under consideration, depending on the geometry of the background namely on the degree of order and chaos of the vacuum supporting the density ρ (of the particles associated with the wavefunction under consideration) the fundamental element which, at a fundamental level, produces an immediate information medium in the behaviour of the particles in EPR-type experiments. The local features of space in our macroscopic domain derives from the fact that, in this domain, the quantum entropy satisfies conditions

$$(\nabla S_Q)^2 \rightarrow (\nabla^2 S_Q) \tag{3.7}$$

for one-body systems and

$$(\nabla_i S_Q)^2 \rightarrow (\nabla_i^2 S_Q) \tag{3.8}$$

for many-body systems.

In summary, in the approach based on the quantum entropy, one can say that the quantum entropy, by producing two quantum corrector terms in the energy, can be indeed interpreted as a sort of intermediary entity between space and the behaviour of quantum particles, and thus between the non-local action of the quantum potential and the behaviour of quantum particles. The introduction of the quantum entropy as the fundamental entity that determines the behaviour of quantum particles leads to an energy conservation law in quantum mechanics (expressed by equations (3.4) and (3.6)) which lets us realize the origin of the property of quantum potential to determine the action of the three-dimensional space as an immediate information medium. The ultimate source which determines the action of the three-dimensional space as an immediate information medium between quantum particles is the geometry of a fundamental vacuum defined by the quantum entropy. The quantum entropy, by producing two quantum corrector terms in the energy, is the fundamental element which gives origin to the non-local action of the quantum potential.

Now, as regards the instantaneous communication between quantum particles in EPR-type experiments and the role of the three-dimensional space as a direct information medium between them, there is another significant element of the geometry of processes. If one imagines to exchange, to invert the roles of the two particles what happens is always the same type of process, namely an instantaneous communication between the two particles. In other words, there is a property of symmetry which characterizes the instantaneous communication between two particles in EPR-type experiments: it occurs both if one intervenes on one and if one intervenes on the other, in both cases the same type of process happens and — we can say — always thanks to space which functions as an immediate information medium. Moreover, if we imagine to film the process of an instantaneous communication between two subatomic particles in EPR-type experiments backwards, namely inverting the sign of time, we should expect to see what really happened. Inverting the sign of time, there is however no guarantee that we obtain something that corresponds to what physically happens. Although the quantum potential ((1.4) for one-body systems and (1.9) for many-body systems) has a space-like, an instantaneous action, however it comes from Schrödinger equation which is not time-symmetric and therefore its expression cannot be considered completely satisfactory just because it can meet problems inverting the sign of time.

On the basis of these considerations, in order to interpret in the correct way, also in symmetric terms in the exchange of t in $-t$, the instantaneous communication between subatomic particles and thus the interpretation of three-dimensional space as an immediate information medium, the author of this book and Sorli recently introduced a research line based on a symmetrized version of the quantum potential (in which a symmetry in time of the processes is satisfied). As regards the contributions in this regard we base our treatment on the references [285, 286, 288].

The symmetrized quantum potential approach can be reviewed by starting from the analysis of the time-symmetric formulation of quantum mechanics developed by Wharton in 2007. Wharton's model consists in applying two consecutive boundary conditions onto solutions of a time-symmetryzed wave equation [293]. In summary, Wharton's proposal is based on the following three postulates:

1. The wavefunction is no longer a solution of the Schrödinger equation, but instead is the solution $|\rangle$ to the time-symmetric equation

$$\begin{pmatrix} H & 0 \\ 0 & -H \end{pmatrix} |C(t)\rangle = i\hbar\frac{\partial}{\partial t}|C(t)\rangle \tag{3.9}$$

where $|C(t)\rangle = \binom{\psi(t)}{\phi(t)}$, $\psi(t)$ is the solution of the standard Schrödinger equation, $\phi(t)$ is the solution to the time-reversed Schrödinger equation.

2. Each measurement Q_M of a wavefunction (at some time t_0) imposes the result of that measurement as an initial boundary condition on $|C_+\rangle = |\psi\rangle + T|\phi\rangle$, and as a final boundary condition on $|C_-\rangle = |\psi\rangle + T|\phi\rangle$ where T is the time-reversal operator. In other words, instead of a collapse postulate, this formulation imposes a boundary condition on the wavefunction at every measurement, equal to the outcome of that measurement.
3. Instead of the standard probability formula, the relative probability of any complete measurement sequence on a wavefunction $|C(t)\rangle$ at times $t_1, t_2, \ldots, t_n$ is

$$P_0 = \prod_{n=1}^{N-1} (C_-(t_n^+))(C_+(t_n^+))(C_+(t_{n+1}^-))(C_-(t_{n+1}^-)) \quad (3.10)$$

where $N > 1$ and each measurement is constrained by the boundary conditions $Q_M \left|C_\pm(t_0^\pm)\right\rangle = q_n | \ C_\pm(t_0^\pm)\rangle$.

This model developed by Wharton constitutes an interesting attempt to build a fully time-symmetric formulation of quantum mechanics, without requiring a time-asymmetric collapse of the wavefunction upon measurement. It can be therefore considered a starting-point in order to interpret in the correct manner both the forward-time and the reversed-time perspectives of a same physical event inside the standard quantum mechanics.

If the non-local geometry of the quantum world is due to the geometrodynamic features of Bohmian quantum potential, to the like-space, instantaneous action of the quantum potential, in order to find the most appropriate and complete mathematical candidate for the state of space as a direct information medium between subatomic particles (in which the symmetry in time needed to interpret also the time-reverse process in the correct manner is assured), the author and Sorli suggested a reformulation of Bohmian quantum mechanics for the time-symmetric equation (3.9), namely a symmetrized reformulation of Bohm's quantum mechanics. Following the references [285, 286, 288], the first step of this approach is, just like in the original Bohmian theory, to decompose the time-symmetric equation (3.9) into two real equations, by expressing the wavefunctions ψ and ϕ in polar form:

$$\psi = R_1 e^{iS_1/\hbar}, \tag{3.11}$$

$$\phi = R_2 e^{iS_2/\hbar} \tag{3.12}$$

where R_1 and R_2 are real amplitude functions and S_1 and S_2 are real phase functions. If we insert (3.11) and (3.12) into (3.9) and separate into real and imaginary parts, we obtain two couples of equations (namely the quantum Hamilton-Jacobi equations and the continuity equations) for the fields R_1, R_2, S_1 and S_2. The real part gives

$$\frac{\partial}{\partial t}\begin{pmatrix} S_1 \\ S_2 \end{pmatrix} + \frac{1}{2m}\begin{pmatrix} (\nabla S_1)^2 \\ (\nabla S_2)^2 \end{pmatrix} - \frac{\hbar^2}{2m}\begin{pmatrix} \dfrac{\nabla^2 R_1}{R_1} \\ -\dfrac{\nabla^2 R_2}{R_2} \end{pmatrix} + \begin{pmatrix} V \\ -V \end{pmatrix} = 0 \tag{3.13}$$

and the imaginary part may be expressed in the form

$$\frac{\partial}{\partial t}\begin{pmatrix} R_1^2 \\ R_2^2 \end{pmatrix} + \nabla \cdot \begin{pmatrix} \dfrac{R_1^2 \nabla S_1}{m} \\ \dfrac{R_2^2 \nabla S_2}{m} \end{pmatrix} = 0. \tag{3.14}$$

In the light of equations (3.13) and (3.14), a symmetryzed extension of Bohmian quantum mechanics emerges where the fundamental entity is a symmetryzed quantum potential at two components of the form

$$Q = -\frac{\hbar^2}{2m}\begin{pmatrix} \dfrac{\nabla^2 R_1}{R_1} \\ -\dfrac{\nabla^2 R_2}{R_2} \end{pmatrix} \tag{3.15}$$

where R_1 is the amplitude function of ψ and R_2 is the amplitude function of ϕ. The symmetryzed quantum potential (3.15) can be considered the starting-point to have a symmetry in time in Bohmian quantum mechanics [285, 286, 288].

The mathematical expression of the symmetrized quantum potential (3.15) indicates clearly that, just like the quantum potential of the original Bohm theory, also the symmetryzed quantum potential (3.15) has an action which is stronger when the mass is more comparable with Planck constant, and the presence of the Laplace operator generates the like-space, non-local, instantaneous action of this potential. The difference with respect to the original Bohm's quantum mechanics lies in the fact that (3.15) has two components, namely depends also on the wavefunction concerning the time-reverse process, and therefore its space-like, non-local, instantaneous action is predicted not only by the forward-time process but also by the time-reverse process (and this means therefore that the process of the instantaneous action between two subatomic particles can be interpreted in the correct way from the point of view of the mathematical symmetry in time). In this picture, the original quantum potential (1.2) can be seen only as a special component of the symmetryzed quantum potential: it is the component that explains the forward-time quantum processes.

The introduction of the symmetryzed quantum potential leads directly to interpret space, in the quantum domain, as a special "state" at two components that determines the following facts. The first component of this special state $Q_1 = -\frac{\hbar^2}{2m}\frac{\nabla^2 R_1}{R_1}$ (that regards the forward-time process and coincides with the original Bohm's quantum potential) determines a non-local, instantaneous communication between the particles into consideration. The second component $Q_2 = \frac{\hbar^2}{2m}\frac{\nabla^2 R_2}{R_2}$, which regards the time-reverse process, reproduces what physically happens if one would imagine to film a quantum process backwards: it allows us to explain quantum phenomena in the most correct and complete way from the point of view of the symmetry in time. According to the symmetryzed quantum potential approach, in EPR-type experiment space acts as an immediate information medium in the sense that the first component of the symmetryzed quantum potential makes physical space an "immediate information medium" which keeps two elementary particles in an immediate contact (while the second component of the symmetryzed quantum potential reproduces the symmetry in time of this communication). Moreover, in this picture, the opposed sign of the second component with respect to the first component constitutes the mathematical translation of the idea that, in a quantum process, time exists only as a measuring system of the numerical order of material changes: the sign of the second component physically means that it is not possible to go backwards in the physical time intended as a numerical order.

It is also interesting to observe that in analogy to what happens in Bohm's original theory, also in this symmetryzed extension the quantum potential (3.15) must not be considered a term ad hoc. It plays a fundamental role in the symmetryzed quantum formalism: in the formal plant of the symmetryzed Bohm's theory it emerges directly from the symmetryzed Schrödinger equation (3.9). Without the term (3.15) the total energy of the physical system under consideration would not be conserved. In fact, equation (3.13) can also be written in the equivalent form

$$\frac{1}{2m}\begin{pmatrix}(\nabla S_1)^2\\(\nabla S_2)^2\end{pmatrix}-\frac{\hbar^2}{2m}\begin{pmatrix}\dfrac{\nabla^2 R_1}{R_1}\\-\dfrac{\nabla^2 R_2}{R_2}\end{pmatrix}+\begin{pmatrix}V\\-V\end{pmatrix}=-\frac{\partial}{\partial t}\begin{pmatrix}S_1\\S_2\end{pmatrix}, \tag{3.16}$$

which can be seen as a real energy conservation law for the forward-time and the reverse-time process in symmetryzed quantum mechanics: here one can easily see that without the symmetryzed quantum potential (3.15) energy would not be conserved. In the light of equation (3.16), we can say also that the reverse-time of a physical process is characterized by a classic potential and a quantum potential which are endowed with an opposed sign with respect to the corresponding potentials characterizing the forward-time process.

It is also interesting to observe that inside this time-symmetric extension of Bohmian quantum mechanics the correspondence principle becomes

$$-\frac{\hbar^2}{2m}\begin{pmatrix} \dfrac{\nabla^2 R_1}{R_1} \\ -\dfrac{\nabla^2 R_2}{R_2} \end{pmatrix} \to \begin{pmatrix} 0 \\ 0 \end{pmatrix}. \tag{3.17}$$

In this classical limit we have the classical Hamilton-Jacobi equation at two components:

$$\frac{\partial}{\partial t}\begin{pmatrix} S_1 \\ S_2 \end{pmatrix} + \frac{1}{2m}\begin{pmatrix} (\nabla S_1)^2 \\ (\nabla S_2)^2 \end{pmatrix} + \begin{pmatrix} V \\ -V \end{pmatrix} = 0 \tag{3.18}$$

which shows us just that the time-reverse of the classical process involves a classic potential which is endowed with an opposed sign with respect to the classic potential characterizing the forward-time process.

Moreover, always following the treatment of [285, 286, 288], in a similar way to the interpretation of the quantum potential suggested by Bohm and Hiley in 1984 as information potential in the behaviour of the particles, also the symmetryzed quantum potential (3.15) can be interpreted as a sort of "information potential": the particles in their movement are guided by the quantum potential just as a ship at automatic pilot can be handled by radar waves of much less energy than that of the ship and this concerns also the time-reverse of this process in the sense that also the time-reverse of this process reproduces what happens as regards the transmission of the information. On the basis of this interpretation, the results of double-slit experiment are explained by saying that the symmetryzed quantum potential (3.15) contains an active information, for example about the slits, and that this information manifests itself in the particles' motions and the time-reverse of these motions can be explained in the same, correct way, namely through the idea of the active information.

Finally, in the case of a many-body system constituted by N particles the symmetryzed quantum potential can be expressed as

$$Q = \sum_{i=1}^{N} -\frac{\hbar^2}{2m_i} \begin{pmatrix} \frac{\nabla_i^2 R_1}{R_1} \\ -\frac{\nabla_i^2 R_2}{R_2} \end{pmatrix} \tag{3.19}$$

where R_1 is the absolute value of the wave-function $\psi = R_1 e^{iS_1/\hbar}$ describing the forward-time process (solution of the standard Schrödinger equation) and R_2 is the absolute value of the wave-function $\phi = R_2 e^{iS_2/\hbar}$ describing the time-reverse process (solution of the time-reverse Schrödinger equation). The symmetryzed quantum potential (3.19) can explain a symmetric and instantaneous communication between subatomic particles and thus can be considered as the fundamental entity which represents the state of the three-dimensional space as an immediate information medium in EPR-type experiments (or, more in general, in each immediate physical phenomenon regarding the quantum domain). In the light of the symmetrized quantum potential (3.19), one can explain non local correlations in many-body systems — and thus EPR experiments — in the correct way (namely also if one would imagine to film back the process of these correlations). The symmetryzed quantum potential (3.19) can be considered the most appropriate candidate to provide a mathematical reality to a three-dimensional space intended as a direct information medium [286, 287]. The first component of the symmetryzed quantum potential (3.19), namely

$$Q_1 = \sum_{i=1}^{N} -\frac{\hbar^2}{2m_i} \frac{\nabla_i^2 R_1}{R_1}, \tag{3.20}$$

which regard the forward-time process and coincides with the original Bohm's quantum potential, is the real physical component (which produces observable effects in the quantum world, such as those obtained by Philippidis, Dewdney, Hiley and Vigier about the classic double-slit experiment, tunnelling, trajectories of two particles in a potential of harmonic oscillator, EPR-type experiments, experiments of neutron-interferometry [46, 294]): it expresses the instantaneous action on quantum particles and thus the immediate action of space on them. The second component

$$Q_2 = \sum_{i=1}^{N} \frac{\hbar^2}{2m_i} \frac{\nabla_i^2 R_2}{R_2} \tag{3.21}$$

is introduced to reproduce in the correct way the time-reverse process of the instantaneous action and thus it guarantees that the quantum world can be interpreted correctly with the idea of space as an immediate information medium if one would imagine to film the process backwards: it must be introduced in order to recover a symmetry in time in quantum processes, to interpret in the correct way quantum processes if one would imagine to film that process backwards. Just like for the one-body systems, the opposed sign of the second component with respect to the physical first component (namely with respect to the original Bohm's quantum potential) can be interpreted as a consequence of the idea of the measurable time as a measuring system of the numerical order of material changes: the mathematical features of the second component of the symmetryzed quantum potential imply that it is not possible to go backwards in the physical time intended as numerical order of physical events.

Moreover, as the author and Sorli showed in the paper [287], both the components (3.20) and (3.21) of the symmetryzed quantum potential can be considered as physical quantities deriving from a more fundamental quantum entropy. The first component can be expressed as

$$Q_1 = \sum_{i=1}^{N} \left[-\frac{\hbar^2}{2m_i} (\nabla_i S_{Q1})^2 + \frac{\hbar^2}{2m_i} (\nabla_i^2 S_{Q1}) \right] \tag{3.22}$$

while the second component can be expressed as

$$Q_2 = \sum_{i=1}^{N} \left[\frac{\hbar^2}{2m_i} (\nabla_i S_{Q2})^2 - \frac{\hbar^2}{2m_i} (\nabla_i^2 S_{Q2}) \right]. \tag{3.23}$$

In equation (3.22) $S_{Q1} = -\frac{1}{2} \ln \rho_1$ is the quantum entropy defining the degree of order and chaos of the vacuum, namely describing the fundamental geometry of the background, for the forward-time processes (where here $\rho_1 = |\psi(\vec{x}_1, \vec{x}_2, \ldots, \vec{x}_N, t)|^2$, $\psi(\vec{x}_1, \vec{x}_2, \ldots, \vec{x}_N, t) = R_1 e^{-iS_1/\hbar}$ being the forward-time may-body wavefunction, solution of the standard Schrödinger equation). In equation (3.23) $S_{Q2} = -\frac{1}{2}\ln\rho_2$ is the quantum entropy, namely describing the fundamental geometry of the background, defining the degree of order and chaos of the vacuum for the time-reverse processes (where here $\rho_2 = |\phi(\vec{x}_1, \vec{x}_2, \ldots, \vec{x}_N, t)|^2$, $\phi(\vec{x}_1, \vec{x}_2, \ldots, \vec{x}_N, t) = R_2 e^{-iS_2/\hbar}$ being the time-reverse many-body wavefunction, solution of the time-reversed Schrödinger equation). Thus the non-local geometry of the quantum world as regards the instantaneous communication of quantum particles in EPR-type experiments, in virtue of its symmetric features, can be associated to the geometry of a fundamental background described by a symmetrized quantum entropy at two components of the form

$$S_Q = \begin{pmatrix} -\dfrac{1}{2}\ln\rho_1 \\ -\dfrac{1}{2}\ln\rho_2 \end{pmatrix}. \tag{3.24}$$

The first component $S_{Q1} = -\frac{1}{2}\ln\rho_1$ of the symmetrized quantum entropy corresponds to the change of the geometric properties of the fundamental background which ultimately reproduces the forward-time processes. The second component $S_{Q2} = -\frac{1}{2}\ln\rho_2$ of the symmetrized quantum entropy corresponds to the deformation of the geometric properties of the fundamental background which allous us to explain the reversed-time processes in agreement with respect to what is observed. The symmetrized quantum potential can be thus written in compact form as

$$Q = \begin{pmatrix} \displaystyle\sum_{i=1}^{N}\left[-\frac{\hbar^2}{2m_i}(\nabla_i S_{Q1})^2 + \frac{\hbar^2}{2m_i}(\nabla_i^2 S_{Q1})\right] \\ \displaystyle\sum_{i=1}^{N}\left[+\frac{\hbar^2}{2m_i}(\nabla_i S_{Q2})^2 - \frac{\hbar^2}{2m_i}(\nabla_i^2 S_{Q2})\right] \end{pmatrix}. \tag{3.25}$$

The energy conservation law for the forward-time process is

$$\sum_{i=1}^{N}\frac{|\nabla_i S_1|^2}{2m_i}-\sum_{i=1}^{N}\frac{\hbar^2}{2m_i}(\nabla_i S_{Q1})^2+V+\sum_{i=1}^{N}\frac{\hbar^2}{2m_i}(\nabla_i^2 S_{Q1})=-\frac{\partial S_1}{\partial t}\tag{3.26}$$

while the energy conservation law for the reversed-time process is

$$\sum_{i=1}^{N}\frac{|\nabla_i S_2|^2}{2m_i}+\sum_{i=1}^{N}\frac{\hbar^2}{2m_i}(\nabla_i S_{Q2})^2-V-\sum_{i=1}^{N}\frac{\hbar^2}{2m_i}(\nabla_i^2 S_{Q1})=-\frac{\partial S_2}{\partial t}.\tag{3.27}$$

The symmetrized quantum entropy can be thus considered as the fundamental entity which determines the non-local features of the quantum geometry. On the basis of the symmetrized energy conservation law expressed by equations (3.26) and (3.27), one can say that the non-local quantum geometry characterized by a symmetric feature as regards the instantaneous correlation between subatomic particles in EPR-type experiments is expressed by the action of the 3D space as an immediate information medium which derives just from the two quantum correctors to the energy of the system under consideration, both for the forward-time processes and for the reversed-time processes, namely from the quantum corrector to the potential energy ($\sum_{i=1}^{N}\frac{\hbar^2}{2m_i}(\nabla_i^2 S_{Q1})$ for the forward-time processes and $-\sum_{i=1}^{N}\frac{\hbar^2}{2m_i}(\nabla_i^2 S_{Q2})$ for the reversed-time processes) and the quantum corrector to the kinetic energy ($-\sum_{i=1}^{N}\frac{\hbar^2}{2m_i}(\nabla_i S_{Q1})^2$ for the forward-time processes and $\sum_{i=1}^{N}\frac{\hbar^2}{2m_i}(\nabla_i S_{Q2})^2$ for the reversed-time processes), while the other two terms $\sum_{i=1}^{N}\frac{|\nabla_i S|^2}{2m_i}$ and V on the left-hand of the two equations (3.26) and (3.27) generate a local feature of space. In other words, the fundamental geometry of the background space associated with the symmetrized quantum entropy given by equation (3.24), by determining two quantum correctors to the energy of the system under consideration (where if one imagines to film the process backwards, one would see what physically happens) can be considered the fundamental entity which, at a fundamental level, determines an immediate information transfer in the behaviour of the particles in EPR-type experiments.

On the basis of its mathematical features, the symmetryzed quantum potential (deriving from the symmetrized quantum entropy) implies that in the quantum domain a timeless three-dimensional space has a crucial role in determining the motion of subatomic particles because the symmetryzed quantum potential produces a like-space and instantaneous action on the particles under consideration and contains an active information about the environment and, on the other hand, implies the concept of time as a numerical order of material changes. In EPR-type experiments (and, more in general, in all immediate physical phenomena regarding the quantum domain) a three-dimensional timeless space acts as an immediate information medium in the sense that the first component of the symmetryzed quantum potential makes physical space an "immediate information medium" which keeps two elementary particles in an immediate contact (while the second component of the symmetrized quantum potential reproduces, from the mathematical point of view, the symmetry in time of this communication and the fact that time exists only as a numerical order of material changes). We can call this peculiar interpretation of quantum non-locality as the "immediate symmetric interpretation" of quantum non-locality.

On the basis of the analysis made in this chapter, the immediate symmetric interpretation of quantum non-locality can be considered a relevant feature of the geometry of space which can be derived by starting from Bohm's approach. The symmetryzed quantum potential is a significant element characterizing the geometry of the quantum world (in the context of the non-local correlations between subatomic particles) by providing a consistent mathematical model to the idea of a three-dimensional timeless space which acts as an immediate information medium.

3.2 The Quantum Geometry in the Symmetrized Quantum Potential Approach

On the basis of the model developed by Anandan in the paper [295], a geometry for quantum theory can be defined by starting from the relations determined by a universal group S, which is the generalization of the group of translation and has a representation in each Hilbert space. The displacement of an object may be characterized by means of any given element $s \in S$, which acts on each of the Hilbert spaces of the particles or fields constituting the object under consideration through the corresponding representation of S. Through this action of s one can associate a certain wavefunction ψ_s to each wavefunction ψ in each of these Hilbert spaces, and the group element s that determines the relation between ψ and ψ_s is independent of the Hilbert space and is therefore universal.

In Anandan's approach, the geometry of the background of quantum processes is based on an invariant geometrical "distance", which corresponds to a gauge field which is the generalization of the translation group. The translation of a material object may be performed by acting on all the quantum states of the particles constituting the object and can be described considering the universal group element $\exp\left(-\frac{i}{\hbar}\hat{p}l\right)$ where $\hat{p}$ is a generator of translation and defines the gauge-covariant transformation $\psi_l(x) = f_l(x)\psi(x)$ where x stands for x^μ and

$$f_l(x) = \exp\left(-\frac{i}{\hbar}\hat{\vec{p}}\cdot\vec{l}\right)\exp\left\{i\frac{q}{\hbar}\int_x^{x+l}\vec{A}(\vec{y},t)\cdot d\vec{y}\right\} \tag{3.28}$$

where the integral is taken along the straight line joining $x = (\vec{x}, t)$ and $(\vec{x}+\vec{l}, t)$ for simplicity. If one applies a gauge transformation of the type $\psi'(x) = u(x)\psi(x)$, where $u(x) = \exp\left\{i\frac{q}{\hbar}\Lambda(x)\right\}$ and $A'_\mu(x) = A_\mu(x) - \partial_\mu\Lambda(x)$, the quantity (3.28) transforms to

$$f'_l(x) = \exp\left(-\frac{i}{\hbar}\hat{\vec{p}}\cdot\vec{l}\right)u(\vec{x}+\vec{l},t)\exp\left\{i\frac{q}{\hbar}\int_x^{x+l}\vec{A}(\vec{y},t)\cdot d\vec{y}\right\}u(\vec{x},t). \tag{3.29}$$

Moreover, the previous equations may be generalized, for an arbitrary gauge field, by replacing eA_μ by $g_0 A^k_\mu T_k$, where T_k generates the gauge group, and $u(x)$ is the corresponding local gauge transformation. Here, the gauge field exponential operator is considered along an arbitrary piecewise-differentiable curve $\hat{\gamma}$ in R^4. The gauge field $g_0 A^k_\mu T_k$ is associated to a geometry defined by a transformation g_γ

on the Hilbert space by $\psi_\gamma(x) = g_\gamma(x)\psi(x)$, where

$$g_\gamma(x) = P\exp\left(-\frac{i}{\hbar}\int_{\bar{\gamma}} \hat{p}_\mu dy^\mu\right) P\exp\left\{-i\frac{g_0}{\hbar}\int_\gamma A^k_\mu(y)T_k dy^\mu\right\} \tag{3.30}$$

with P denoting path ordering, and γ is a curve in space-time that is congruent to $\hat{\gamma}$ while $\bar{\gamma}$ is the curve γ traversed in the reversed order. In equation (3.30), the curve γ begins at x and ends at $x+l$, where l^μ is a fixed vector (independent of x^μ), and $\bar{\gamma}$ begins at $x+l$ and ends in x. In Anandan's model, the quantity g_γ given by equation (3.30) can be considered as a quantum distance that replaces the classical space-time distance along the curve γ [310].

Now, in the recent paper [296] (see also [297] for a review of this concept) the author of this book and Sorli generalized Anandan's geometric approach based on the geometrical distance (3.30) in the picture of the symmetryzed quantum potential approach by considering the symmetrized quantum distance given by the following relation:

$$g_\gamma(x) = \begin{pmatrix} P\exp\left(-\frac{i}{\hbar}\int_{\bar{\gamma}} \hat{p}_\mu dy^\mu\right) P\exp\left\{-i\frac{g_0}{\hbar}\int_\gamma A^k_\mu(y)T_k dy^\mu\right\} \\ P\exp\left(\frac{i}{\hbar}\int_{\bar{\gamma}} \hat{p}_\mu dy^\mu\right) P\exp\left\{i\frac{g_0}{\hbar}\int_\gamma A^k_\mu(y)T_k dy^\mu\right\} \end{pmatrix}. \tag{3.31}$$

In equation (3.31), the second component

$$g_\gamma(x) = P\exp\left(\frac{i}{\hbar}\int_{\bar{\gamma}} \hat{p}_\mu dy^\mu\right) P\exp\left\{i\frac{g_0}{\hbar}\int_\gamma A^k_\mu(y)T_k dy^\mu\right\} \tag{3.32}$$

regards the time-reverse processes, namely is introduced to guarantee a symmetry in time of the processes of translation if one would imagine to see the process backwards.

In the light of Anandan's results, if one makes a local gauge transformation the symmetrized quantum distance (3.31) transforms as

$$g'_\gamma(x) = \begin{pmatrix} P\exp\left(-\frac{i}{\hbar}\int_{\bar{\gamma}} \hat{p}'_\mu dy^\mu\right) u(x+l)\exp\left\{-i\frac{g_0}{\hbar}\int_\gamma A^k_\mu(y^\mu)T_k dy^\mu\right\} u^+(x) \\ P\exp\left(\frac{i}{\hbar}\int_{\bar{\gamma}} \hat{p}'_\mu dy^\mu\right) u(x+l)P\exp\left\{i\frac{g_0}{\hbar}\int_\gamma A^k_\mu(y)T_k dy^\mu\right\} u^+(x) \end{pmatrix} \tag{3.33}$$

where $\hat{p}'_0 = H'$, $\hat{p}'_i = \hat{p}_i = i\hbar\frac{\partial}{\partial x^i}$. Moreover, if one takes into account that

$$P\exp\left(-\frac{i}{\hbar}\int_{\bar{\gamma}} \hat{p}_\mu dy^\mu\right) = \left\{P\exp\left(-\frac{i}{\hbar}\int_\gamma \hat{p}_\mu dy^\mu\right)\right\}^+ \tag{3.34}$$

one obtains

$$\langle\psi|g_\gamma|\psi\rangle = \begin{pmatrix} \int d^3x \left[P\exp\left(-\frac{i}{\hbar}\int_\gamma \hat{p}_\mu dy^\mu\right)\psi(x)\right]^+ P\exp\left(-ig_0\int_\gamma A^k_\mu T_k dy^\mu\right)\psi(x) \\ \int d^3x \left[P\exp\left(\frac{i}{\hbar}\int_\gamma \hat{p}_\mu dy^\mu\right)\psi(x)\right]^+ P\exp\left(ig_0\int_\gamma A^k_\mu T_k dy^\mu\right)\psi(x) \end{pmatrix} \tag{3.35}$$

which indicates that, since the integrand is gauge invariant, also the expectation value is gauge invariant (for both the components).

According to the approach developed in [296, 297], the ultimate entity which describes the geometry of the quantum world associated with the gauge field which is the generalization of the translation group is the symmetrized quantum distance (3.33). In this approach, as it has been shown in [296, 297], the fundamental parameter which describes the geometrical properties of the background space of quantum processes, expressing a deformation of the geometrical properties of space with respect to the Euclidean geometry characteristic of classical physics, is the quantum entropy. In this picture, the action of the gauge field $g_0A^k_\mu T_k$ — which is the generalization of the group of translation and determines the symmetrized quantum distance (3.33) — emerges from the quantum entropy. Thus, the non-local quantum geometry associated with the 3D space which acts as an immediate information medium can be described by introducing the following notions of quantum distances:

$$L_{quantum} = \begin{pmatrix} \dfrac{1}{\sqrt{(\nabla S_{Q1})^2 - \nabla^2 S_{Q1}}} \\ \dfrac{1}{\sqrt{-(\nabla S_{Q2})^2 + \nabla^2 S_{Q2}}} \end{pmatrix}$$

$$= \begin{pmatrix} P\exp\left(-\frac{i}{\hbar}\int_{\bar{\gamma}} \hat{p}_\mu dy^\mu\right) P\exp\left\{-i\frac{g_0}{\hbar}\int_{\gamma} A_\mu^k(y)T_k dy^\mu\right\} \\ P\exp\left(\frac{i}{\hbar}\int_{\bar{\gamma}} \hat{p}_\mu dy^\mu\right) P\exp\left\{i\frac{g_0}{\hbar}\int_{\gamma} A_\mu^k(y)T_k dy^\mu\right\} \end{pmatrix} \tag{3.36}$$

for one-body systems and

$$L_{quantum} = \begin{pmatrix} \dfrac{1}{\sqrt{\sum_{i=1}^N ((\nabla_i S_{Q1})^2 - \nabla_i^2 S_{Q1})}} \\ \dfrac{1}{\sqrt{\sum_{i=1}^N (-(\nabla_i S_{Q2})^2 + \nabla_i^2 S_{Q2})}} \end{pmatrix}$$

$$= \begin{pmatrix} P\exp\left(-\frac{i}{\hbar}\int_{\bar{\gamma}} \hat{p}_\mu dy^\mu\right) P\exp\left\{-i\frac{g_0}{\hbar}\int_{\gamma} A_\mu^k(y)T_k dy^\mu\right\} \\ P\exp\left(\frac{i}{\hbar}\int_{\bar{\gamma}} \hat{p}_\mu dy^\mu\right) P\exp\left\{i\frac{g_0}{\hbar}\int_{\gamma} A_\mu^k(y)T_k dy^\mu\right\} \end{pmatrix} \tag{3.37}$$

for many-body systems. Inside the symmetrized quantum potential approach, the equations (3.36) and (3.37) define typical quantum-entropic lengths that can be used to evaluate the strength of quantum effects and, therefore, the degree of non-local correlation and the modification of the geometrical properties of space with respect to the Euclidean geometry characteristic of classical physics. In particular, the first component of (3.36) and (3.37) describes the geometry of the fundamental background which reproduces the forward-time processes while the second component of (3.36) and (3.37) characterizes the geometry of the fundamental background which allows us to explain the time-reverse processes in agreement with what physically occurs. Once the quantum-entropic lengths (3.36) and (3.37) become non-negligible the system into consideration goes into a quantum regime, characterized by a non-local geometry. In this picture, Heisenberg's uncertainty principle expresses our incapability to perform a classical measurement to distances smaller than the quantum-entropic lengths. In other words, the size of a measurement has to be bigger than the quantum-entropic lengths

$$\Delta L \geq L_{quantum} = \begin{pmatrix} \dfrac{1}{\sqrt{(\nabla S_{Q1})^2 - \nabla^2 S_{Q1}}} \\ \dfrac{1}{\sqrt{-(\nabla S_{Q2})^2 + \nabla^2 S_{Q2}}} \end{pmatrix} \tag{3.38}$$

(for one-body systems)

$$\Delta L \geq L_{quantum} = \begin{pmatrix} \dfrac{1}{\sqrt{\sum_{i=1}^{N}\left((\nabla_i S_{Q1})^2 - \nabla_i^2 S_{Q1}\right)}} \\ \dfrac{1}{\sqrt{\sum_{i=1}^{N}\left(-(\nabla_i S_{Q2})^2 + \nabla_i^2 S_{Q2}\right)}} \end{pmatrix} \tag{3.39}$$

(for many-body systems). In the non-local quantum geometry corresponding to the 3D space which acts as a direct information medium and is associated with a gauge field which is the generalisation of the universal group of translation, equations (3.36) and (3.37) may be considered as the fundamental definitions of distances. They indicate that there is an equivalence between the effect of the gauge field associated with $g_0 A_\mu^k T_k$, where T_k generates the gauge group of translation, and the action of the quantum entropy. On one hand, equations (3.36) and (3.37) mean that the gauge field associated with $g_0 A_\mu^k T_k$ determines a 3D non-Euclidean space which acts as an immediate information medium and which is described by a symmetrized quantum distance describing the geometrical properties of a vacuum associated with a symmetrized quantum entropy. On the other hand, the action of the gauge field of generating a deformation of the geometry in the quantum regime derives just from the symmetrized quantum entropy, which represents the ultimate parameter defining the geometry of the quantum world inside the symmetrized quantum potential approach.

3.3 The Symmetrized Quantum Potential Approach in the Relativistic de Broglie-Bohm Theory in Curved Space-time

If at the fundamental level of quantum processes, non-local correlations are due to a background space which acts as a direct information medium between the particles under consideration, inside a complete physical theory it is legitimate consider the possibility that, also in other domains of physical reality, space functions as a direct information medium. The next important step is to apply the symmetrized quantum potential approach in the context of A. Shojai's and F. Shojai's Bohmian model regarding the motion of a spinless particle in a relativistic curved space-time (developed in the papers [96, 97, 157–169]). In this regard, one considers a wavefunction at two components $|C(t)\rangle = \binom{\psi(t)}{\phi(t)}$, where $\psi(t)$ is the solution to the standard Klein-Gordon equation, $\phi(t)$ is the solution to the time-reversed Klein-Gordon equation. By making the usual decomposition of these wavefunctions in polar form

$$\psi = R_1 e^{iS_1/\hbar}, \tag{3.40}$$

$$\phi = R_2 e^{iS_2/\hbar} \tag{3.41}$$

where R_1 and R_2 are real amplitude functions and S_1 and S_2 are real phase functions, if one substitutes into the time-symmetric Klein-Gordon equation (in the picture of F. Shojai's and A. Shojai's model) and separates real and imaginary parts, one obtains a symmetrized quantum Hamilton-Jacobi equation that, by imposing that it be Poincarè invariant and have the correct non-relativistic limit, assumes the following form

$$\partial_\mu \begin{pmatrix} S_1 \\ S_2 \end{pmatrix} \partial^\mu \begin{pmatrix} S_1 \\ S_2 \end{pmatrix} = m^2c^2 \exp \frac{\hbar^2}{m^2c^2} \begin{pmatrix} \dfrac{\left(\nabla^2 - \dfrac{1}{c^2}\dfrac{\partial^2}{\partial t^2}\right)|\psi|}{|\psi|} \\ \dfrac{\left(-\nabla^2 + \dfrac{1}{c^2}\dfrac{\partial^2}{\partial t^2}\right)|\phi|}{|\phi|} \end{pmatrix}, \tag{3.42}$$

with the symmetrized quantum potential defined as

$$Q = \frac{\hbar^2}{m^2c^2} \begin{pmatrix} \dfrac{\left(\nabla^2 - \dfrac{1}{c^2}\dfrac{\partial^2}{\partial t^2}\right)|\psi|}{|\psi|} \\ \dfrac{\left(-\nabla^2 + \dfrac{1}{c^2}\dfrac{\partial^2}{\partial t^2}\right)|\phi|}{|\phi|} \end{pmatrix}, \tag{3.43}$$

and the symmetrized continuity equation

$$\partial_\mu \left(\begin{pmatrix} \rho_1 \\ \rho_2 \end{pmatrix} \partial^\mu \begin{pmatrix} S_1 \\ S_2 \end{pmatrix} \right) = 0 \tag{3.44}$$

where $\rho_1 = R_1^2$, $\rho_2 = R_2^2$ are the ensemble of particles regarding forward-time processes and time-reverse processes respectively.

Here, in analogous way to the original F. Shojai's and A. Shojai's model, the extension to the case of a particle moving in a curved background can be achieved by changing the ordinary differentiating ∂_μ with the covariant derivative ∇_μ and by replacing the Lorentz metric with the curved metric $g_{\mu\nu}$. In this way one obtains that the equations of motion for a particle (of spin 0) in a curved background in the symmetrized quantum potential approach can be written as:

$$\nabla_\mu \left(\begin{pmatrix} \rho_1 \\ \rho_2 \end{pmatrix} \nabla^\mu \begin{pmatrix} S_1 \\ S_2 \end{pmatrix} \right) = 0, \tag{3.45}$$

$$g^{\mu\nu} \nabla_\mu \begin{pmatrix} S_1 \\ S_2 \end{pmatrix} \nabla_\nu \begin{pmatrix} S_1 \\ S_2 \end{pmatrix} = m^2 c^2 \exp Q \tag{3.46}$$

where

$$Q = \frac{\hbar^2}{m^2 c^2} \begin{pmatrix} \dfrac{\left(\nabla^2 - \dfrac{1}{c^2} \dfrac{\partial^2}{\partial t^2} \right)_g |\psi|}{|\psi|} \\ \dfrac{\left(-\nabla^2 + \dfrac{1}{c^2} \dfrac{\partial^2}{\partial t^2} \right)_g |\phi|}{|\phi|} \end{pmatrix} \tag{3.47}$$

is the symmetrized quantum potential. The symmetrized quantum potential (3.47) at two components is introduced in order to reproduce in the correct way also the time-reverse process of the motion of a spinless particle in a relativistic curved space-time where the quantum and gravitational aspects of matter are highly coupled and the quantum potential emerges as the conformal degree of freedom of the space-time metric. The symmetrized quantum potential (3.47) can be considered the opportune candidate to express the state of space in the relativistic curved space-time: its instantaneous action physically means that in the relativistic curved space-time the high coupling between quantum and gravitational aspects of matter are produced by a background space which acts as an immediate information medium (where the second component, of opposed sign with respect to the first component, allows the recover of a symmetry in time in the analysis of the processes, a correct interpretation of the processes if one would imagine to film them backwards).

Moreover, in analogy to the symmetrized Bohmian non-relativistic quantum mechanics, also in the symmetrized extension of de Broglie-Bohm theory in curved space-time both the components and of the symmetryzed quantum potential (3.47) can be considered as physical quantities deriving from an opportune quantum entropy defining the geometrical properties of the background, namely indicating a degree of order and chaos of the background space which determines the processes. The first component can be expressed as

$$Q_1 = -\frac{\hbar^2}{m^2c^2}\left(\nabla_\mu S_{Q1}\right)^2 + \frac{\hbar^2}{m^2c^2}\left(\left(\nabla^2 - \frac{1}{c^2}\frac{\partial^2}{\partial t^2}\right)_g S_{Q1}\right) \tag{3.48}$$

while the second component can be expressed as

$$Q_2 = \frac{\hbar^2}{m^2c^2}\left(\nabla_\mu S_{Q2}\right)^2 - \frac{\hbar^2}{m^2c^2}\left(\left(\nabla^2 - \frac{1}{c^2}\frac{\partial^2}{\partial t^2}\right)_g S_{Q2}\right), \tag{3.49}$$

where $S_{Q1} = -\frac{1}{2}\ln\rho_1$ is the quantum entropy defining the degree of order and chaos of the vacuum of processes for the forward-time processes and $S_{Q2} = -\frac{1}{2}\ln\rho_2$ is the quantum entropy defining the degree of order and chaos of the vacuum for the time-reverse processes. Thus, the non-local geometry of the quantum world in the relativistic curved space-time for spinless particles (where there is an instantaneous high coupling between quantum and gravitational aspects of matter), can be associated to a symmetrized quantum entropy at two components of the form

$$S_Q = \begin{pmatrix} -\dfrac{1}{2}\ln\rho_1 \\ -\dfrac{1}{2}\ln\rho_2 \end{pmatrix} \tag{3.50}$$

where the first component $S_{Q1} = -\frac{1}{2}\ln\rho_1$ corresponds to the change of the geometric properties of physical space for the forward-time processes while the second component $S_{Q2} = -\frac{1}{2}\ln\rho_2$ corresponds to the change of the geometric properties of physical space for the reversed-time processes, which explains in the correct way the processes if one would imagine to see them backwards. The symmetrized quantum potential can be thus written in compact form as

$$Q=\begin{pmatrix} -\frac{\hbar^2}{m^2c^2}(\nabla_\mu S_{Q1})^2+\frac{\hbar^2}{m^2c^2}\left(\left(\nabla^2-\frac{1}{c^2}\frac{\partial^2}{\partial t^2}\right)_g S_{Q1}\right) \\ \frac{\hbar^2}{m^2c^2}(\nabla_\mu S_{Q2})^2-\frac{\hbar^2}{m^2c^2}\left(\left(\nabla^2-\frac{1}{c^2}\frac{\partial^2}{\partial t^2}\right)_g S_{Q2}\right) \end{pmatrix}. \tag{3.51}$$

In the symmetrized quantum potential approach, where the fundamental entity describing the geometrical properties of the fundamental background is the symmetrized quantum entropy, the equations of motion for a spinless particle in a curved background assume the following form:

$$\tilde{g}^{\mu\nu}\frac{1}{c}\begin{pmatrix} \frac{\partial S_{Q1}}{\partial t} \\ \frac{\partial S_{Q2}}{\partial t} \end{pmatrix}=\tilde{g}^{\mu\nu}\left[-\left(p_\mu\tilde{\nabla}^\mu\begin{pmatrix} S_{Q1} \\ S_{Q2} \end{pmatrix}\right)+\frac{1}{2}\nabla_\mu p^\mu\right] \tag{3.52}$$

and

$$\tilde{g}^{\mu\nu}\tilde{\nabla}_\mu\begin{pmatrix} S_1 \\ S_2 \end{pmatrix}\tilde{\nabla}_\nu\begin{pmatrix} S_1 \\ S_2 \end{pmatrix}=m^2c^2 \tag{3.53}$$

where

$$\tilde{g}_{\mu\nu}=\frac{M^2}{m^2}g_{\mu\nu} \tag{3.54}$$

is a symmetrized conformal transformation that indeed is determined by the symmetrized quantum entropy (3.50), $\tilde{\nabla}_\mu$ represents the covariant differentiation with respect to the metric $\tilde{g}_{\mu\nu}$,

$$M^2=m^2\exp\begin{bmatrix} -\frac{\hbar^2}{m^2c^2}(\nabla_\mu S_{Q1})^2+\frac{\hbar^2}{m^2c^2}\left(\left(\nabla^2-\frac{1}{c^2}\frac{\partial^2}{\partial t^2}\right)_g S_{Q1}\right) \\ \frac{\hbar^2}{m^2c^2}(\nabla_\mu S_{Q2})^2-\frac{\hbar^2}{m^2c^2}\left(\left(\nabla^2-\frac{1}{c^2}\frac{\partial^2}{\partial t^2}\right)_g S_{Q2}\right) \end{bmatrix} \tag{3.55}$$

is the variable quantum mass.

On the basis of its mathematical features, the symmetryzed quantum potential (deriving from the symmetrized quantum entropy) implies that in the relativistic curved space-time a background space which acts as a direct information medium has a crucial role in determining the motion of spinless particles in such a way that there is an instantaneous tied link between quantum and gravitational aspects of matter, as a consequence of the symmetrized quantum entropy describing the degree of order and chaos of the vacuum supporting the density of the particles, of the fact that the ensemble of particles associated with the wavefunction of the physical system determines a modification of the ultimate geometry. In this approach, the geometrization of the quantum aspects of matter is reproduced in the correct way also if one imagines to film the processes backwards, in a picture based on the idea that the density of particles associated with a given wavefunction determines a modification of the geometry of the fundamental background of the processes. The effects of gravity on geometry and the quantum effects on the geometry of space-time are highly coupled because they are both determined by the fundamental background space described by the quantum entropy, they are both produced by the degree of order and chaos of the vacuum supporting the density of particles associated with the wavefunction under consideration and the second component of the symmetrized quantum potential — determined by the second component of the symmetrized quantum entropy — can reproduce in the correct way also the time-reverse of the processes. The symmetrized quantum entropy associated with the fundamental background is the fundamental entity which generates itself a curvature which may have a large influence on the classical contribution to the curvature of the space-time, thus producing an instantaneous coupling between gravitational and quantum effects of matter.

Moreover, in the context of this symmetrized entropic approach to F. Shojai's and A. Shojai's model of relativistic spinless particles in a curved space-time, the geometrical properties of the vacuum can be characterized by introducing an appropriate symmetrized quantum length associated with the conformal metric (3.54) and determined by the symmetrized quantum entropy:

$$L_{quantum} = \begin{pmatrix} \dfrac{1}{\sqrt{(\nabla_\mu S_{Q1})^2 - \left(\nabla^2 - \dfrac{1}{c^2}\dfrac{\partial^2}{\partial t^2}\right)_g S_{Q1}}} \\ \dfrac{1}{\sqrt{-(\nabla_\mu S_{Q2})^2 + \left(\nabla^2 - \dfrac{1}{c^2}\dfrac{\partial^2}{\partial t^2}\right)_g S_{Q2}}} \end{pmatrix}. \quad (3.56)$$

The symmetrized quantum length (3.56) can be used to evaluate the strength of quantum effects and, therefore, the modification of the geometry — introduced in a relativistic curved space-time regime — with respect to the Euclidean geometry characteristic of classical physics. It is the physical parameter that, in the vacuum described by the symmetrized quantum entropy, measures the degree of non-locality in spinless systems in a relativistic curved space-time regime. In particular, the first component of (3.56) describes the geometry of the background for the forward-time processes while the second component of (3.56) is introduced to characterize the time-reverse processes in a compatible way with what is observed if one would imagine to film the processes bakwards. Once the symmetrized quantum length (3.56) becomes non-negligible the spinless particle into consideration goes into a quantum regime where the quantum and gravitational effects are highly related. In this picture, Heisenberg's uncertainty principle derives from the fact that we are unable to perform a classical measurement to distances smaller than this symmetrized quantum length, and also the time-reverse process turns out to be in agreement with what occurs. A quantum regime where there is a coupling between gravitational and quantum effects is entered when the symmetrized quantum length (3.56) must be taken under consideration.

On the basis of equation (3.56), it becomes thus permissible the following re-reading of the geometry of the background space regarding the motion of a spinless particle in the context of F. Shojai's and A. Shojai's model. The symmetrized quantum entropy can be considered as the fundamental entity which provides a measure of the geometrical properties of a regime where the quantum behaviour of matter becomes highly related with its gravitational effects. In virtue of the link between the symmetrized quantum potential and the symmetrized quantum entropy, the geometrical properties of the background space which correspond with a regime where the quantum behaviour of matter is highly coupled with gravitational effects are determined by the symmetrized quantum entropy and the condition (3.56) about the quantum length indicates when this coupling between quantum and gravitational effects must be taken into consideration and provides a measure of the degree of non-local correlations in this regime.

3.4 Towards a Symmetrized Extension of Bohm's Relativistic Quantum Field Theory

By considering the classic model of a Bohmian relativistic bosonic quantum field theory developed by Bohm, Hiley, Bell and Kaloyerou in [69, 145, 146], a symmetrized approach can be developed if one starts by writing the time-symmetric Schrödinger equation for the wave functional at two components $C(\{\varphi(\vec{x})\},t) = \begin{pmatrix} \Psi(\{\varphi(\vec{x})\},t) \\ \Phi(\{\varphi(\vec{x})\},t) \end{pmatrix}$ (here we assume $\hbar = c = 1$):

$$\begin{pmatrix} H & 0 \\ 0 & -H \end{pmatrix} \begin{pmatrix} \Psi(\{\varphi(\vec{x})\},t) \\ \Phi(\{\varphi(\vec{x})\},t) \end{pmatrix} = i\frac{\partial}{\partial t} \begin{pmatrix} \Psi(\{\varphi(\vec{x})\},t) \\ \Phi(\{\varphi(\vec{x})\},t) \end{pmatrix}, \tag{3.57}$$

where $\{\varphi(\vec{x})\} = \varphi_1(\vec{x}), \varphi_2(\vec{x}), \varphi_3(\vec{x}), \ldots$ are Bose-Einstein fields in the Schrödinger representation,

$$H = \sum_k \int d^3x \left[-\frac{1}{2}\frac{\delta^2}{\delta\varphi_k^2(\vec{x})} + \frac{1}{2}\left|\nabla\varphi_k(\vec{x})\right|^2 \right] + V(\{\varphi(\vec{x})\}) \tag{3.58}$$

is the Hamiltonian in the Schrödinger representation. If one expresses both the components of the wave functional in the polar representation as follows:

$$\begin{aligned} \Psi(\{\varphi(\vec{x})\},t) &= R_1(\{\varphi(\vec{x})\},t)e^{iS_1(\{\varphi(\vec{x})\},t)} \\ \Phi(\{\varphi(\vec{x})\},t) &= R_2(\{\varphi(\vec{x})\},t)e^{iS_2(\{\varphi(\vec{x})\},t)} \end{aligned} \tag{3.59}$$

where R_1 and R_2 are real amplitude functions and S_1 and S_2 are real phase functions, and inserts them into the Schrödinger equation (3.57), one obtains two coupled partial differential functional equations:

$$\frac{\partial}{\partial t}\begin{pmatrix} S_1 \\ S_2 \end{pmatrix} + \frac{1}{2}\sum_k \int d^3x \left[-\frac{1}{2}\frac{\delta^2}{\delta\varphi_k^2(\vec{x})} + \frac{1}{2}\left|\nabla\varphi_k(\vec{x})\right|^2 \right] + \begin{pmatrix} V \\ -V \end{pmatrix} + \begin{pmatrix} Q_1 \\ Q_2 \end{pmatrix} = 0, \tag{3.60}$$

$$\frac{\partial}{\partial t}\begin{pmatrix} R_1^2 \\ R_2^2 \end{pmatrix} + \sum_k \int d^3x \frac{\delta}{\delta\varphi_k(\vec{x})} J_k(\{\varphi(\vec{x})\}, t) = 0 \tag{3.61}$$

where

$$Q_1(\{\varphi(\vec{x})\}, t) = -\frac{1}{2}\sum_k \int d^3x \frac{1}{R_1}\frac{\delta^2 R_1(\{\varphi(\vec{x})\}, t)}{\delta\varphi_k^2(\vec{x})} \tag{3.62}$$

is the quantum potential regarding the forward-time process,

$$Q_2(\{\varphi(\vec{x})\}, t) = \frac{1}{2}\sum_k \int d^3x \frac{1}{R_2}\frac{\delta^2 R_2(\{\varphi(\vec{x})\}, t)}{\delta\varphi_k^2(\vec{x})} \tag{3.63}$$

is the quantum potential regarding the time-reverse process and, as usual,

$$J_k(\{\varphi(\vec{x})\}, t) = R^2 \frac{\delta S}{\delta\varphi_k(\vec{x})} \tag{3.64}$$

is the generalized current density in the field space. In the symmetrized extension of relativistic bosonic quantum field theory we have thus a symmetrized quantum potential at two components of the form

$$Q(\{\varphi(\vec{x})\}, t) = \begin{pmatrix} -\frac{1}{2}\sum_k \int d^3x \frac{1}{R_1}\frac{\delta^2 R_1(\{\varphi(\vec{x})\}, t)}{\delta\varphi_k^2(\vec{x})} \\ \frac{1}{2}\sum_k \int d^3x \frac{1}{R_2}\frac{\delta^2 R_2(\{\varphi(\vec{x})\}, t)}{\delta\varphi_k^2(\vec{x})} \end{pmatrix} \tag{3.65}$$

where the second component is introduced to reproduce in the correct way also the time-reverse of the processes depending of Bose-Einstein fields in a quantum relativistic regime. The symmetrized quantum potential (3.65) means that, in relativistic bosonic quantum field theory, processes linked with Bose-Einstein fields are produced by a background space which acts as an immediate information medium.

In this picture, each component of the symmetryzed quantum potential can be considered as a physical quantity deriving from an opportune quantum entropy of fields indicating a degree of order and chaos of the background space which determines the processes. The non-local geometry of the quantum world in relativistic bosonic quantum field theory can be associated to a vacuum defined by a symmetrized quantum entropy at two components of the form

$$S_Q = \begin{pmatrix} -\frac{1}{2}\ln\rho_1 \\ -\frac{1}{2}\ln\rho_2 \end{pmatrix} \tag{3.66}$$

where the first component $S_{Q1} = -\frac{1}{2}\ln\rho_1$ is the quantum entropy for fields corresponding to the change of the geometric properties of physical space for the forward-time processes while the second component $S_{Q2} = -\frac{1}{2}\ln\rho_2$ is the quantum entropy for fields corresponding to the change of the geometric properties of physical space for the reversed-time processes (and has the function to reproduce in the correct way the processes if one would imagine to film them backwards). The symmetrized quantum potential can be thus written in compact form as

$$Q(\{\varphi(\vec{x})\},t) = -\frac{1}{2}\begin{pmatrix} \sum_k \int d^3x \frac{1}{R_1}\left[\left(\frac{\delta S_{Q1,f}}{\delta\varphi_k(\vec{x})}\right)^2 - \frac{\delta^2 S_{Q1,f}}{\delta\varphi_k^2(\vec{x})}\right] \\ \sum_k \int d^3x \frac{1}{R_2}\left[-\left(\frac{\delta S_{Q2,f}}{\delta\varphi_k(\vec{x})}\right)^2 + \frac{\delta^2 S_{Q2,f}}{\delta\varphi_k^2(\vec{x})}\right] \end{pmatrix} \tag{3.67}$$

and thus the quantum Hamilton-Jacobi equation (3.61) becomes:

$$
\begin{aligned}
&\frac{1}{2}\begin{pmatrix} \sum_k \int d^3x \left[-\frac{1}{2}\frac{\delta^2}{\delta\varphi_k^2(\vec{x})} + \frac{1}{2}\left|\nabla\varphi_k(\vec{x})\right|^2 \right] + V \\ \sum_k \int d^3x \left[-\frac{1}{2}\frac{\delta^2}{\delta\varphi_k^2(\vec{x})} + \frac{1}{2}\left|\nabla\varphi_k(\vec{x})\right|^2 \right] - V \end{pmatrix} \\
&-\frac{1}{2}\begin{pmatrix} \sum_k \int d^3x \frac{1}{R_1}\left[\left(\frac{\delta S_{Q1,f}}{\delta\varphi_k(\vec{x})}\right)^2 - \frac{\delta^2 S_{Q1,f}}{\delta\varphi_k^2(\vec{x})} \right] \\ \sum_k \int d^3x \frac{1}{R_2}\left[-\left(\frac{\delta S_{Q2,f}}{\delta\varphi_k(\vec{x})}\right)^2 + \frac{\delta^2 S_{Q2,f}}{\delta\varphi_k^2(\vec{x})} \right] \end{pmatrix} \\
&= -\begin{pmatrix} \frac{\partial S_1}{\partial t} \\ \frac{\partial S_2}{\partial t} \end{pmatrix}
\end{aligned} \tag{3.68}
$$

which can be considered as the fundamental equation of motion, in the symmetrized approach of bosonic field theory, which evidences the primary role of the fundamental background expressed by the symmetrized quantum entropy in determining the processes in this regime.

In this picture, the symmetrized quantum potential (deriving from the symmetrized quantum entropy) physically indicates that in the relativistic bosonic quantum field theory, a background space which acts as a direct information medium has a crucial role in determining the behaviour of systems, as a consequence of the symmetrized quantum entropy of fields describing the degree of order and chaos of the vacuum supporting the density of the fields, of the fact that the fields associated with the wave functional of the physical system determines a modification of the geometry.

Moreover, in order to evaluate the strength of quantum effects in this regime one can also introduce the symmetrized quantum-entropic length given by relation

$$L_{quantum,Bose-Einstein} = \left(\begin{array}{c} \dfrac{1}{\sqrt{\sum_k \int d^3x \dfrac{1}{R_1}\left(\left(\dfrac{\delta S_{Q1,f}}{\delta \varphi_k(\vec{x})}\right)^2 - \dfrac{\delta^2 S_{Q1,f}}{\delta \varphi_k^2(\vec{x})}\right)}} \\ \dfrac{1}{\sqrt{\sum_k \int d^3x \dfrac{1}{R_2}\left(-\left(\dfrac{\delta S_{Q2,f}}{\delta \varphi_k(\vec{x})}\right)^2 + \dfrac{\delta^2 S_{Q2,f}}{\delta \varphi_k^2(\vec{x})}\right)}} \end{array} \right). \quad (3.69)$$

The symmetrized quantum-entropic length (3.69) can be used to describe the modification of the geometry — introduced by Bose-Einstein fields in a quantum relativistic regime — with respect to the Euclidean geometry characteristic of classical fields and consequently to provide a measure of the degree of non-locality in this regime (and, of course, the second component is introduced in order to reproduce in the correct way also the time-reverse processes). It shows how the vacuum supporting the space-temporal distribution of the fields under consideration deforms the geometry of the processes inside a Bohmian approach to relativistic bosonic quantum field theory in a picture where the processes are reproduced in the correct way also if one imagines to film them backwards. Once the symmetrized quantum-entropic length (3.69) becomes non-negligible we have Bose-Einstein fields in a quantum relativistic regime. And here Heisenberg's uncertainty principle can be seen as a consequence of the fact that we are unable to perform a classical measurement to distances smaller than this symmetrized quantum-entropic length corresponding with Bose-Einstein fields in a quantum-relativistic regime.

In analogous way, as regards Nikolic's treatment of Bohmian particle trajectories in relativistic bosonic quantum field theory provided in the papers [148–151], the symmetrized quantum potential responsible of the geometrodynamics of the processes assumes the form

$$Q = -\begin{pmatrix} \dfrac{1}{2|\Psi|} \\ \dfrac{1}{2|\Phi|} \end{pmatrix} \int d^3x \begin{bmatrix} \left(\dfrac{\delta S_{Q1,f}}{\delta\phi(\vec{x})}\right)^2 - \dfrac{\delta^2 S_{Q1,f}}{\delta\phi^2(\vec{x})} \\ -\left(\dfrac{\delta S_{Q2,f}}{\delta\phi(\vec{x})}\right)^2 + \dfrac{\delta^2 S_{Q2,f}}{\delta\phi^2(\vec{x})} \end{bmatrix} \tag{3.70}$$

where $S_{Q1} = -\frac{1}{2}\ln\rho_1$ is the quantum entropy for fields which describes the geometric properties of the physical vacuum for the forward-time processes and $S_{Q2} = -\frac{1}{2}\ln\rho_2$ is the quantum entropy for fields which describes the geometric properties of physical space for the reversed-time processes. Thus, the laws describing the causal evolution of the field $\phi(\vec{x})$, for forward-time processes and time-reverse processes respectively, are the following

$$\begin{aligned}&(\partial_0^2 - \nabla^2 + m^2c^2)\phi(x)\\ &\quad = J(\phi(x)) + \frac{\delta}{\delta\phi(\vec{x})}\left(\frac{1}{2|\Psi|}\int d^3x\left[\left(\frac{\delta S_{Q1,f}}{\delta\phi(\vec{x})}\right)^2 - \frac{\delta^2 S_{Q1,f}}{\delta\phi^2(\vec{x})}\right]\right)_{\phi(\vec{x})=\phi(x)}\end{aligned} \tag{3.71}$$

$$\begin{aligned}&(\partial_0^2 - \nabla^2 + m^2c^2)\phi(x)\\ &\quad = J(\phi(x)) + \frac{\delta}{\delta\phi(\vec{x})}\left(\frac{1}{2|\Phi|}\int d^3x\left[-\left(\frac{\delta S_{Q2,f}}{\delta\phi(\vec{x})}\right)^2 + \frac{\delta^2 S_{Q2,f}}{\delta\phi^2(\vec{x})}\right]\right)_{\phi(\vec{x})=\phi(x)}.\end{aligned} \tag{3.72}$$

These two equations represent the two fundamental laws of motion which express the dependence of the behaviour of bosonic fields with the fundamental background associated to the symmetrized quantum entropy, in other words indicate how the symmetrized quantum entropy influences the evolution of the bosonic fields, in particular that this backgouund space acts as an immediate information medium.

Moreover, the geometrical properties of the fundamental background associated to the symmetrized quantum entropy can be characterized by introducing an opportune symmetrized quantum-entropic length given by relation

$$L_{quantum,Bose-Einstein} = \begin{pmatrix} \dfrac{1}{\sqrt{\int d^3x \dfrac{1}{|\Psi|}\left(\left(\dfrac{\delta S_{Q1,f}}{\delta\phi(\vec{x})}\right)^2 - \dfrac{\delta^2 S_{Q1,f}}{\delta\phi^2(\vec{x})}\right)}} \\ \dfrac{1}{\sqrt{\int d^3x \dfrac{1}{|\Phi|}\left(-\left(\dfrac{\delta S_{Q2,f}}{\delta\phi(\vec{x})}\right)^2 + \dfrac{\delta^2 S_{Q2,f}}{\delta\phi^2(\vec{x})}\right)}} \end{pmatrix}. \quad (3.73)$$

The symmetrized quantum-entropic length (3.73) allows us to measure, in the symmetrized extension of Nikolic's model (which reproduces in the correct way also the time-reverse of the processes) the modification of the geometrical properties and the corresponding degree of non-locality — introduced by bosonic fields in a quantum relativistic regime — with respect to the Euclidean geometry characteristic of classical fields. Once the quantum length (3.73) becomes non-negligible we have bosonic fields in a quantum relativistic regime.

Finally, as regards Nikolic's model of a Bohmian covariant interpretation for the many-fingered-time Tomonaga-Schwinger equation for relativistic quantum field theory (developed in the paper [152]), a symmetrized approach considers the symmetrized quantum potential expressed by relation

$$Q = -\frac{1}{2\,|q(\vec{s})|^{1/2}} \begin{pmatrix} \dfrac{1}{R_1} \\ \dfrac{1}{R_2} \end{pmatrix} \int d^3x \begin{bmatrix} \left(\dfrac{\delta S_{Q1,f}}{\delta\phi(\vec{s})}\right)^2 - \dfrac{\delta^2 S_{Q1,f}}{\delta\phi^2(\vec{s})} \\ -\left(\dfrac{\delta S_{Q2,f}}{\delta\phi(\vec{s})}\right)^2 + \dfrac{\delta^2 S_{Q2,f}}{\delta\phi^2(\vec{s})} \end{bmatrix}. \quad (3.74)$$

In equation (3.74) R_1 and R_2 are real amplitude functions of the wave functional ψ_1 and ψ_2 regarding the forward-time and reverse-time processes respectively, solutions of the time-symmetric Tomonaga-Schwinger equation

$$\begin{pmatrix} \hat{\mathrm{H}} & 0 \\ 0 & -\hat{\mathrm{H}} \end{pmatrix} \begin{pmatrix} \Psi_1 \\ \Psi_2 \end{pmatrix} = i\frac{\delta}{\delta T(\vec{x})} \begin{pmatrix} \Psi_1 \\ \Psi_2 \end{pmatrix} \tag{3.75}$$

where $\hat{\mathrm{H}}(\vec{x})$ the Hamiltonian density operator, $\delta T(\vec{x})$ represents an infinitesimal change of the timelike Cauchy hypersurface Σ where the dynamical field $\phi(\vec{x})$ is defined. Moreover, $S_{Q1} = -\frac{1}{2}\ln\rho_1$ is the quantum entropy for fields which describes the geometric properties of the physical vacuum for the forward-time processes and $S_{Q2} = -\frac{1}{2}\ln\rho_2$ is the quantum entropy for fields which describes the geometric properties of physical space for the reversed-time processes. $q(\vec{s})$ is the induced metric on Σ which makes manifestly covariant the theory. Here, the fundamental equation of motion is the covariant quantum many-fingered-time symmetrized Klein-Gordon equation, which assumes the following form:

$$\left(\left(\frac{\partial}{\partial\tau(\vec{s})}\right)^2 - \nabla^i\nabla_i + m^2\right)\Phi[\vec{s}, X]$$
$$= \left[\frac{1}{|q(\vec{s})|^{1/2}}\left(\begin{array}{c} \dfrac{\partial\left(\dfrac{1}{2R_1\,|q(\vec{s})|^{1/2}} \times \int d^3x \left[\left(\dfrac{\delta S_{Q1,f}}{\delta\phi(\vec{s})}\right)^2 - \dfrac{\delta^2 S_{Q1,f}}{\delta\phi^2(\vec{s})}\right]\right)}{\partial\phi(\vec{s})} \\ \dfrac{\partial\left(\dfrac{1}{2R_2\,|q(\vec{s})|^{1/2}} \times \int d^3x \left[-\left(\dfrac{\delta S_{Q2,f}}{\delta\phi(\vec{s})}\right)^2 + \dfrac{\delta^2 S_{Q2,f}}{\delta\phi^2(\vec{s})}\right]\right)}{\partial\phi(\vec{s})} \end{array}\right)\right]_{\phi=\Phi} \tag{3.76}$$

which puts in evidence the action of the background associated with the symmetrized quantum entropy in this regime. The geometrical properties of the background associated with the symmetrized quantum entropy, which determine the behaviour of bosonic fields in the many-fingered-time Tomonaga-Schwinger regime for relativistic quantum field theory, can be characterized by introducing a symmetrized quantum-entropic distance of the form:

$$L_{quantum,MFT-TS} = \left(\frac{\frac{1}{\sqrt{\frac{1}{2R_1 |q(\vec{s})|^{1/2}} \int d^3x \left[\left(\frac{\delta S_{Q1,f}}{\delta \phi(\vec{s})} \right)^2 - \frac{\delta^2 S_{Q1,f}}{\delta \phi^2(\vec{s})} \right]}}}{\frac{1}{\sqrt{\frac{1}{2R_2 |q(\vec{s})|^{1/2}} \int d^3x \left[- \left(\frac{\delta S_{Q2,f}}{\delta \phi(\vec{s})} \right)^2 + \frac{\delta^2 S_{Q2,f}}{\delta \phi^2(\vec{s})} \right]}}} \right). \tag{3.77}$$

The symmetrized quantum-entropic length (3.77) allows us to evaluate the modification of the geometrical properties and the corresponding degree of non-locality — introduced by bosonic fields in a three-dimensional many-fingered-time background in a quantum relativistic regime — with respect to the Euclidean geometry characteristic of classical fields in a picture where one can reproduce in the correct way the processes also if one imagines to film them backwards. Once the symmetrized quantum length (3.77) becomes non-negligible we have fields in a quantum relativistic regime of Tomonaga-Schwinger many-fingered-time background in a picture where the processes are correctly explained with the idea of a background space as an immediate information medium.

3.5 The Symmetrized Quantum Potential for Gravity in the Context of Wheeler-deWitt Equation

As regards the quantum gravity domain, since 2012 the author of this book and Sorli developed a toy mathematical model of quantum gravity in the context of WDW equation in which the idea of stage of processes as a direct, immediate information medium can be embedded. This model consists in introducing the considerations made in chapter 3.1 inside Wheeler-DeWitt equation and thus to build a symmetryzed version of the Bohmian approach to Wheeler-DeWitt equation. In this chapter, by following the reference [288], we want to analyse the fundamental results of this model (see also the book [297] for a review and analysys of these results).

As we have seen in chapter 1.7, in the context of WDW equation one can derive a fundamental entity describing the geometry of space, namely the "quantum potential for gravity" Q_G, given by equation (1.349), which is the crucial element that guides the behaviour of the universe. Just like the quantum potential of non-relativistic quantum mechanics, also the quantum potential of gravity (1.349) has an instantaneous, like-space action. As a consequence, if the original Bohm's quantum potential (1.9) can be associated — in the context of quantum non-locality in the subatomic world — with the idea of space as an immediate information medium between subatomic particles, in analogous way the quantum potential for gravity (1.349) — just in virtue of its non-local, instantaneous action — can be associated with the idea of a background space in the quantum gravity domain as an immediate information medium. The quantum potential for gravity (1.349) can be thus considered a appropriate candidate to characterize the special state of space in the quantum gravity domain as an immediate, direct information medium.

Moreover, taking into account that, just like it happens in the original Bohm's pilot wave theory, also the original Bohm's approach to quantum gravity cannot be considered completely convincing from the point of view of the mathematical symmetry in time in the sense that the standard quantum laws regarding WDW equation (and thus also Bohm's approach to WDW equation which derives from it) are not time-symmetric and therefore if one inverts the sign of time, the author and Sorli developed a symmetrized extension of WDW equation and a time-symmetric reformulation of the Bohmian approach to WDW equation. In this picture, the filming of a process backwards in the quantum gravity and the quantum cosmology corresponds correctly to what physically happens.

The time-symmetric extension of WDW equation has the form

$$\begin{pmatrix} H & 0 \\ 0 & -H \end{pmatrix} C = 0 \tag{3.78}$$

where

$$H = \left[(8\pi G) G_{abcd} p^{ab} p^{cd} + \frac{1}{16\pi G} \sqrt{g} (2\Lambda - {}^{(3)}R) \right] \tag{3.79}$$

and $C = \binom{\Psi}{\Phi}$, Ψ is the solution of the standard WDW equation, Φ is the solution of the time-reversed WDW equation.

The time-symmetric reformulation of the Bohmian approach to WDW equation in the light of the symmetryzed WDW equation (3.78) is obtained in analogous way to the program followed in paragraph 3.1 regarding the time-symmetric extension of Bohm's non-relativistic quantum mechanics. One starts by decomposing the time-symmetric WDW equation (3.78) into two real equations, by expressing the wave-functionals ψ and Φ in polar form:

$$\Psi = R_1 e^{iS_1}, \tag{3.80}$$

$$\Phi = R_2 e^{iS_2} \tag{3.81}$$

where R_1 and R_2 are real amplitude functionals and S_1 and S_2 are real phase functionals. If we substitute (3.80) and (3.81) into (3.78) and separate into real and imaginary parts we obtain the following quantum Hamilton-Jacobi equation for quantum general relativity

$$(8\pi G) G_{abcd} \begin{pmatrix} \dfrac{\delta S_1}{\delta g_{ab}} \dfrac{\delta S_1}{\delta g_{cd}} \\ \dfrac{\delta S_2}{\delta g_{ab}} \dfrac{\delta S_2}{\delta g_{cd}} \end{pmatrix} - \frac{1}{16\pi G} \sqrt{g} \begin{pmatrix} 2\Lambda - {}^{(3)}R \\ -2\Lambda + {}^{(3)}R \end{pmatrix} + \begin{pmatrix} Q_{G1} \\ Q_{G2} \end{pmatrix} = 0 \tag{3.82}$$

where

$$Q_{G1} = \hbar^2 N g G_{abcd} \frac{1}{R_1} \frac{\delta^2 R_1}{\delta g_{ab} \delta g_{cd}}, \tag{3.83}$$

$$Q_{G2} = -\hbar^2 N g G_{abcd} \frac{1}{R_2} \frac{\delta^2 R_2}{\delta g_{ab} \delta g_{cd}}. \tag{3.84}$$

In this way a symmetrized extension of Bohmian version of WDW equation emerges which is characterized by the following symmetryzed quantum potential for gravity at two components

$$Q_G = \begin{pmatrix} \hbar^2 N g G_{abcd} \dfrac{1}{R_1} \dfrac{\delta^2 R_1}{\delta g_{ab} \delta g_{cd}} \\ \\ -\hbar^2 N g G_{abcd} \dfrac{1}{R_2} \dfrac{\delta^2 R_2}{\delta g_{ab} \delta g_{cd}} \end{pmatrix}. \tag{3.85}$$

The first component (3.83) coincides with the original "quantum potential for gravity" (1.349): it can explain the forward-time process of the space-like, instantaneous action of quantum gravity. The second component (3.84) is considered in order to explain also the time-reverse of a process in the quantum gravity and cosmology domain through the instantaneous action (this second component guarantees that one could see what really happens if one would imagine to film the process backwards). In the light of its features, the symmetryzed quantum potential for gravity can be considered a good mathematical candidate for the state of space in the quantum gravity domain (in the context of WDW equation) that expresses a direct, immediate information medium. As it has been shown in the paper [288], the symmetryzed quantum potential for gravity implies that also in the WDW equation regime a fundamental stage of physical processes exists which acts as an immediate information medium.

Moreover, each component of the symmetryzed quantum potential for gravity can be considered as a physical quantity deriving from an opportune quantum entropy of gravity indicating a degree of order and chaos of the background space which determines the processes in the quantum gravity domain. The non-local geometry of the quantum gravity in WDW equation regime can be associated with a symmetrized quantum entropy of the form

$$S_Q = q \ln \begin{pmatrix} \rho_1 \\ \\ \rho_2 \end{pmatrix} \tag{3.86}$$

where $\rho_1 = R_1^2$, $\rho_2 = R_2^2$ are the densities of the particles associated with the wave functionals of the universe Ψ_1 and Ψ_2 regarding the forward-time and reverse-time processes respectively. With the introduction of the symmetrized quantum entropy (3.86), the symmetrized quantum potential for the gravitational field (3.85) can be expressed in the following entropic way:

$$Q_G = \begin{pmatrix} \hbar^2 N g G_{abcd} \dfrac{1}{R_1} \dfrac{\delta^2 R_1}{\delta g_{ab} \delta g_{cd}} S_{Q1} \\ -\hbar^2 N g G_{abcd} \dfrac{1}{R_2} \dfrac{\delta^2 R_2}{\delta g_{ab} \delta g_{cd}} S_{Q2} \end{pmatrix}. \tag{3.87}$$

On the basis of equation (3.87), the symmetrized quantum entropy for the gravitational field can be considered as the fundamental physical entity which determines the action of the quantum potential for the gravitational field. It indicates that the action of the quantum potential for gravity in the behaviour of the universe as a whole is instantaneously influenced by a vacuum characterized by a certain degree of order and chaos in a picture where one can reproduce in the correct way also the time-reverse of the processes.

The symmetrized quantum potential may be extended also to F. Shojai's and A. Shojai's model as well as Pinto-Neto's model of quantum cosmology and Farag Ali's and Das' model. In F. Shojai's and A. Shojai's model, one has a symmetrized quantum potential for gravity (3.87) which is determined by a symmetrized quantum entropy for the gravitational field indicating the degree of order and chaos of the gravitational field, and a symmetrized quantum potential for matter which is associated with a symmetrized quantum entropy for matter in terms of relation

$$Q_m = \hbar^2 \frac{N\sqrt{H}}{2} \begin{pmatrix} \left[\left(\dfrac{\delta S_{Q1,m}}{\delta \varphi_1} \right)^2 - \dfrac{\delta^2 S_{Q1,m}}{\delta \varphi_1^2} \right] \\ \left[-\left(\dfrac{\delta S_{Q2,m}}{\delta \varphi_2} \right)^2 + \dfrac{\delta^2 S_{Q2,m}}{\delta \varphi_2^2} \right] \end{pmatrix} \tag{3.88}$$

where $S_{Q1,m}$ and $S_{Q2,m}$ are the forward-time and time-reverse components of the symmetrized quantum entropy. Here, one has a fundamental vacuum which is described by the symmetrized quantum entropy for the gravitational field and by the symmetrized quantum entropy for matter. As a consequence, in this picture one has a quantum corrector tensor at two components $S^{\mu\nu}$ — which describes the modification, the deformation of the geometry of the physical space produced by matter and gravity in WDW equation's regime — in the sense that it depends on both the components of the symmetrized quantum entropy for gravity and on both the components of the symmetrized quantum entropy for matter, where the second components allow us to explain the processes correctly — with the idea of a non-local background which acts as a direct information medium — also if one imagines to film them backwards.

In Pinto-Neto's approach, the quantum potential reads

$$Q = -\begin{pmatrix} \dfrac{\hbar^2}{R_1} \\ \dfrac{\hbar^2}{R_2} \end{pmatrix} kG_{abcd}q \begin{pmatrix} \left(\dfrac{\delta^2}{\delta q_{ab}\delta q_{cd}}\right) S_{q1} \\ -\left(\dfrac{\delta^2}{\delta q_{ab}\delta q_{cd}}\right) S_{q2} \end{pmatrix}$$

$$- \begin{pmatrix} \dfrac{\hbar^2}{2R_1\sqrt{q}} \\ \dfrac{\hbar^2}{2R_2\sqrt{q}} \end{pmatrix} \begin{bmatrix} \left(\dfrac{\delta S_{Q1,m}}{\delta \varphi_1}\right)^2 - \dfrac{\delta^2 S_{Q1,m}}{\delta \varphi_1^2} \\ \left(-\dfrac{\delta S_{Q2,m}}{\delta \varphi_2}\right)^2 + \dfrac{\delta^2 S_{Q2,m}}{\delta \varphi_2^2} \end{bmatrix}. \tag{3.89}$$

Equation (3.89) indicates that the symmetrized quantum entropy associated to the geometrical properties of the vacuum of processes is the fundamental physical parameter which, with its peculiar features, makes the geometry of the ordinary space of our level of physical reality a non-degenerate four-geometry (which can be Euclidean or hyperbolic), or a degenerate four-geometry indicating the presence of special vector fields and the breaking of the space-time structure as a single entity (in a wider class of possibilities).

Finally, in Farag Ali's and Das' model regarding the link between the quantum potential and the cosmological constant in the context of a quantum corrected Raychaudhuri equation, one can think to extend the symmetrized approach, by considering a time-symmetric

Raychaudhuri equation of the form

$$\frac{d\theta}{d\lambda} = -\frac{1}{3}\theta^2 - R_{cd}u^c u^d + \frac{\hbar^2}{m^2}q^{ab}\left[\left[\begin{pmatrix} (\partial_\mu S_{Q1})^2 - \left(\nabla^2 - \frac{1}{c^2}\frac{\partial^2}{\partial t^2}\right) S_{Q1} \\ -(\partial_\mu S_{Q2})^2 + \left(\nabla^2 - \frac{1}{c^2}\frac{\partial^2}{\partial t^2}\right) S_{Q2} \end{pmatrix}\right]\right] + \frac{\varepsilon_1 \hbar^2}{m^2} q^{ab} R_{;ab}. \tag{3.90}$$

Here, if one considers a Gaussian form $\Psi \approx \exp(-r^2/L_0^2)$ or, in the context of a scalar field theory with an interaction of strength g, the wavefunction $\Psi = \Psi_0 \tanh(r/L_0\sqrt{2})$ (for $g > 0$) and $\Psi = \sqrt{2}\Psi_0 \sec h(r/L_0)$ (for $g < 0$) where L_0 is the characteristic length scale in the problem which is of the order of the Compton wavelength, the cosmological constant emerges directly from the symmetrized quantum entropy on the basis of relation

$$\frac{\hbar^2}{m^2}q^{ab}\left[\begin{pmatrix} \left[(\partial_\mu S_Q)^2 - \left(\nabla^2 - \frac{1}{c^2}\frac{\partial^2}{\partial t^2}\right) S_Q\right] \\ \left[-(\partial_\mu S_{Q1})^2 + \left(\nabla^2 - \frac{1}{c^2}\frac{\partial^2}{\partial t^2}\right) S_{Q2}\right] \end{pmatrix}\right] = \frac{1}{L_0^2} = (mc/\hbar)^2 \tag{3.91}$$

In this way, the mass m of gravitons (or axions), with gravity (or Coulomb field) following the Yukawa type of force law (1.375) may be regarded as the result of the fundamental symmetrized vacuum described by the symmetrized quantum entropy.

In summary, on the basis of the treatments provided in chapters 3.1, 3.2, 3.3, 3.4 and 3.5 one can therefore conclude that a symmetryzed Bohm's version of non-relativistic quantum mechanics, a symmetryzed extension of de Broglie-Bohm-theory in curved space-time, a symmetrized extension of relativistic quantum field theory and a symmetrized Bohm's version of WDW equation constitute relevant approaches which allow us to obtain significant results as regards the fundamental geometry of space generating the processes in their respective level of reality: both the wavefunctions of subatomic particles and the wave-functionals in relativistic quantum field theory regime and the wave functionals of the gravitational field in the quantum gravity domain in WDW equation regime determine a space medium, a special state of physical reality (represented, respectively, by an opportune symmetryzed quantum potential) which acts as a direct, immediate information medium in its respective domain, and which derives from a vacuum defined by an opportune symmetrized quantum entropy.

Chapter 4

The Quantum Potential ... and the Quantum Vacuum

4.1 The Space-time Arena and the Quantum Vacuum

If one assumes non-locality as the ultimate plot which characterizes physical world in quantum mechanics, one can think that there are two possible ways in order to develop a fundamental theory which describes the geometrodynamics and the arena of physical processes, exploring the actual behaviour of matter at the fundamental level:

a) the quantum geometry is assumed as primary and non-local, and therefore it is necessary to introduce additional hypotheses about its deep structure, or
b) the space-time manifold must be considered an emergence of the deepest processes situated at the level of quantum gravity. The germinal theory which introduced this type of research can be considered Sacharov's original proposal of deducing gravity as "metric elasticity" of quantum vacuum in which the action of space-time is interpreted as the effect of quantum fluctuations of the vacuum in a curved space [298]. Successively, a lot of research can be inserted in this context. In particular, we can mention Haisch's and Rueda's model regarding the interpretation of inertial mass and gravitational mass as effects of an electromagnetic quantum vacuum [299], Puthoff's polarizable vacuum

model of gravitation [300] and the most recent one by Consoli on ultra-weak excitations in a condensed manifold as a model for gravity and Higgs mechanism [301–303].

The image of the physical world suggested by the approaches a) and b) implies that the entire connected and local structure of both space-time and quantum information reveals themselves as the explicit order of a hidden, implicit order, in other words of a quantum vacuum which acts as a "fabric of reality" at a subquantum level, fundamentally discrete and non-commutative [141]. At the beginning of chapter 1.5 we have underlined that quantum field theory constitutes the most general syntax to build physical theories and, in particular, to describe interactions. Quantum field theory emerges as a realistic interpretation of quantum mechanics, showing that the quantum reality is fundamentally non-local in the sense that each physical event can be correlated with the others by an ultimate vacuum. In fact, a crucial consequence of contemporary quantum field theories (such as the quantum electrodynamics, the Weinberg-Salam-Glashow theory of electroweak interactions and the quantum chromodynamics of strong interactions) lies just in the existence of a unified quantum vacuum as a fundamental medium subtending the observable forms of matter, energy and space-time. In the light of the predictions of quantum field theories, the notion of an "empty" space devoid of any physical properties can be replaced with that of a quantum vacuum state, defined to be the ground state of a collection of quantum fields. In the picture of the physical vacuum, one can say that the physical space in which elementary particles and stellar objects move has origin in a fundamental quantum vacuum endowed with fundamental quantum fields that exhibit zero-point fluctuations everywhere in space, even in regions which are devoid of matter and radiation. In other words, contemporary quantum field theories imply that the foundation and origin of our tangible physical world is something totally abstract, comprising of discrete quantum fields that possess intrinsic, spontaneous, unpredictable quantum fluctuations. These zero-point fluctuations of the quantum fields, as well as other "vacuum phenomena" of quantum field theory, generate an enormous vacuum energy density.

The physical vacuum acts as a real quantum arena (a kind of quantum fluid) filling out the whole universal space and can be seen as an ever-changing collection of virtual particles which are created alone (photons) or in massive particle-antiparticle pairs, both of which are jumping in and out of existence within the constraints of the Heisenberg uncertainty principle [304]. The real particles such as electrons, positrons, photons, hadrons etc. as well as all macroscopic bodies can be considered as quantum wave-like excitations of this medium endowed with certain quantum numbers which have the role to ensure their relative stability. The fundamental particles involved at the basis of our daily physical reality are excitations of their respective underlying quantum fields possessing propagating states of discrete energies, and it is these which constitute the primary reality. The particles may be seen as bubbles guided by ocean waves. In other words, the ocean waves represent pilot-waves guiding the bubbles. Waves may be created and destroyed, but the ocean — namely the quantum vacuum — is eternal [305, 306].

As we have already mentioned in chapter 1.7.3, on the basis of the quantum field theories, various contributions to the vacuum energy density exist: the fluctuations characterizing the zero-point field, the fluctuations characterizing the quantum chromodynamic level of subnuclear physics, the fluctuations linked with the Higgs field, as well as perhaps other contributions from possible existing sources outside the Standard Model (for instance, Grand Unified Theories, string theories, etc...). On the other hand, since there is no structure within the Standard Model which suggests any relations between the different contributions to the quantum vacuum energy density, it is permissible to assume that the total vacuum energy density is, at least, as large as any of these individual contributions. Several authors recently made a detailed analysys of the role of the different contributions to the vacuum energy density, such as, for example, Rugh and Zinkernagel who studied the connection between the vacuum concept in quantum field theory and the conceptual origin of the cosmological constant problem [307], and Timashev who examined the possibility of considering the physical vacuum as a unified system governing the processes taking place in microphysics and macrophysics, which manifests itself on all space-time scales, from subnuclear to cosmological [308].

In this chapter, we will show that Bohm's quantum potential approach can be considered the real epistemological foundation of the idea of a unified quantum vacuum as a fundamental medium subtending the observable forms of matter, energy and space-time. We will see how, by assuming the non-local geometry associated with the quantum potential as the ultimate plot of quantum processes, the behaviour of a subatomic particle (such as, for example, the electron) in a given quantum experiment derives from elementary processes of creation/annihilation of quanta corresponding to elementary fluctuations of the energy density of a three-dimensional (3D) timeless non-local quantum vacuum. We will explore how the idea of a 3D timeless non-local quantum vacuum allows us to interpret the real background of physics as well as the meaning of "motion" of a subatomic particle.

4.2 Bohm's Quantum Potential Approach as the Epistemological Foundation of the Quantum Vacuum

As shown clearly by Hiley in [47], what one can desume from the analysis of Bohm's original 1952 papers is that the quantum potential implies a universal interconnection of things that could no longer be questioned, enables the global properties of quantum phenomena to be focused on the particle aspect where the 'particle' is not independent of the background. In this regard, the ultimate visiting card of Bohm's work has been the crucial role of non-locality to be introduced *ab initio* in the structural corpus of the physical theory, and the quantum potential, even in the plurality of the mathematical treatments, is in the core of such structure, the non-local *trait d'union* between the post-classical features of quantum physics and the most advanced perspectives of field theory [141, 144, 309].

In this way, in his research of 70s and 80s Bohm suggested the revolutionary view that quantum non-locality does not look like a field at all, but it is a plot "written" in the informational structure of a pre-space that Bohm and Hiley called just "Implicate Order". This deeper descriptive level of physical reality is revealed only partially, depending on the information the observer chooses to extract from the system, and gives to quantum mechanics its characteristic "contextuality" [310]. From a general epistemological point of view, the theory of Implicate/Explicate order can be considered as the first real attempt to realize the J.A. Wheeler program of *It from Bit* (or *QBit*), the possibility to describe the emergent features of space-time-matter as expressions, constrained and conveyed, of an informational matrix "at the bottom of the world" [311].

Here we claim the view according to which the fundamental, deep arena subtended by the geometrodynamic features of the quantum potential and which determines the processes regarding subatomic particles, is a three-dimensional (3D) timeless non-local quantum vacuum defined by elementary processes of creation/annihilation of quanta corresponding to elementary fluctuations of the energy density. In this regard, according to the author, the real starting point and epistemological foundation of the view of a 3D timeless non-local quantum vacuum as fundamental arena of processes can be drawn from the following famous excerpt of Bohm's "duel" with M. Pryce, a defender of the orthodox vision, broadcasted by BBC in 1962: "We wondered what actually an electron does. What should it do while it passes from the source to the slit? That's the point. Well, I could propose, for example, that the electron is not a particle in the sense it is currently meant, but an event. I assume such event happens in a generic medium — a "field" — we can suppose in this field there's an impulse. A wave moves forward and converges in a point so producing a very strong impulse and then diverges and scatters away. Let's imagine these impulses in a series all reaching a line there producing a series of intense pulses. The impulses will be very close one to the other, and so they will look like a particle. In most cases, all that will behave just like a particle and will behave differently when goes through the two slits, because each impulse will come out according to the way the incident wave passes the slits. The result is that we are looking at something it's neither a wave nor a particle. If you wonder how the electron has actually passed the slit and if it has really passed one slit or the other, I would reply that probably is not that kind of thing which can pass a slit or the other one. Actually, it is something which forms and dissolves continuously and that can be the way it really acts." [312]. According to the author of this book, as regards the double slit experiment, in the light of these last two sentences one can easily find the real most ultimate, crucial key in order to throw new light as regards the problem of the meaning of "motion" of a subatomic particle, of the existence of different descriptive levels of physical reality and thus of the possible deeper origin, at a fundamental level, of physical processes described by the geometrodynamic features of the quantum potential. Bohm's sentences listed above suggest clearly that the electron intended as a wave or a corpuscle (satisfying the well known laws of quantum theory) has not a primary physical existence but its "physical appearance" actually emerges from more elementary processes of formation and dissolving of quanta of a fundamental informational arena. According to the author, the idea of a fundamental quantum vacuum (in particular, in our model, of a fundamental 3D timeless non-local quantum vacuum) as the ultimate arena of the universe, which determines the behaviour of subatomic particles in terms of zero-point fluctuations corresponding to elementary processes of creation/annihilation, somewhat draws inspiration from (and indeed can be considered the natural development of) the physics contained in the sentences of Bohm's 1962 words at BBC listed above. In particular, in this regard, the key idea of our approach is to introduce Bohm's quantum potential into a 3D isotropic quantum vacuum characterized by a Planckian metric and defined by elementary processes of creation/annihilation of quanta.

4.3 From Bohm's View to Kastner's Possibilist Transactional Interpretation and Chiatti's and Licata's Approach about Transactions

Once one realizes that each subatomic particle, such as the electron, is something which forms and dissolves continuously and thus that its behaviour is generated by fluctuations corresponding to elementary processes of creation/annihilation of quanta of an informational vacuum, the fundamental problem becomes to describe mathematically and characterize conceptually this ultimate vacuum. Assuming non-locality as a fundamental feature of this vacuum, a first step in this regard could be to consider the possibilist transactional approach recently proposed R. Kastner in the references [313–316].

The transactional interpretation was originally introduced by John G. Cramer in a series of papers in the 1980s [317–320] on the basis of previous results obtained by theories of Wheeler-Feynman and of Hoyle-Narlikar. According to Cramer's approach, the behaviour of a subatomic particle such as the electron is a time-symmetric process, in which a given "absorber" generates confirmation waves in response to an emitted offer wave: a system emits a field in the form of half-retarded, half-advanced solutions to the wave equation, and the response of the absorber allows the creation of a process that transfers energy from the emitter to the absorber. Cramer's theory leads to introduce a vacuum which gives form to the anticipated waves; these inform a particle about the global situation in its space-time region and in this way one can explain the quantum behaviours, and in particular what happens in EPR-type experiments, in terms of information — the transaction events — coming from the future owed to the relativistic nature of the vacuum.

In the recent papers [313–316], Ruth Kastner extended Cramer's approach into a more general and fundamental theory, the possibilist transactional interpretation, which can explain the nature of the process leading to an actualized transaction. In Kastner's approach, space-time is not a pre-existing substance, a structured container for events, but rather unfolds as an emergent manifold from a more fundamental collective structure, a "pre-spacetime" characterized by "transactional processes" (constituted by emission and absorption of quanta) involving de Broglie waves. In Kastner's model, the behaviour of a subatomic particle such as the electron is the emergence of realized transactions resulting in transfers of energy from an emitter to an observer, which must be considered as pre-spacetime objects at the micro-level. Kastner's view implies that the transactions are processes that somewhat transcend the space-time structure, in other words are the expression of the non-local feature of the processes characterizing subatomic particles. In summary, in this approach space-time of classical physics, of the ordinary quantum mechanics as well as of quantum field theory with the enormous richeness of its manifestations derive from a network of non-local transactions, which therefore turn out to become the logical plot of the physical world.

Kastner's approach impies that each quantum particle unfolds from actualizations of transactions, and thus from emissions and absorptions of quanta, originating the related events taking place in the space-time arena. An important feature of the transactions is that the actualization of a transaction turns out to be an a-spatiotemporal process. In Kastner's possibilist transactional approach, the apparent four-dimensional space-time universe we perceive is not something "already there", there is no "space-time" without actualized transactions; rather, space-time with its dynamics and locality are emergent features which crystallize from an indeterminate (but real) non-local pre-spacetime of dynamical possibility. As a consequence, in the light of Kastner's theory, as regards the evolution of a subatomic particle in a given experiment, only the "now" is the fundamental empirical realm and that the changing properties of the physical system under consideration are associated with electromagnetic signals that transfer energy from what we are observing to our sense organs (by way of actualized transactions). In summary, according to the possibilist transactional interpretation, the now is the ultimate, primary reality of the universe, and the evolution of the particles may be recorded as changes of the now in the sense that it emerges from non-local transactional processes resulting in transfers of energy from an emitter to an observer.

Once one realizes that the non-local behaviour of subatomic particles can be associated to transactions — involving de Broglie waves and resulting in transfers of energy from an emitter to an observer — which must be considered as pre-spacetime objects at the micro-level, the next fundamental step is to describe mathematically and characterize conceptually the fundamental, ultimate informational vacuum represented by the network of these transactional processes. In this regard, an interesting way is provided by the approach of the archaic vacuum developed by Chiatti and Licata in the references [321–324].

By following the papers [321–324], in Chiatti's and Licata's model, the physical vacuum is not only an eigenstate of minimal energy but above all provides the network of all the possible transactions of the field modes generating an "undivided oneness". In the picture proposed by Chiatti and Licata, the physical vacuum must be considered as a radical non-local and event symmetric state: as a fundamental arena of the universe it reveals itself as an archaic, atemporal manifold that constrains and conveys the dynamical processes we observe in nature. Chiatti's and Licata's approach is based on the idea that the only truly existent "things" in the physical world are the events of creation and destruction (or, in other words, physical manifestation and demanifestation) of certain qualities [321]. Such properties' measurement is all that we know of the physical world from an operational view point [322]. In this picture, any other construction in physics — like the continuous space-time notion itself or the evolution operators — has the role to causally connect the measured properties: they are "emergent" with respect to the network of events. By making use of the language of quantum field theory the events of creation and destruction constitute the "interaction vertices", while the different sets of manifested/demanifested qualities in the same vertex represent the "quanta".

In the formal structure of Chiatti's and Licata's transactional theory, the probability of the occurrence of a creation/destruction event for a quantum Q in a point event x is associated to the probability amplitude $\psi_Q(x)$, which can be a spinor of any degree. Each component $\psi_{Q,i}(x)$ of this spinor obeys the Klein-Gordon quantum relativistic equation

$$(-\hbar^2\partial^\mu\partial_\mu + m^2c^2)\psi_{Q,i}(x) = 0, \tag{4.1}$$

where $\hbar$ is Planck's reduced constant and m is the mass of the quantum. At the non-relativistic limit, this equation becomes a pair of Schrodinger equations [325]:

$$-\frac{\hbar^2}{2m}\nabla^2\psi_{Q,i}(x) = i\hbar\frac{\partial}{\partial t}\psi_{Q,i}(x) \tag{4.2}$$

for creation events

$$-\frac{\hbar^2}{2m}\nabla^2\psi_{Q,i}(x) = -i\hbar\frac{\partial}{\partial t}\psi^*_{Q,i}(x) \tag{4.3}$$

for destruction events.

Equations (4.2) and (4.3) imply that there are no causal propagations from the future. In fact, equation (4.2) has only retarded solutions, which classically correspond to a material point with impulse $\mathbf{p}$ and kinetic energy $\mathbf{E} = \mathbf{p}\cdot\mathbf{p}/2m > 0$. Equation (4.3) has only advanced solutions, which correspond to a material point with kinetic energy $\mathbf{E} = -\mathbf{p}\cdot\mathbf{p}/2m < 0$. From the point of view of the dynamical laws for the probability amplitudes of these events, the creation of quality Q is associated with the initial condition for $\psi_{Q,i}(x)$ in equation (4.2); the destruction of quality Q is associated with the "initial", actually the final, condition for $\psi^*_{Q,i}(x)$ in equation (4.3).

As Chiatti showed in [321], in this approach the transactions, from a general point of view, can be defined in the following way. One has at time $t = t_1$ the event of the creation-destruction of a quality $Q(|Q\rangle\langle Q|)$ and at time $t = t_2$ the event of the creation-destruction of a quality $R(|R\rangle\langle R|)$. The link between these two processes is determined by a time evolution operator S which acts in the following way: $|Q\rangle$ is transported from S into $|Q'\rangle$ and projected onto $\langle R|$, $|R\rangle$ is transported by S^+ into $|R'\rangle$ and projected onto $\langle Q|$. The amplitudes product $\langle R|S|Q\rangle\langle Q|S^+|R\rangle = |\langle R|S|Q\rangle|^2$ provides directly the probability of the entire process. The result already seen with the non-relativistic expressions (4.2) and (4.3) can be obtained by moving to the representation of the coordinates and by substituting bras and kets with wavefunctions. Here, the propagation operator S^+ is as "real" as the S, in the sense that the initial condition $|Q\rangle$ and the final one $\langle R|$ are connected in a non-local way, in the light of the well-known non-local phenomena EPR and GHZ [326], confirmed by solid experimental evidence [327–330].

In Chiatti's and Licata's approach, the two extreme events of a transaction, which correspond to two reductions of the two state vectors which describe the evolution of the quantum process in the two directions of time, are also called "**R** processes" (where **R** stands for *reduction*) and here can be considered as the only real physical processes. They are constituted of interaction vertices in which real elementary particles are created or destroyed. Both the probability amplitudes in the two directions of time and the time evolution operators which act on them are mathematical entities whose only purpose is to describe the causal connection between the extreme events of the transaction, namely between **R** processes. The connection between **R** processes is something possible by an immediate, instantaneous action in the sense that is associated to the transformation of the same aspatial and atemporal "substratum". According to this approach, therefore, the history of the Universe and thus the behaviour of matter, considered at the ultimate level, are interpreted as a complete network of past, present and future **R** processes unfolding from the same invariant atemporal substratum. It looks as if every quantum process regarding subatomic particles (as well as time itself) emerge from this invariant atemporal substratum and re-absorbed within it.

Moreover, in this picture, Chiatti and Licata called this substratum which rules the actual behaviour of matter "archaic vacuum" because it must be distinguished from the quantum field theory traditional dynamic vacuum and containes all the self-consistency logical constraints which rule "the fabric of reality". In this atemporal substratum, at a fundamental level only the transactions between field modes exist and the quantum-mechanical wave-function simply emerges as a statistical coverage of a great amount of elementary transitions. This atemporal vacuum generates and conveys the dynamical processes regarding subatomic particles by means of **R** processes whose patterns influence the vacuum activity, in a quantum feedback.

4.4 From Transactions ... to a Three-dimensional Timeless Non-local Quantum Vacuum ... to the Emergence of the Geometrodynamics of Standard Quantum Theory from the Non-local Quantum Vacuum

If, in the transactional approach, the quantum behaviour of matter derives from a network of **R** processes, transactional processes defining an archaic, atemporal vacuum and if, in the light of contemporary quantum field theories, the unified quantum vacuum, as a fundamental medium subtending the observable forms of matter, energy and space-time, is characterized by zero-point fluctuations which generate an enormous vacuum energy density, how can the archaic vacuum energy density be described? And, in summary, what is the actual, physical link and the relationship, both from the mathematical and from the conceptual point of view, between archaic vacuum, vacuum energy density, non-locality and the geometrodynamic features of the quantum potential?

In order to provide an answer to these questions, before all we observe that as regards the quantum vacuum energy density which should give rise to the macroscopic space-time, the Planck energy density

$$\rho_{pE} = \sqrt{\frac{c^{14}}{\hbar^2 G^4}} \approx 4,641266 \cdot 10^{113}\, J/m^3 \cong 10^{97}\, Kg/m^3 \qquad (4.4)$$

is generally considered as the origin of the dark energy, if the dark energy is assumed to be owed to an interplay between quantum mechanics and gravity. However the observations are compatible with a dark energy given by relation

$$\rho_{DE} \cong 10^{-26} Kg/m^3. \qquad (4.5)$$

The dark energy (4.5) turns out to be 123 orders of magnitude larger than (4.4), thus generating the so-called "cosmological constant problem". An interesting explanation for the actual value (4.5) of the dark energy was provided by Santos in the papers [331, 332]. Santos' model invokes fluctuations of the quantum vacuum as the ultimate events which determine a curvature of space and, on the basis of plausible hypotheses within quantized gravity, establishes a relation between the two-point correlation function of the vacuum fluctuations

$$C(|\vec{r}_1 - \vec{r}_2|) = \frac{1}{2} \langle vac | \hat{\rho}(\vec{r}_1, t)\hat{\rho}(\vec{r}_2, t) + \hat{\rho}(\vec{r}_2, t)\hat{\rho}(\vec{r}_1, t) | vac \rangle \quad (4.6)$$

and the space curvature. In Santos' approach the dark energy ρ_{DE} emerges as the effect of the quantum vacuum fluctuations on the curvature of space-time according to equation

$$\rho_{DE} \cong 70G \int_0^\infty C(s) s ds \quad (4.7)$$

where $C(s)$ is a two-point correlation function of vacuum density fluctuations. Santos' model of quantized general relativity implies that the quantum vacuum fluctuations actually generate a curvature of space-time similar to the curvature produced by a "dark energy" density.

In the recent paper *Space-time curvature of general relativity and energy density of a three-dimensional quantum vacuum* [333], by following the philosophy that is at the basis of Santos' approach [331, 332], as well as of Haisch's and Rueda's model of inertial mass and gravitational mass as effects of the electromagnetic quantum vacuum [299], Puthoff's polarized vacuum model of gravitation [300] and Consoli's approach based on ultra-weak excitations in a condensed manifold in order to describe gravitation and Higgs mechanism [301–303], the author of this book and Sorli proposed a model of a 3D quantum vacuum in which the fluctuations of the quantum vacuum energy density generate a curvature of space-time similar to the curvature produced by a "dark energy" density and produce a shadowing of the gravitational space which determines the motion of the other material objects present in the region under consideration. Although this approach is in its germinal stages of development, it suggests the relevant perspective to interpret the curvature of space-time characteristic of general relativity as a mathematical value of a more fundamental actual energy density of quantum vacuum which has a concrete physical meaning.

On the basis of the treatment made in [333], in this model the consideration of a Planckian metric as starting-point leads directly to define, as fundamental arena of the universe, a 3D isotropic quantum vacuum composed by elementary packets of energy having the size of Planck volume and whose most universal property is the energy density. In this picture, the ordinary space-time we perceive is generated by this 3D isotropic quantum vacuum, which is assumed to be characterized by processes analogous to Chiatti's and Licata's transactions. By following the treatment of the paper [333], let us review now the fundamental ideas of this approach, in order to find which are the fundamental connections with the quantum behaviour of matter, non-locality and the geometrodynamic features of the quantum potential.

In the free space, in the absence of matter, the energy density of the 3D quantum vacuum is at its maximum and is given by the Planck energy density

$$\rho_{pE} = \frac{m_p \cdot c^2}{l_p^3} = \frac{2,1767 \cdot 10^{-8} \cdot (3 \cdot 10^8)^2}{(1,6161 \cdot 10^{-35})^3}$$
$$= \frac{19,5903 \cdot 10^8}{4,220896 \cdot 10^{-105}} = 4,641266 \cdot 10^{113} \frac{Kg}{ms^2}. \tag{4.8}$$

In the ordinary space-time we perceive, the appearance of matter endowed with a given mass m corresponds to a more fundamental diminishing of the quantum vacuum energy density. The energy density of quantum vacuum at the centre of this object is given by relation:

$$\rho_{qvE} = \rho_{pE} - \frac{m \cdot c^2}{V} \tag{4.9}$$

where V is the volume of the massive object under consideration. The variation of the quantum vacuum energy density $\Delta\rho_{qvE} = \rho_{qvE} - \rho_{pE}$ generates thus the appearance of a massive particle of mass m in a given region of the ordinary space-time on the basis of equation

$$m = \frac{V\Delta\rho_{qvE}}{c^2}. \tag{4.10}$$

Taking account of Santos' results, in the approach developed in [333] the quantized metric of the 3D quantum vacuum condensate is

$$d\hat{s}^2 = \hat{g}_{\mu\nu} dx^\mu dx^\nu \tag{4.11}$$

whose coefficients (in polar coordinates) are defined by equations

$$\hat{g}_{00} = -1 + \hat{h}_{00}, \quad \hat{g}_{11} = 1 + \hat{h}_{11}, \quad \hat{g}_{22} = r^2 + (1 + \hat{h}_{22}),$$
$$\hat{g}_{33} = r^2 \sin^2 \vartheta (1 + \hat{h}_{33}), \quad \hat{g}_{\mu\nu} = \hat{h}_{\mu\nu} \quad \text{for} \quad \mu \neq \nu \tag{4.12}$$

where multiplication of every term times the unit operator is implicit and, at the order $O(r^2)$, one has $\langle \hat{h}_{\mu\nu} \rangle = 0$ except $\langle \hat{h}_{00} \rangle = \frac{8\pi G}{3} \left(\frac{\Delta\rho_{qvE}}{c^2} + \frac{35Gc^2}{2\pi\hbar^4 V} \left(\frac{V}{c^2} \Delta\rho_{qvE}^{DE} \right)^6 \right) r^2$ and

$$\langle \hat{h}_{11} \rangle = \frac{8\pi G}{3} \left(-\frac{\Delta\rho_{qvE}}{2c^2} + \frac{35Gc^2}{2\pi\hbar^4 V} \left(\frac{V}{c^2} \Delta\rho_{qvE}^{DE} \right)^6 \right) r^2. \tag{4.13}$$

In equations (4.13), $\Delta\rho_{qvE}^{DE}$ are opportune fluctuations of the quantum vacuum energy density which give rise to the dark energy density on the basis of relation

$$\rho_{DE} \cong 70 \frac{G}{4\pi} \left(\frac{V}{c^2} \Delta\rho_{qvE}^{DE} \right)^2 \frac{1}{l} \cdot \frac{1}{l^3} \tag{4.14}$$

where

$$l = \frac{\hbar}{\left(\frac{V}{c^2} \Delta\rho_{qvE}^{DE} \right) c}. \tag{4.15}$$

The quantized metric (4.11) is associated with an underlying microscopic geometry expressed by equations

$$\Delta x \geq \frac{\hbar}{2\Delta p} + \frac{\Delta p}{2\hbar} (2\pi^2/3)^{2/3} l^{2/3} l_P^{4/3}, \tag{4.16}$$

which is the uncertainty in the measure of the position,

$$\Delta t \geq \frac{\hbar}{2\Delta E} + \frac{\Delta E T_0^2}{2\hbar}, \tag{4.17}$$

which is the time uncertainty and

$$\Delta L \cong \frac{(2\pi^2/3)^{1/3} l^{1/3} l_p^{2/3} T_0 E}{2\hbar} \tag{4.18}$$

which indicates in what sense the curvature of a region of size L can be related to the presence of energy and momentum in it. Equations (4.16)–(4.18) are obtained from the quantum uncertainty principle [334] and from the hypotheses of space-time discreteness at the Planck scale in the picture of Ng's treatment [335–338] in which the structure of the space-time foam can be inferred from the accuracy in the measurement of a distance l — in a spherical geometry over the amount of time $T = 2l/c$ it takes light to cross the volume — given by

$$\delta l \geq (2\pi^2/3)^{1/3} l^{1/3} l_P^{2/3}. \tag{4.19}$$

In the light of the quantized metric (4.11), the quantum Einstein equations

$$\hat{G}_{\mu\nu} = \frac{8\pi G}{c^4} \hat{T}_{\mu\nu} \tag{4.20}$$

(where the quantum Einstein tensor operator $\hat{G}_{\mu\nu}$ is expressed in terms in terms of the operators $\hat{h}_{\mu\nu}$) may be obtained directly: this means that the curvature of space-time characteristic of general relativity may be considered as a mathematical value which emerges from the quantized metric (4.11) and thus from fluctuations of the quantum vacuum energy density (on the basis of equations (4.12) and (4.13)).

Now, the fundamental point of our approach based on a 3D quantum vacuum defined by a quantized metric and by a variable energy density, which allows us to open a connection with the ideas illustrated in the previous paragraphs of this chapter, lies in the fact that the quantized metric (4.11) of the 3D quantum vacuum condensate (whose coefficients are directly associated with the fluctuations of the quantum vacuum energy density on the basis of equations (4.12) and (4.13)) is strictly tied to the quantum behaviour of subatomic particles deriving from transactional processes. In this regard, on the basis of the results obtained in the recent paper *Perspectives about quantum mechanics in a model of a three-dimensional quantum vacuum where time is a mathematical dimension* [339], in our model, in analogy with Chiatti's and Licata's approach, the elementary fluctuations of the quantum vacuum energy density can be physically associated to the events of preparation of the initial state (creation of a particle or object from the 3D quantum vacuum) and of detection of the final state (annihilation or destruction of a particle or object from the 3D quantum vacuum). The creation and annihilation of an elementary quantum can be considered as the two only primary extreme physical events of the 3D quantum vacuum and each corresponds to a peculiar reduction of a state vector (which is constituted of interaction vertices in which real elementary particles are created or destroyed). In our model, they are also called "**RS** processes" where **RS** stands for *state reduction.* These two quantum events of reduction, these two **RS** processes correspond, in our level of physical reality, to the evolution of the quantum process in the two directions of time (namely forward-time evolution and time-reversed evolution respectively, which can be associated with the opportune time evolution operators S and S^+ respectively of the Chiatti-Licata model, and where the time-reversed evolution is considered in order to reproduce correctly the things if one films the processes backwards).

Moreover, always following the treatment of [339], the model of 3D quantum vacuum developed by the author of this book and Sorli foresees that the behaviour and evolution of a particle or object — which is originated by fluctuations of the 3D quantum vacuum energy density — is generated by appropriate waves of the vacuum associated with the wavefunction which describes the amplitude of creation or destruction events associated with the corresponding fluctuations of the quantum vacuum energy density. These waves of the vacuum act in a non-local way through an appropriate quantum potential of the vacuum (which, so to speak, guides the occurring of the processes of creation or annihilation in the 3D quantum vacuum). The quantum potential of the vacuum can be considered as the primary mathematical reality which emerges from the very real extreme primary physical realities, namely from the processes of creation and annihilation of elementary quanta. The primary physical reality of the processes of creation and annihilation as well as the non-local action of the quantum potential which is associated with the amplitudes of them, make the 3D quantum vacuum, as a fundamental medium subtending the observable forms of matter, energy and space-time, a timeless background in which each **RS** process constitutes a self-connection.

The mathematical formalism of the **RS** processes and the quantum potential of the vacuum is the following. One starts by considering the probability amplitudes $\psi_{Q,i}(x)$ and $\phi_{Q,i}(x)$ which describe the probability of the occurrence of a creation event and destruction event, respectively, for a quantum particle Q of mass (4.10) in a point event x. The generic spinor $C = \binom{\psi_{Q,i}}{\phi_{Q,i}}$ satisfies a time-symmetric extension of Klein-Gordon quantum relativistic equation of the form

$$\begin{pmatrix} H & 0 \\ 0 & -H \end{pmatrix} C = 0 \tag{4.21}$$

where $H = (-\hbar^2\partial^\mu\partial_\mu + m^2c^2)$. Equation (4.21) corresponds to the two following equations

$$(-\hbar^2\partial^\mu\partial_\mu + m^2c^2)\psi_{Q,i}(x) = 0 \tag{4.22}$$

for creation events

$$(\hbar^2\partial^\mu\partial_\mu - m^2c^2)\phi_{Q,i}(x) = 0 \tag{4.23}$$

for destruction events which can also be conveniently written as

$$\left(-\hbar^2\partial^\mu\partial_\mu + \frac{V^2}{c^2}(\Delta\rho_{qvE})^2\right)\psi_{Q,i}(x) = 0 \tag{4.24}$$

for creation events

$$\left(\hbar^2\partial^\mu\partial_\mu - \frac{V^2}{c^2}(\Delta\rho_{qvE})^2\right)\phi_{Q,i}(x) = 0 \tag{4.25}$$

for destruction events respectively, where $m = \frac{V\Delta\rho_{qvE}}{c^2}$ is the mass of the quantum particle under consideration. At the non-relativistic limit, equation (4.21) becomes a pair of Schrödinger equations:

$$-\frac{\hbar^2}{2m}\nabla^2\psi_{Q,i}(x) = i\hbar\frac{\partial}{\partial t}\psi_{Q,i}(x) \tag{4.26}$$

for creation events

$$-\frac{\hbar^2}{2m}\nabla^2\phi_{Q,i}(x) = -i\hbar\frac{\partial}{\partial t}\phi^*_{Q,i}(x) \tag{4.27}$$

for destruction events, which read respectively

$$-\frac{\hbar^2 c^2}{2V\Delta\rho_{qvE}}\nabla^2\psi_{Q,i}(x) = i\hbar\frac{\partial}{\partial t}\psi_{Q,i}(x) \tag{4.28}$$

$$-\frac{\hbar^2 c^2}{2V\Delta\rho_{qvE}}\nabla^2\phi_{Q,i}(x) = i\hbar\frac{\partial}{\partial t}\phi^*_{Q,i}(x). \tag{4.29}$$

In a similar way to Chiatti's and Licata's transactional approach, the creation of a quantum particle Q of mass (4.10) is associated with the initial condition for $\psi_{Q,i}(x)$ in the equation (4.28), which has only retarded solutions, which classically corresponds to a material point with impulse $\mathbf{p}$ and kinetic energy $\mathrm{E} = \mathbf{p}\cdot\mathbf{p}/2m > 0$; the destruction of the quantum particle Q is associated with the "initial", actually the final, condition for $\psi^*_{Q,i}(x)$ in equation (4.29), which has only advanced solutions, which correspond to a material point with kinetic energy $\mathrm{E} = -\,\mathbf{p}\cdot\mathbf{p}/2m < 0$. In general, however, the two conditions are different and therefore generate different solutions for the two equations, which are not necessarily mutual complex conjugates.

The non-locality of the 3D quantum vacuum is determined by a quantum potential of the 3D quantum vacuum which can be introduced as the fundamental element which guides the occurring of the processes of creation and annihilation of quanta in the different regions of the 3D quantum vacuum. In this regard, by writing the two components of the spinor in polar form

$$\psi_{Q,i} = |\psi_{Q,i}| \exp\left(\frac{iS^{\psi}_{Q,i}}{\hbar}\right), \tag{4.30}$$

$$\phi_{Q,i} = |\phi_{Q,i}| \exp\left(\frac{iS^{\phi}_{Q,i}}{\hbar}\right) \tag{4.31}$$

and decomposing the real and imaginary parts of the Klein-Gordon equation (4.21), for the real part one obtains a couple of quantum Hamilton-Jacobi equations that, by imposing the requirement that they are Poincarè invariant and have the correct non-relativistic limit, assume the following form

$$\partial_\mu \begin{pmatrix} S^{\psi}_{Q,i} \\ S^{\phi}_{Q,i} \end{pmatrix} \partial^\mu \begin{pmatrix} S^{\psi}_{Q,i} \\ S^{\phi}_{Q,i} \end{pmatrix} = \frac{V^2}{c^2} \left(\Delta\rho_{qvE}\right)^2 \exp\begin{pmatrix} Q^{\psi}_{Q,i} \\ -Q^{\phi}_{Q,i} \end{pmatrix}, \tag{4.32}$$

while the imaginary part gives the continuity equation

$$\partial_\mu \left(\rho \partial^\mu \begin{pmatrix} S^{\psi}_{Q,i} \\ S^{\phi}_{Q,i} \end{pmatrix}\right) = 0 \tag{4.33}$$

where ρ is the ensemble of particles associated with the spinor under consideration and

$$Q_{Q,i} = \frac{\hbar^2 c^2}{V^2(\Delta\rho_{qvE})^2} \begin{pmatrix} \dfrac{\left(\nabla^2 - \dfrac{1}{c^2}\dfrac{\partial^2}{\partial t^2}\right)|\psi_{Q,i}|}{|\psi_{Q,i}|} \\ -\dfrac{\left(\nabla^2 - \dfrac{1}{c^2}\dfrac{\partial^2}{\partial t^2}\right)|\phi_{Q,i}|}{|\phi_{Q,i}|} \end{pmatrix} \tag{4.34}$$

is the quantum potential of the vacuum. In the non-relativistic limit, equation (4.32) becomes

$$\frac{c^2}{2V(\Delta\rho_{qvE})} \begin{pmatrix} |\nabla S^{\psi}_{Q,i}|^2 \\ |\nabla S^{\phi}_{Q,i}|^2 \end{pmatrix} + Q_{Q,i} + \begin{pmatrix} V \\ -V \end{pmatrix} = -\frac{\partial}{\partial t} \begin{pmatrix} S^{\psi}_{Q,i} \\ S^{\phi}_{Q,i} \end{pmatrix} \tag{4.35}$$

and equation (4.33) becomes

$$\frac{\partial}{\partial t}\begin{pmatrix}|\psi_{Q,i}|^2\\|\phi_{Q,i}|^2\end{pmatrix}+\nabla\cdot\begin{pmatrix}\frac{|\psi_{Q,i}|^2\,\nabla S^{\psi}_{Q,i}}{m}\\\frac{|\phi_{Q,i}|^2\,\nabla S^{\phi}_{Q,i}}{m}\end{pmatrix}=0 \qquad (4.36)$$

where

$$Q_{Q,i}=-\frac{\hbar^2 c}{2V(\Delta\rho_{qvE})}\begin{pmatrix}\frac{\nabla^2|\psi_{Q,i}|}{|\psi_{Q,i}|}\\-\frac{\nabla^2|\phi_{Q,i}|}{|\phi_{Q,i}|}\end{pmatrix} \qquad (4.37)$$

is the non-relativistic quantum potential of the vacuum.

Equations (4.34) and (4.37), both in the relativistic domain and in the non-relativistic domain, both for the processes of creation and for the processes of annihilation, indicate that the quantum potential of the vacuum has a non-local, instantaneous action. In other words, the processes of creation and annihilation of quanta in the different regions of the 3D quantum vacuum are choreographed by the quantum potential of the vacuum in a non-local, instantaneous manner. In sum, **RS** processes are characterized by a non-local connection as a consequence of the instantaneous action of the quantum potential guiding the evolution of the occurring of the processes of creation or annihilation of quanta (corresponding to opportune changes of the quantum vacuum energy density) in the different regions of the 3D quantum vacuum. The first component of the quantum potential regards the processes of creation of quanta, the second component regards the processes of annihilation of quanta in the 3D quantum vacuum.

In the light of the mathematical formalism of the **RS** processes of the 3D quantum vacuum constituted by equations (4.21)–(4.37), the following re-reading of the quantum behaviour of matter and of the quantum geometrodynamics of subatomic particles becomes permissible. The non-local action of the quantum potential of the vacuum may be considered as the ultimate structure which rules, at a fundamental level, the behaviour and the quantum geometrodynamics of subatomic particles: the behaviour of the matter in the universe can be seen as an undivided network of **RS** processes that take place in the 3D timeless quantum vacuum. This means, in other words, that the quantum potential of the vacuum may be considered as the real deep origin of the quantum geometrodynamics, which itself gives origin to the original quantum potential of Bohm's approach as expression of an upper level of reality.

In order to characterize in detail these aspects (and in particular the link between the quantum geometrodynamics ruled by Bohm's original approach and the quantum potential of the 3D quantum vacuum), according to the author, the crucial point is to take into consideration Sbitnev's view of the physical vacuum intended as a super-fluid medium, which contains pairs of particles-antiparticles giving rise a Bose-Einstein condensate. The discussion we make here is based on the paper [340].

By assuming that the **RS** processes of creation/annihilation of particles-antiparticles with opposite orientations of spins act as a super-fluid medium, the 3D quantum vacuum can be characterized by the following Einstein energy-momentum tensor

$$T^{\mu\nu} = (\varepsilon + p)u^{\mu}u^{\nu} + p\eta^{\mu\nu}. \tag{4.38}$$

In equation (4.38) ε and p are functions per unit volume expressed in units of pressure and the metric tensor $\eta^{\mu\nu}$ has the spacelike signature $(-,+,+,+)$. The energy-momentum tensor (4.38) leads directly to formulate the following conservation law

$$\partial_{\mu}(VT^{\mu\nu}/n) = 0 \tag{4.39}$$

where n is the number of the **RS** processes of virtual sub-particles characterizing the vacuum medium.

Now, by following the philosophy that underlines Sbitnev's hydrodynamic picture provided in [72], from equations (4.38)–(4.39) one obtains the first Fick's law in the relativistic limit in the form

$$j_{\mu} = -\frac{D}{c^2}\partial_{\mu}(\Delta\rho_{qvE}) \tag{4.40}$$

where D is a diffusion coefficient having the dimension of $length^2/time$. Equation (4.40) suggests that the diffusion flux vector emerges as a result of scattering of the sub-particles of the **RS** processes characterizing the vacuum on each other. In particular, it turns out to be proportional to the negative value of the gradient of the fluctuations of the quantum vacuum energy density. Now, the motion of the virtual particles constituting the 3D quantum vacuum generates in space-time a virtual radiation with frequency ω, which is linked to the diffusion coefficient through relation:

$$D = \frac{c^2}{\omega}. \tag{4.41}$$

Thus, taking account that the zero-point of the radiation is $E = \frac{\hbar\omega}{2}$, one has

$$D = \frac{\hbar c^2 n}{2\Delta\rho_{qvE} V}. \tag{4.42}$$

On the basis of equation (4.42), the diffusion coefficient associated with the scattering of the sub-particles of the **RS** processes characterizing the vacuum on each other turns out to depend of the number of the **RS** processes of virtual sub-particles characterizing the volume V of the vacuum medium into consideration and of the corresponding fluctuations of the quantum vacuum energy density. On the basis of equations (4.41) and (4.42), the frequency of the virtual radiation determined by the motion of the virtual particles of the **RS** processes corresponding to the elementary fluctuations of the energy density, may be expressed as

$$\omega = \frac{2\Delta\rho_{qvE} V}{\hbar n}. \tag{4.43}$$

In the light of equation (4.43), each elementary fluctuation of the quantum vacuum energy density in a given volume produces an oscillation of space at a peculiar frequency. This means that each material object given by mass (4.12) corresponds to oscillations of the 3D quantum vacuum given by equation (4.43).

Moreover, always following the treatment of [340], the collisions between the virtual particles of the vacuum give rise to a pressure p

which can be expressed as

$$p = p_1 + p_2 \tag{4.44}$$

where

$$p_1 = -\frac{D^2}{c^2}\left[\nabla^2 \Delta\rho_{qvE} - \frac{1}{c^2}\frac{\partial^2}{\partial t^2}\Delta\rho_{qvE}\right] \tag{4.45}$$

derives from Fick's law and

$$p_2 = \frac{D^2}{2\Delta\rho_{qvE}c^2}\left[(\nabla\Delta\rho_{qvE})^2 - \frac{1}{c^2}\left(\frac{\partial}{\partial t}\Delta\rho_{qvE}\right)^2\right] \tag{4.46}$$

is the pressure corresponding to the average momentum transfer per unit area per unit time. In this way, the quantum potential of the vacuum for creations events, namely associated with the **RS** processes of creation of quanta of the 3D quantum vacuum, may be written as:

$$\begin{aligned} Q = V\frac{p_1 + p_2}{n} &= -\frac{D^2}{c^2}\left[\nabla^2 \Delta\rho_{qvE} - \frac{1}{c^2}\frac{\partial^2}{\partial t^2}\Delta\rho_{qvE}\right] \\ &+ \frac{D^2}{2\Delta\rho_{qvE}c^2}\left[(\nabla\Delta\rho_{qvE})^2 - \frac{1}{c^2}\left(\frac{\partial}{\partial t}\Delta\rho_{qvE}\right)^2\right] \\ &= -\frac{\hbar^2 c^2 n}{2\Delta\rho_{qvE}V}\frac{\partial_\mu \partial^\mu R}{R} \end{aligned} \tag{4.47}$$

where R is the square root of the density distribution of the virtual particles in the vacuum (and, of course, D is the diffusion coefficient given by (4.42)). The quantum potential of the vacuum (4.47) describes the ultimate geometry of the vacuum generated by the pressures that arise between ensembles of virtual particles populating the vacuum.

Now, on the basis of equations (4.38) and (4.47) one has

$$\left(\frac{\varepsilon + p}{p}\gamma\right)\partial^\nu u_\nu u^\nu - \partial^\nu\left(\frac{\varepsilon + p}{p}\gamma\right) + \partial^\nu Q = 0, \tag{4.48}$$

where

$$\gamma = \left(1 - \frac{v^2}{c^2}\right)^{-1/2}. \tag{4.49}$$

The solution of (4.48) is

$$\frac{\Delta\rho_{qvE}V}{c^2 n}v^2 - \frac{\Delta\rho_{qvE}V}{c^2 n}c^2 + 2Q = 2C, \tag{4.50}$$

C being an integration constant having the dimension of energy.

Here, if one introduces the scalar field $S + \hbar\omega\gamma^{-1}t$, where S characterizes a degree of mobility of the virtual particles near the 4-point $(t, \vec{r})$, taking into account that the velocity can be expressed as

$$v^2 = \frac{c^4 n^2}{(\Delta\rho_{qvE})^2 V^2}\partial_\mu S \partial^\mu S + \frac{2Ec^3 n^2}{\gamma(\Delta\rho_{qvE})^2 V^2}\partial_0 S + \frac{E^2 c^2 n^2}{\gamma^2(\Delta\rho_{qvE})^2 V^2}, \tag{4.51}$$

one gets

$$\partial_\mu S \partial^\mu S + \frac{2E}{c\gamma}\partial_0 S + \frac{E^2}{c^2\gamma^2} - \frac{(\Delta\rho_{qvE})^2 V^2}{c^2 n^2} - \hbar^2\frac{\partial_\mu\partial^\mu R}{R} = 2\frac{(\Delta\rho_{qvE})V}{c^2 n}C. \tag{4.52}$$

At the relativistic limit one has $v \to c$, $\gamma \to \infty$ and thus $\frac{2E}{c\gamma}\partial_0 S + \frac{E^2}{c^2\gamma^2} \to 0$; in this regime, equation (4.52) becomes therefore

$$\partial_\mu S \partial^\mu S - \frac{(\Delta\rho_{qvE})^2 V^2}{c^2 n^2} - \hbar^2\frac{\partial_\mu\partial^\mu R}{R} = 2\frac{(\Delta\rho_{qvE})V}{c^2 n}C \tag{4.53}$$

which is a quantum Hamilton-Jacobi equation extracted from the Klein-Gordon type equation for the energy density of quantum vacuum (for creation events)

$$\partial_\mu\partial^\mu\psi + \frac{(\Delta\rho_{qvE})^2 V^2}{\hbar^2 c^2 n^2}\psi + 2\frac{(\Delta\rho_{qvE})V}{\hbar^2 c^2 n}C\psi = 0 \tag{4.54}$$

for the wavefunction $\psi = R\exp(iS/\hbar)$ of the quantum vacuum (for creation events). The Klein-Gordon equation of ordinary quantum mechanics may be obtained directly as a special case of (4.54) if one substitutes $m = \frac{\Delta\rho_{qvE}V}{c^2n}$ for the mass of the particle.

At the non-relativistic limit one has $v \to 0$, $\gamma \to 1$ and thus $\frac{E^2}{c^2} - \frac{(\Delta\rho_{qvE})^2V^2}{c^2n^2} = 0$; in this way, after some mathematical manipulations one obtains

$$\frac{\partial}{\partial t}S + \frac{c^2n}{2(\Delta\rho_{qvE_0})V}(\nabla S)^2 - \hbar^2\frac{c^2n}{2(\Delta\rho_{qvE_0})V}\frac{\nabla^2 R}{R} = C, \qquad (4.55)$$

where $\Delta\rho_{qvE_0}$ are the fluctuations of the quantum vacuum energy density at rest. Equation (4.55) is the quantum Hamilton-Jacobi equation which can be drawn from the Schrödinger-type equation for the energy density of quantum vacuum (for creation events)

$$i\hbar\frac{\partial\psi}{\partial t} = -\hbar^2\frac{c^2n}{2(\Delta\rho_{qvE_0})V}\nabla^2\psi - C\psi \qquad (4.56)$$

for the wavefunction $\psi = R\exp(iS/\hbar)$ describing the quantum vacuum (for creation events). Even here, the Schrödinger equation of ordinary quantum mechanics may be obtained directly as a special case of (4.56) if one replaces $m = \frac{\Delta\rho_{qvE}V}{c^2n}$ for the mass of the particle.

On the basis of equations (4.44)–(4.56), the following re-reading of the fundamental geometrodynamics of the quantum processes becomes natural. The quantum behaviour of matter represents the result, at an upper level, of a more fundamental geometry of a 3D quantum vacuum acting as a super-fluid medium, which consists of enormous amount of **RS** processes of creation/annihilation of virtual pairs of particles-antiparticles (thus constituting an organized Bose ensemble) and which is defined by the energy-momentum tensor (4.38). In other words, the level of ordinary quantum mechanics described by the standard Klein-Gordon and Schrödinger equations emerges from the deep, ultimate level of the 3D quantum vacuum described by the non-local action of the quantum potential of the vacuum. In fact, appropriate Hamilton-Jacobi equations — associated with Klein-Gordon and Schrödinger equations respectively — can be obtained directly as a result of the pressure due to the collisions of the virtual particles of the medium, which is linked with the fluctuations of the quantum vacuum energy density. In this picture, therefore, the fundamental point is that the quantum behaviour of matter as we know it from standard quantum theory derives ultimately from the quantum potential of the 3D quantum vacuum which describes the geometry of the vacuum through the pressures that are originated by the collisions between the virtual particles-antiparticles populating the vacuum and corresponding to the **RS** processes.

As a consequence, also the non-local correlations between the quantum particles associated with the ordinary quantum potential can be considered as an effect of the quantum potential of the quantum vacuum (4.47). On the basis of the approach of the 3D quantum vacuum analised in this chapter, the quantum-entropic (Bell) length (2.10) characterizing ordinary non-relativistic quantum mechanics (and based on the two quantum correctors to the energy of the system) may be generalized into a more general Bell length of the 3D quantum vacuum (intended as the ultimate parameter describing the fundamental geometry which is responsible of the non-local correlations between elementary regions of physical space) [340]. In fact, since Bohm's quantum potential emerges from the fundamental quantum potential of the vacuum, by comparing equations (4.47) and (2.2), one may make the positions

$$-\frac{\hbar^2}{2m}(\nabla S_Q)^2 = \frac{D^2}{2\Delta\rho_{qvE}c^2}\left[(\nabla\Delta\rho_{qvE})^2 - \frac{1}{c^2}\left(\frac{\partial}{\partial t}\Delta\rho_{qvE}\right)^2\right] \quad (4.57)$$

for the quantum corrector to the kinetic energy and

$$\frac{\hbar^2}{2m}(\nabla^2 S_Q) = -\frac{D^2}{c^2}\left[\nabla^2\Delta\rho_{qvE} - \frac{1}{c^2}\frac{\partial^2}{\partial t^2}\Delta\rho_{qvE}\right] \quad (4.58)$$

for the quantum corrector to the potential energy. In virtue of equations (4.57) and (4.58), the term $\frac{D^2}{2\Delta\rho_{qvE}c^2}[(\nabla\Delta\rho_{qvE})^2 - \frac{1}{c^2}(\frac{\partial}{\partial t}\Delta\rho_{qvE})^2]$ can be considered the origin of the quantum corrector to the kinetic energy and the term $-\frac{D^2}{c^2}[\nabla^2\Delta\rho_{qvE} - \frac{1}{c^2}\frac{\partial^2}{\partial t^2}\Delta\rho_{qvE}]$ can be considered the origin of the quantum corrector to the potential energy, of the physical system into consideration. In other words, in our approach, on the basis of the results obtained in [340], the quantum correctors to the energy of the system derive from more fundamental entities linked with the elementary fluctuations of the energy density of the 3D quantum vacuum. Equations (4.57) and (4.58) lead directly to express the quantities appearing in the Bell length (2.10) in terms of the parameters of the geometry of the 3D quantum vacuum as:

$$(\nabla S_Q)^2 = \frac{D^2 V}{c^4\hbar^2 n}\left[-(\nabla\Delta\rho_{qvE})^2 + \frac{1}{c^2}\left(\frac{\partial}{\partial t}\Delta\rho_{qvE}\right)^2\right] \tag{4.59}$$

and

$$(\nabla^2 S_Q) = -\frac{2D^2\Delta\rho_{qvE}V}{c^4\hbar^2 n}\left[\nabla^2\Delta\rho_{qvE} - \frac{1}{c^2}\frac{\partial^2}{\partial t^2}\Delta\rho_{qvE}\right] \tag{4.60}$$

respectively. In this way, the Bell length (2.10) of ordinary non-relativistic quantum mechanics may be considered as an emergent structure, which derives from the more general Bell length of the 3D quantum vacuum

$$L_{quantum} = \frac{c^2\hbar}{D\sqrt{\begin{array}{l}\frac{V}{n}\left[-(\nabla\Delta\rho_{qvE})^2 + \frac{1}{c^2}\left(\frac{\partial}{\partial t}\Delta\rho_{qvE}\right)^2\right.\\ \qquad \left. - \Delta\rho_{qvE}(\nabla^2\Delta\rho_{qvE} - \frac{1}{c^2}\frac{\partial^2}{\partial t^2}\Delta\rho_{qvE})\right]\end{array}}}. \tag{4.61}$$

The quantum length of the vacuum (4.61) can be considered as the ultimate parameter which generates the non-local correlations in the quantum domain. The non-local correlations between subatomic particles in the ordinary quantum theory is ultimately determined by the quantum length (4.61), which is the ultimate parameter characterizing the non-local and timeless features of the 3D quantum vacuum defined by **RS** processes of creation/annihilation of virtual particles-antiparticles organized in Bose ensembles and corresponding to fluctuations of the quantum vacuum energy density.

The maximum value of the Bell length of the 3D quantum vacuum (4.61), which implies the maximum de-localization of a quantum system, is 1, which means

$$\frac{c^2\hbar}{D\sqrt{\frac{V}{n}\left[-\left(\nabla\Delta\rho_{qvE}\right)^2+\frac{1}{c^2}\left(\frac{\partial}{\partial t}\Delta\rho_{qvE}\right)^2 -\Delta\rho_{qvE}\left(\nabla^2\Delta\rho_{qvE}-\frac{1}{c^2}\frac{\partial^2}{\partial t^2}\Delta\rho_{qvE}\right)\right]}}=1 \qquad (4.62)$$

namely

$$\frac{2\Delta\rho_{qvE}V}{n\sqrt{\frac{V}{n}\left[-\left(\nabla\Delta\rho_{qvE}\right)^2+\frac{1}{c^2}\left(\frac{\partial}{\partial t}\Delta\rho_{qvE}\right)^2 -\Delta\rho_{qvE}\left(\nabla^2\Delta\rho_{qvE}-\frac{1}{c^2}\frac{\partial^2}{\partial t^2}\Delta\rho_{qvE}\right)\right]}}=1 \qquad (4.63)$$

namely

$$\frac{2\Delta\rho_{qvE}V^{1/2}}{n^{1/2}\sqrt{\left[-\left(\nabla\Delta\rho_{qvE}\right)^2+\frac{1}{c^2}\left(\frac{\partial}{\partial t}\Delta\rho_{qvE}\right)^2 -\Delta\rho_{qvE}\left(\nabla^2\Delta\rho_{qvE}-\frac{1}{c^2}\frac{\partial^2}{\partial t^2}\Delta\rho_{qvE}\right)\right]}}=1. \qquad (4.64)$$

In this way, it is possible to obtain a simple relation satisfied by the number of virtual particles-antiparticles of the **RS** processes of the 3D quantum vacuum in the condition of maximum entanglement and de-localization in a quantum system:

$$n^{1/2}=\frac{2\Delta\rho_{qvE}V^{1/2}}{\sqrt{\left[-\left(\nabla\Delta\rho_{qvE}\right)^2+\frac{1}{c^2}\left(\frac{\partial}{\partial t}\Delta\rho_{qvE}\right)^2 -\Delta\rho_{qvE}\left(\nabla^2\Delta\rho_{qvE}-\frac{1}{c^2}\frac{\partial^2}{\partial t^2}\Delta\rho_{qvE}\right)\right]}}. \qquad (4.65)$$

Equation (4.65) expresses the mathematical condition which must be satisfied by the number of **RS** processes in a given volume of physical space in order to generate the maximum grade of non-locality in a quantum system having the mass $m=\frac{\Delta\rho_{qvE}V}{c^2 n}$ produced by the fluctuations of the quantum vacuum energy density corresponding to the same **RS** processes [340].

4.5 About the Behaviour of Subatomic Particles in the Three-dimensional Timeless Non-local Quantum Vacuum

In virtue of the fact that the quantum geometrodynamics of Bohm's quantum potential of ordinary quantum theory emerges from a more fundamental non-local geometry of a quantum potential of a 3D quantum vacuum defined by **RS** processes of creation/annihilation of quanta corresponding to elementary fluctuations of the quantum vacuum energy density, we can throw new light as regards what is the actual behaviour of a subatomic particle. As the author of this book showed in the paper [341], in the model presented in this chapter, the evolution of each subatomic particle (such as the electron in a double-slit interference) emerges from opportune elementary **RS** processes of creation/annihilation of quanta described by the quantum potential of the vacuum. The behaviour of a subatomic particle described by Bohm's approach of quantum mechanics turns out to be ultimately generated by a more fundamental evolution of elementary fluctuations of the quantum vacuum energy density corresponding to elementary **RS** processes of creation/annihilation of quanta in the context of a non-local and timeless geometry associated with a quantum potential of the vacuum.

Moreover, one can say that each elementary **RS** process of creation/annihilation of quanta in the 3D quantum vacuum has indeed a corpuscular and a wave nature. For example, electron can be seen as the result of opportune elementary **RS** processes of creation/annihilation of quanta which can be associated with opportune waves of the vacuum which evolve according to the general equations (4.24) and (4.25) (which become equations (4.28) and (4.29) respectively in the non-relativistic limit). The wave associated to the electron of the ordinary quantum level can be considered as the result of the waves of the vacuum. Because of its origin from the elementary fluctuations of the quantum vacuum energy density, electron can be therefore associated with appropriate waves of the vacuum which guide it in the different regions of the 3D quantum vacuum through the action of the quantum potential of the vacuum.

In this approach, the behaviour of a quantum particle as we know it from ordinary quantum mechanics actually derives from the equations describing the fundamental geometry of the 3D timeless non-local quantum vacuum. In particular, if we consider creation events, in the relativistic domain the fundamental equation of motion is the first of the quantum Hamilton-Jacobi equations (4.32), namely:

$$\partial_\mu S_{Q,i}\partial^\mu S_{Q,i} = \frac{V^2}{c^2}(\Delta\rho_{qvE})^2 \exp Q_{Q,i}, \tag{4.66}$$

where

$$Q_{Q,i} = \frac{\hbar^2 c^2}{V^2(\Delta\rho_{qvE})^2} \frac{\left(\nabla^2 - \frac{1}{c^2}\frac{\partial^2}{\partial t^2}\right)|\psi_{Q,i}|}{|\psi_{Q,i}|} \tag{4.67}$$

is the quantum potential of the vacuum. In the non-relativistic limit, equation (4.66) becomes

$$\frac{c\,|\nabla S_{Q,i}|^2}{2V\,(\Delta\rho_{qvE})} + Q_{Q,i} + V = -\frac{\partial S_{Q,i}}{\partial t} \tag{4.68}$$

where

$$Q_{Q,i} = -\frac{\hbar^2 c}{2V(\Delta\rho_{qvE})} \frac{\nabla^2\,|\psi_{Q,i}|}{|\psi_{Q,i}|} \tag{4.69}$$

is the (non-relativistic) quantum potential of the vacuum. Equations (4.66)–(4.69) allow us to provide the following interpretation of the evolution of subatomic particles: electrons and other elementary particles have precise positions at every time and follow precise trajectories which emerge from the evolution laws regarding creation events of quanta of the 3D quantum vacuum (4.66) and (4.68) and the appearance of these quanta in the various regions of this fundamental background is ruled by the corresponding quantum potentials (4.67) and (4.69) which indicate just the quantum force exerting on these quanta. In this way the 3D timeless non-local quantum vacuum model proposed in this chapter implies that the quantum potential of ordinary non-relativistic quantum mechanics (which is given by (1.4) for a one-body system) can be considered an emergent structure, which derives from the more fundamental quantum potential of the quantum vacuum of creation events (4.69) (and an analogous result regards the quantum potential of Klein-Gordon's relativistic quantum mechanics which derives from the more fundamental quantum potential of the 3D quantum vacuum (4.67)) [341].

4.6 Unifying Perspectives of the Non-local Quantum Potential of the Vacuum

The geometry of the 3D quantum vacuum described by the non-local quantum potential of the vacuum which guides the occurring of the **RS** processes of creation/annihilation of quanta and emerges from the pressures that arise between ensembles of virtual particles populating the vacuum, has also the merit to provide unifying perspectives of gravity, electromagnetic fields and quantum behaviour of matter as different aspects of the same fluctuations of the quantum vacuum energy density.

Taking account of equation (4.47), the frequency (4.43) associated with each elementary fluctuation of the quantum vacuum energy density is determined by the quantum potential of the vacuum on the basis of relation:

$$\omega = -\frac{\hbar c^2}{Q}\frac{\partial_\mu \partial^\mu R}{R} \tag{4.70}$$

where R is the square root of the density distribution of the virtual particles in the vacuum. The frequency (4.70) and thus the quantum potential of the vacuum may be considered to be the origin of the electromagnetic effects of the 3D quantum vacuum. In fact, the electromagnetic field inside a cavity of perfect reflecting can be seen as an expansion of infinite different modes of the fundamental 3D quantum vacuum, where each mode corresponds to an independent oscillation defined by frequency (4.70) determined by a specific **RS** process of creation/annihilation of quanta in correspondence to elementary fluctuations of the 3D quantum vacuum and thus by a specific behaviour of the quantum potential of the vacuum.

In the light of the results obtained in the paper [340], the electric and magnetic fields deriving from the zero-point fluctuations of the 3D quantum vacuum, in the stochastic electrodynamics (SED) approximation, can be written as

$$\vec{E}^{zp}_{\tau}(\vec{r},t) = \sum_{\lambda=1}^{2} \int d^3k(\Delta\rho_{qvE} V/n\pi^2)^{1/2} \times \hat{\varepsilon}(\vec{k},\lambda)\cos\left[\vec{k}\cdot\vec{r} - \frac{\hbar^2 c}{Q}\frac{\partial_\mu\partial^\mu R}{R}t - \theta(\vec{k},\lambda)\right], \quad (4.71)$$

$$\vec{B}^{zp}(\vec{r},t) = \sum_{\lambda=1}^{2} \int d^3k(\Delta\rho_{qvE} V/n\pi^2)^{1/2} \times [\hat{k}\times\hat{\varepsilon}(\vec{k},\lambda)]\cos\left[\vec{k}\cdot\vec{r} - \frac{\hbar^2 c}{Q}\frac{\partial_\mu\partial^\mu R}{R}t - \theta(\vec{k},\lambda)\right]. \quad (4.72)$$

According to relations (4.71) and (4.72), the electromagnetic radiations, which are expressed in expansion of plane waves, where the sum is over two polarization states, $\hat{\varepsilon}$ is a unit vector, $\vec{k}$ is the polarization vector such that $|\vec{k}| = \omega/c$ and $\theta(\vec{k},\lambda)$ is a random variable uniformly distributed in the interval $(0, 2\pi)$ and independently for each wave vector $\vec{k}$ and polarization index λ, are determined directly by the specific behaviour of the quantum potential of the vacuum. Now, if one considers the transformation from a stationary frame to a Rindler frame (characterized by a constant proper acceleration) which experiences an asymmetric event horizon leading to a non-zero electromagnetic energy and momentum flux and which is defined by the following velocity and Lorentz factor respectively

$$\frac{v}{c} = \tanh\left(\frac{a\tau}{c}\right), \quad (4.73)$$

$$\gamma_\tau = \cosh\left(\frac{a\tau}{c}\right), \quad (4.74)$$

where α is the object's proper acceleration and τ its proper time, one obtains the following expressions for the electric and magnetic zero-point fluctuations

$$\vec{E}_{\tau}^{zp}(0,\tau) = \sum_{\lambda=1}^{2} \int d^3k(\Delta\rho_{qvE}V/n\pi^2)^{1/2} \left\{ \hat{x}\hat{\varepsilon}_x + \hat{y}\cosh\left(\frac{a\tau}{c}\right) \right.$$
$$\times \left[\hat{\varepsilon}_y - \tanh\left(\frac{a\tau}{c}\right)\left(\hat{k}\times\hat{\varepsilon}\right)_z\right] + \hat{z}\cosh\left(\frac{a\tau}{c}\right)$$
$$\left. \times \left[\hat{\varepsilon}_z + \tanh\left(\frac{a\tau}{c}\right)\left(\hat{k}\times\hat{\varepsilon}\right)_y\right]\right\} \cdot$$
$$\cos\left[k_x\frac{c^2}{a}\cosh\left(\frac{a\tau}{c}\right) - \frac{\hbar c^3}{aQ}\frac{\partial_\mu\partial^\mu R}{R}\sinh\left(\frac{a\tau}{c}\right) - \theta\left(\vec{k},\lambda\right)\right], \tag{4.75}$$

$$\vec{B}_{\tau}^{zp}(0,\tau) = \sum_{\lambda=1}^{2} \int d^3k(\Delta\rho_{qvE}V/n\pi^2)^{1/2} \{ \hat{x}(\hat{k}\times\hat{\varepsilon})_x + \hat{y}\cosh\left(\frac{a\tau}{c}\right)$$
$$\times \left[\left(\hat{k}\times\hat{\varepsilon}\right)_y - \tanh\left(\frac{a\tau}{c}\right)(\hat{\varepsilon})_z\right]$$
$$\left. + \hat{z}\cosh\left(\frac{a\tau}{c}\right)\left[\hat{\varepsilon}_z + \tanh\left(\frac{a\tau}{c}\right)\left(\hat{k}\times\hat{\varepsilon}\right)_y\right]\right\} \cdot$$
$$\cos\left[k_x\frac{c^2}{a}\cosh\left(\frac{a\tau}{c}\right) - \frac{\hbar c^3}{aQ}\frac{\partial_\mu\partial^\mu R}{R}\sinh\left(\frac{a\tau}{c}\right) - \theta\left(\vec{k},\lambda\right)\right], \tag{4.76}$$

where $\hat{\varepsilon}_x$ is the scalar projection of the $\hat{\varepsilon}$ unit vector along the x−direction, and similarly for $\hat{\varepsilon}_y$ and $\hat{\varepsilon}_z$. On the basis of equations (4.71) and (4.72) (as well as of equations (4.75) and (4.76) for Rindler frames), electric and magnetic fields can be interpreted as two different kinds of polarization of the 3D quantum vacuum generated by the frequencies of the radiation associated with the motion of the virtual particles of the **RS** processes, namely by the fluctuations of the quantum vacuum energy density, and thus by the fundamental geometry of the quantum potential of the vacuum.

Now, if in the approach proposed by Haisch, Rueda and Puthoff in the papers [299, 342–346], the inertial mass and the gravitational mass are seen as effects of the electromagnetic quantum vacuum (inertia emerges as a kind of acceleration-dependent electromagnetic quantum vacuum drag force acting upon electromagnetically interacting elementary particles whilst gravitational mass — as manifest in weight — results from what may in a limited sense be viewed as acceleration of the electromagnetic quantum vacuum past a fixed object), in our approach, in the light of the electric and magnetic properties of the quantum vacuum expressed by relations (4.71) and (4.72) (and, in a Rindler frame, by relations (4.75) and (4.76)), the explanation of the weak equivalence principle provided by Haisch, Rueda and Puthoff may be derived, re-read, understood and justified in a simple, suggestive and unifying way. The view presented in this chapter suggests a picture where — contrary to Haisch's, Rueda's and Puthoff's model, characterized by an ad hoc high-frequency approximation — the frequencies defined by equation (4.70), and thus the non-local action of the quantum potential of the vacuum, represent the primary ontology. If Haisch, Rueda and Puthoff introduce a mesoscopic level as origin of the inertial and gravitational mass, here this mesoscopic level can be seen as a secondary ontological level which emerges from the primary existence of an enormous amount of **RS** processes of the 3D quantum vacuum and, therefore, from the fundamental non-local geometry of the quantum potential of the vacuum which rules their evolution. The virtual particles-antiparticles with opposite orientations of spins, associated with the **RS** processes, are characterized by motions which generate a virtual radiation defined by specific frequencies (given by (4.70)). Our approach of the 3D non-local quantum vacuum allows us to obtain the mesoscopic level considered by Haisch's, Rueda's and Puthoff's model as an emergent order on the basis of the consideration that the frequencies of the electromagnetic zero-point fields analysed by these three authors derive from more fundamental specific frequencies associated with the motions of the virtual particles of the **RS** processes, and thus from the fundamental geometry of the quantum potential of the vacuum, in agreement with the fundamental relation (4.70).

Here, the inertial mass of an object emerges from the interacting fraction of an energy density characterizing electromagnetic properties of the 3D quantum vacuum, which are determined by the frequencies associated with opportune **RS** processes of creation/annihilation, and corresponding to elementary fluctuations of the quantum vacuum energy density, according to equation

$$m_i = \left[4\frac{V^4}{\hbar^2\pi^2 n^3 c^5}\int \eta(\rho)(\rho_{pE}-\rho)^3 d\rho\right] \tag{4.77}$$

where c is the speed of light, $\eta(\omega)$ is the spectral factor, interacting with the body, which physically measures the relative strength of the interaction between the oscillations produced by the motions of the virtual particles of the **RS** processes and the massive object, an interaction which acts to oppose the acceleration. As a consequence, the explanation of the weak equivalence principle provided by Haisch, Rueda and Puthoff gets a new simple, suggestive and more unifying re-reading: here, the equivalence principle does not need to be independently postulated but derives directly as a consequence of the **RS** processes, and thus of the elementary fluctuations of the energy density, of the fundamental geometry of the same 3D quantum vacuum. In summary, in our approach of the 3D quantum vacuum, the physical basis for the principle of equivalence is the fact that accelerating through the regions of quantum vacuum characterized by the frequencies (4.70) of the radiation field determined by the motion of the virtual particles of the **RS** processes, is identical to remaining fixed in a gravitational field and having the electromagnetic proprieties of the 3D quantum vacuum fall past on curved geodesics. In both situations the observer will experience an asymmetry in the radiation pattern of the electromagnetic properties of the quantum vacuum which results in a force — either the inertia reaction force or weight — which becomes then the same thing within this more general perspective.

Moreover, as shown by the author of this book in [347], another important merit of the 3D quantum vacuum model presented in this chapter lies in the possibility to provide a new suggetive key of reading of the dark matter. The 3D quantum vacuum allows us to provide a unifying approach suggesting that the real explanation for the dark matter lies in the fluctuations of the 3D quantum vacuum energy density and thus on the fundamental geometry associated with the quantum potential of the vacuum. In order to explain dark matter, by following the paper [347], we make a generalization of the treatment made in chapter 4.4, by adding to the energy-momentum tensor (4.38) a term $\Pi^{\mu\nu}$ describing the viscosity of the vacuum:

$$T^{\mu\nu} = (\varepsilon + p)\, u^{\mu}u^{\nu} + p\eta^{\mu\nu} + \Pi^{\mu\nu}. \tag{4.78}$$

According to equation (4.78), the 3D quantum vacuum acts — in general — as a non-perfect fluid medium characterized by a given viscosity. The consideration of the 3D quantum vacuum as a non-perfect superfluid medium finds its justification in the fact that the energy-momentum tensor (4.78) can reproduce the existence of (other) large scale dynamical phenomena caused by forms of matter beyond the usual baryonic component.

If one considers the case of an incompressible vacuum, the conservation law of the energy-momentum tensor (4.78) takes the form

$$\partial_{\mu}(VT^{\mu\nu}/n) = \partial_{\mu}\left(\frac{V(\varepsilon+p)}{n}\gamma u^{\mu}u^{\nu}\right) + \partial^{\nu}Q + \partial_{\mu}\left(\frac{V\mu(t)}{n}\right)\pi^{\mu\nu} = 0 \tag{4.79}$$

where

$$\pi^{\mu\nu} = c(\partial^{\mu}u^{\nu} + \partial^{\nu}u^{\mu}) - c\frac{2}{3}\partial^{\mu}u_{\mu}\eta^{\mu\nu} \tag{4.80}$$

and the parameter $\mu(t)$, a fluctuating quantity about zero, describes the dispersion of the viscosity average. Fluctuations of the viscosity about zero are responsible of the exchanging energy of the orbital rotation with the zero-point fluctuations of the physical vacuum on the ultra-low frequencies. In agreement with the treatment provided by Sbitnev in [72], from equation (4.79) the following relativistic form of the Navier-Stokes-type equation for the states of the 3D quantum vacuum as non-perfect fluid may be obtained

$$\frac{V(\varepsilon+p)}{n}\gamma\left(\frac{1}{c}\frac{\partial v}{\partial t} + (\vec{v}\cdot\nabla\vec{v})\right) + \nabla Q - \frac{v}{c}\frac{\partial Q}{\partial t}$$
$$+\frac{\partial(V\mu(t)/n)}{\partial t}\cdot(\pi^{0,i} - v\pi^{0,0}) + \frac{V\mu(t)}{n}(\partial_{\mu}\pi^{\mu,i} - \vec{v}\partial_{\mu}\pi^{\mu,0}) = 0, \tag{4.81}$$

whose non-relativistic limit is

$$\frac{(\Delta\rho_{qvE_0})V}{c^2 n}\left(\frac{\partial\vec{V}}{\partial t}+\left(\vec{V}\cdot\nabla\vec{V}\right)\right)=\nabla Q+\frac{V\mu(t)}{n}\nabla^2\vec{V} \tag{4.82}$$

where $\vec{V} = c\vec{v}$ is the real velocity of the non-relativistic fluid associated with the fluctuations of the quantum vacuum. By defining $\vec{\omega} = \nabla\times\vec{V}$ the vorticity of the vacuum, the curl operator of equation (4.82) yields

$$\frac{\partial\vec{\omega}}{\partial t}+(\vec{\omega}\cdot\nabla)\vec{V}=\frac{\mu(t)c^2}{\Delta\rho_{qvE_0}}\nabla^2\vec{\omega} \tag{4.83}$$

where the kinetic viscosity $\frac{\mu(t)c^2}{\Delta\rho_{qvE_0}}$ has the physical dimension of the diffusion constant (4.41).

Now, a interesting perspective opened by the relativistic hydrodynamics based on equations (4.78)–(4.83), which characterizes the state of the 3D quantum vacuum in general (namely when the viscosity term must be taken into consideration), is represented by the description of vortices arising in such a non-perfect fluid vacuum. In fact, by considering a vortex tube having the cross-section oriented along the z-axis and its centre placed in the coordinate origin of the plane (x;y), equation(4.83) may be expressed as

$$\frac{\partial\omega}{\partial t}=\frac{\mu(t)c^2}{\Delta\rho_{qvE_0}}\left(\frac{\partial^2\omega}{\partial r^2}+\frac{1}{r}\frac{\partial\omega}{\partial r}\right). \tag{4.84}$$

The solution of equation (4.84) reads

$$\omega(r,t)=\frac{\Gamma}{\Sigma(t)}\cdot\exp\left(-\frac{r^2}{\Sigma(t)}\right) \tag{4.85}$$

and thus the orbital speed becomes

$$V(r,t)=\frac{\Gamma}{2r}\left(1-\exp\left[-\frac{r^2}{\Sigma(t)}\right]\right) \tag{4.86}$$

where Γ is an integration constant and the denominator is

$$\Sigma(t) = 4\left(\int_0^t \frac{\mu(t)c^2}{\Delta\rho_{qvE_0}} dt' + \sigma^2\right), \tag{4.87}$$

σ being an arbitrary constant such that the denominator is always positive. In particular, if, for sake of simplicity, one makes the choice

$$\mu(t) = \mu\cos(\Omega t) \tag{4.88}$$

one obtains

$$\Sigma(t) = 4\left(\frac{\mu c^2}{\Omega\Delta\rho_{qvE_0}}\right)\sin(\Omega t) + \sigma^2 \tag{4.89}$$

where Ω is an oscillation frequency and $\mu = \hbar/2$. The vorticity (4.85) and the orbital speed (4.86) turn out to oscillate about some average values limited by the constant σ [70, 71]. The dynamics of the vortices described by equations (4.84)–(4.89) implies that the vortices seem to be characterized by pulsations, as a consequence of the annihilation-creation processes of the virtual particle–antiparticle pairs (and thus of the fundamental geometry corresponding to the quantum potential of the vacuum, in the light of frequency (4.70) and of the presence of the quantum potential in equations (4.81) and (4.82)). Moreover, because of the fact that averaged in time the viscosity coefficient vanishes, but its dispersion is not zero, the vortices can live infinitely long and their radius trembles because of the exchange of the vortex energy with the quantum vacuum fluctuations.

The formalism of the states of the quantum vacuum depending of the viscosity, based on equations (4.78)– (4.89), has also the merit to provide a description of the motion of the spiral galaxies. Simulation of the vortex exhibits a spiral form which manifests a similarity with that of a galaxy, which has a core that does not exceed the radius $r^* \approx 5,29 \cdot 10^{-11} m$ where the point of inflection of the orbital velocity takes place and we see agglomeration of the slowly rotating matter. The stars in the galaxy are subjected to a rotation around the galactic core with almost constant speed even being located far from the galactic core. Really, the orbital speed of stars and gas, given by equation (4.86), has origin inside the core and reaches almost constant level outside the core in agreement with the results obtained in [348].

As regards the features of the viscosity, one can assume the existence of a wide spectrum of the viscosity coefficients and that these viscosity coefficients are discrete with equidistant position of each component and are condensed in the point $\Omega = 0$:

$$\Sigma_n(t) = 4\left(\frac{\mu c^2}{\Omega_n^2 \Delta\rho_{qvE_0}} \sin(\Omega_n t) + \sigma_n^2\right). \tag{4.90}$$

In the light of equation (4.90), the strongest contribution to the vorticity gives modes with frequencies close to zero. Moreover, equation (4.90) leads directly to the following relations for the vorticity and the orbital speed:

$$\omega(r,t) = \frac{\Gamma}{N}\sum_{n=1}^{N} \frac{1}{\Sigma_n(t)} \exp\left(-\frac{r^2}{\Sigma_n(t)}\right), \tag{4.91}$$

$$V(r,t) = \frac{\Gamma}{2rN}\sum_{i=1}^{N}\left(1 - \exp\left[-\frac{r^2}{\Sigma_n(t)}\right]\right). \tag{4.92}$$

Here, if one chooses $\Gamma = 10^{27} m^2/s$, $\Omega_n = 10^{-11} s^{-1}$, and $N = 25$ (in such a way that the parameter $\sigma_n = 4c/\Omega_n$ ranges from 10000 to 300000 light years, which covers the diameter of the ordinary spiral galaxies), relations (4.91) and (4.92) provide an explanation into a hydrodynamical picture of the rotating motion of the spiral galaxies reproducing the observed flattening of the orbital speeds. More precisely, the thermalization of the vorticity and the angular velocities of spiral galaxies turn out to be the result of the energy exchange of baryonic matter with the 3D quantum vacuum in regime of ultra-low frequencies (which, in terms of wavelengths, cover just almost entirely the diameter of the ordinary spiral galaxies). In other words, fluctuations of the viscosity about zero describes exchanging energy of the orbital rotation with the zero-point fluctuations of the 3D quantum vacuum on the ultra-low frequencies (ultimately associated with specific behaviours of the quantum potential of the vacuum).

In regime of ultra-low frequencies, exchange of the energy of the rotating galactic matter with the quantum vacuum fluctuations, ultimately determined by the geometry of the non-local quantum potential of the vacuum, because of the term $\Sigma_n(t)$, generates an effect that looks as a breathing of the galaxy. The orbital speed (4.92) is characterized by small fluctuations in time, corresponding to the breathing of the galaxy. In other words, the breathing of the galaxy manifests itself in variations of the vorticity. On the ultra-low frequencies variations of the vorticity take place because of the virtual particle-antiparticle pairs, a phenomenon that, according to the standard cosmological model, is believed to be derived from dark matter (see, for example, the reference [349]). In this way, the quantum vacuum fluctuations (and therefore the fundamental geometry of the quantum potential of the vacuum) on the ultra-low frequencies can explain the observed flattening of the orbital speeds of the spiral galaxies (on radii far exceeding the radius of the galactic core) without invoking the idea of dark matter. To sum up, the 3D quantum vacuum model here considered introduces the suggestive perspective that dark matter does not exist as a primary physical reality but derives just from opportune quantum vacuum energy density fluctuations, associated to virtual pairs of particles-antiparticles, in a 3D quantum vacuum characterized by a fluctuating viscosity (and thus from the fundamental geometry associated with specific behaviours of the quantum potential of the vacuum). In this regard, we can find a good agreement with (and provide a new key of reading of) the results obtained by Hajdukovic as regards dark matter. In fact, Hajdukovic explicitly declares: "Assuming that a particle and its antiparticle have the gravitational charge of the opposite sign, the physical vacuum may be considered as a fluid of virtual gravitational dipoles, that allow the gravitational polarization of the quantum vacuum. Following this hypothesis [...] we have revealed the first indications that what we call dark matter may be consequence of the gravitational repulsion between matter and antimatter and the corresponding gravitational polarization of the quantum vacuum by the existing baryonic matter. [...] The gravitational vacuum polarization could be an alternative to dark matter in the explanation of several phenomena such as the galactic rotational curves. [..] Dark matter does not exist but is an illusion created by the polarization of the quantum vacuum by the gravitational field of the baryonic matter. Hence, for the first time, the quantum vacuum fluctuations, well established in quantum field theory but mainly neglected in astrophysics and cosmology, are related to the problem of dark matter." [350]. The description of the motion of spiral galaxies as well as the explanation of the origin of dark matter as phenomena deriving from the 3D quantum vacuum energy density — and thus from the fundamental geometry associated with specific behaviours of the non-local quantum potential of the vacuum — may be considered as other important results of the 3D quantum vacuum model analysed in this chapter.

Conclusions

Despite 20th century theoretical physics has opened important perspectives in the exploration of new territories (such as the meaning of matter at the Planck scale and the role of the quantum information), it is characterized by significant foundational problems which the inner unity of physics knowledge depend on. In the light of the topics of quantum physics analysed in this book, relevant foundational problems regard the arena of the universe and the fundamental geometry which are subtended by the quantum potential.

Bohm's quantum potential emerges as the crucial element which allows us to portray a geometrodynamic picture in the different contexts of quantum physics. It is a geometric entity, it contains an information woven into space-time, both in the non-relativistic domain and in the relativistic domain and in relativistic quantum field theory and in the quantum gravity domain and in quantum cosmology. However, if one takes into consideration the recent developments about the geometrodynamic features of the quantum potential in the different contexts of quantum physics, the following fundamental questions become natural. What must be considered the real, ultimate arena of quantum processes? What is the real background of physics? What is space? What is the geometry that rules the behaviour of a subatomic particle or a physical field at the most fundamental level? What is matter? What is the meaning of "moving"? These are some of the foundational questions that one must answer in order to develop a novel satisfactory picture of the subatomic world and find a new unifying ontology in quantum physics.

The treatment made in this book suggests that the quantum potential introduces relevant perspectives about the fundamental arena of quantum processes, that a significant link exists between the geometrodynamic action of the quantum potential and a fundamental quantum vacuum, in the different contexts of quantum physics. A primary quantum vacuum defined by specific geometrical properties emerges as the ultimate arena which rules and determines the behaviour of quantum systems.

In this regard, already Bohm's implicate order and Hiley's pre-space can be considered as non-local monistic models in which "the whole is prior to its parts, and thus the cosmos is viewed as fundamental, with metaphysical explanation which is drawn from the One" [351]. In Bohm's and Hiley's view non-locality is a characteristic subtended of space-time and quantum particles are seen as vibration modes of the global field which is the dynamical expression of a fundamental level, of the deep geometrical structure associated with the implicate order. Since the first attempts to develop models about the underlying fundamental level of quantum reality represented by the implicate order, there has been a considerable amount of mathematical work exploring a possible deeper structure than space-time of which Bohm was unaware. These attempts have become better known to the physics community under the term 'non-commutative geometry'. In the light of these theoretical constructions, the space-time of the classical world would be some statistical approximation and not all quantum processes can be projected into this space without producing the familiar paradoxes, including non-separability and non-locality. According to the research of the last Bohm and Hiley's research, quantum domain is to be regarded as a structure or order evolving in space-time, but space-time is to be regarded as a higher order abstraction arising from this process involving events and abstracted notions of space or space-like points. Moreover, Hiley's monistic approach of pre-space extends Bohm's interpretation of quantum mechanics to the relativistic regime and allows a sort of unification between space-time geometry and material processes. In fact, on the basis of Hiley's view, things do not take place in a background space-time intended as the fundamental arena: Hiley's research has the aim to "start from something more primitive from which both geometry and material process unfold together" [136].

In Hiley's approach, all the macroscopic manifestations of the physical reality which come into light in the measurement outcomes are ultimately generated from something more "primitive" and fundamental, namely the "holomovement", which is characterized just by two intertwined and correlated aspects: the "implicate order" (described algebraically) and all the physics constructions derived from it, such as space-time geometry, which represent the "explicate order". In other words, in Hiley's view, the holomovement can be regarded the primary physical reality, the primitive elementary process, which contains as its constituting parts implicate order and explicate order, where the latter expresses aspects of the former.

In the light of the research of the last Bohm as well as of Hiley's results, from the perspective of the approach of the implicate order, particles and pilot waves cannot be considered fundamental physical realities but are emergent from the implicate order. In this view, the space-time as well as the matter manifestations of the classical world represent only statistical approximations, abstractions of higher order which arise from a structure process of more fundamental nature, and here it is just this fundamental arena the key of quantum non-locality. The space and time structures are themselves derived from the fundamental process. To sum up, Bohm's and Hiley's research on implicate order and pre-space imply that the mutual interaction between implicate order and explicate order determined by the holomovement has to be taken as fundamental whilst space-time, fields and matter are indeed secondary entities which emerge from this fundamental process. And the quantum potential represents here just an active information medium which underlies this fundamental process.

The treatment made in this book allows us to throw new light on the fundamental vacuum, on the characterization of the ultimate arena responsible of quantum processes and which determines the geometrodynamic properties of the quantum potential, in the different domains of quantum physics. The picture here proposed as regards the fundamental, ultimate arena of quantum processes can be considered as an alternative way to formulate Bohm's implicate order and Hiley's pre-space. In the light of the analysis made in this book, in particular, three are the main ideas which introduce novel perspectives and scenarios as regards the fundamental stage of quantum physics: the quantum entropy, the symmetrized quantum potential and the three-dimensional timeless non-local quantum vacuum defined by **RS** (reduction-state) processes of creation/annihilation of quanta in correspondence to elementary fluctuations of the quantum vacuum energy density.

If Hiley's approach of holomoviment implies that the geometry of space-time is derived from the algebra of process characterizing the implicate order, if in Hiley's approach the crucial idea is that, in the quantum world, there is not *a priori* given manifold, in the sense that the fundamental reality is the algebra defining the implicate order and the geometry is then derived from the algebra [155, 352], instead the considerations made in this book allow us to turn over the scenarios as regards the role of algebra and geometry at the level of quantum processes: the geometry must be considered as fundamental whilst algebraical constructions are emergent. The approaches here proposed suggest that the ultimate arena of quantum processes, the ultimate reality of the quantum world is the non-local geometry of a fundamental quantum vacuum defined by a quantum entropy, a symmetrized quantum potential and **RS** processes of creation/annihilation of quanta corresponding to elementary fluctuations of the quantum vacuum energy density.

Before all, in each regime of quantum physics, both in the non-relativistic domain, in the relativistic domain, in the relativistic quantum field theory, in quantum gravity and in quantum cosmology, a fundamental stage of processes exists, which is defined by a quantum entropy which indicates the degree of order and chaos of the vacuum supporting the density of the particles or of the fields associated to the system into consideration, and all the physical laws can be derived from this arena. The quantum entropy is the physical entity which shows how the vacuum determines a deformation of the geometrical properties of the background of the processes into consideration, thus producing the direct action of a non-local correlation.

Moreover, in the different contexts of quantum physics, a fundamental non-local geometry associated to a fundamental vacuum can be considered which is expressed by a symmetrized quantum potential at two components (where the second component allows us to reproduce in the correct way also the time-reverse of the processes). This symmetrized quantum potential is derived itself from a symmetrized quantum entropy, which is the fundamental entity describing the geometrical properties of the fundamental background which determines the processes into consideration. The crucial result of the symmetrized quantum potential approach as regards the fundamental geometry of space generating the processes in their respective level of quantum reality, is that both the wavefunctions of subatomic particles and the wave functionals in relativistic quantum field theory regime and the wave functionals of the gravitational field in the quantum gravity domain determine a space medium, a special state of physical reality (represented, respectively, by an opportune symmetryzed quantum potential) which acts as a direct, immediate information medium in its respective domain.

On the other hand, by starting from Bohm's 1962 words at BBC evidencing that the electron intended as a wave or a corpuscle (satisfying the well known laws of quantum theory) has not a primary physical existence but its "physical appearance" actually emerges from more elementary processes of formation and dissolving, one arrives naturally to Kastner's transactional interpretation and then to Chiatti's and Licata's transactions as fundamental processes of an archaic vacuum as "fabric of reality". Hence, in order to reproduce the energy density of quantum vacuum, the step is brief and the approach of the three-dimensional timeless non-local quantum vacuum characterized by elementary **RS** processes of creation/annihilation of quanta proposed by the author in a series of recent papers allows us to "close the circle". According to this picture, the quantum geometrodynamics of Bohm's quantum potential of ordinary quantum theory, which determines the processes regarding subatomic particles, emerges from a more fundamental arena, namely from the non-local geometry of a quantum potential of a three-dimensional quantum vacuum defined by **RS** processes of creation/annihilation of quanta corresponding to fluctuations of the quantum vacuum energy density. The geometry of the three-dimensional quantum vacuum described by the non-local quantum potential of the vacuum which guides the occurring of the **RS** processes of creation/annihilation of quanta and is produced by the pressures that arise between ensembles of virtual particles-antiparticles populating the vacuum, allows us also to provide unifying perspectives of gravity, electromagnetic fields and quantum behaviour of matter as different aspects of the same fluctuations of the quantum vacuum energy density [340].

By using again a usual Bohmian terminology, we can define the reality constituted by the **RS** processes and their evolution through the non-local action of the quantum potential of the vacuum as the fundamental background from whose differentiation the foreground constituted by the events of a given subatomic particle or system (governed by the well-known laws of quantum theory) emerges. By attributing the status of primary physical reality to the arena of **RS** processes and the frequencies of vibration associated with them (ultimately linked with the elementary fluctuations of the quantum vacuum energy density and, therefore, with the non-local quantum potential of the vacuum) it follows that the ordinary spacetime we perceive is, so to speak, materialized by **RS** processes. In analogous way, also the background associated with the de Broglie-Bohm pilot-wave theory but also of Bohm's implicate order and Hiley's pre-space can be seen as manifolds deriving from this more fundamental arena represented by the three-dimensional timeless non-local quantum vacuum, are someway "materialized" by **RS** processes and by the frequencies of vibration associated with them (ultimately linked with the elementary fluctuations of the quantum vacuum energy density and the fundamental geometry associated with the non-local quantum potential of the vacuum) [339]. The philosophy of this model as regards the link between our level of physical reality (regarding measurement processes), the implicate order and the three-dimensional timeless non-local quantum vacuum actually someway goes beyond the program that Bohm had already sketched out when he studied the relationship between implicate and explicate order, suggesting that the real origin and explanation of Bohm's implicate order lies just in the three-dimensional timeless non-local quantum vacuum. In this way, the problem regarding the existence, in quantum physics, of different descriptive levels of physical reality, whether formal analogies exist between them or there is a deeper meaning, gets a new significance. If each subatomic particle (such as the electron in the famous double-slit interference experiment) is indeed the evolution of opportune **RS** processes of creation/annihilation of quanta corresponding to elementary fluctuations of the quantum vacuum energy density (and thus to the fundamental geometry associated with specific behaviours of the quantum potential of the vacuum), both the ordinary quantum mechanics and Bohm's implicate order and Hiley's pre-space emerge directly from the three-dimensional timeless non-local quantum vacuum. A new suggestive paradigm is thus derived from Bohm's quantum potential approach, in which the wavefunction of a quantum particle represents the emergence of innate quantum fluctuations at the core of a more fundamental three-dimensional non-local quantum vacuum. The frequencies of vibration of elementary processes of creation/annihilation of quanta and, therefore, the fundamental geometry corresponding to opportune behaviours of the quantum potential of the vacuum must be considered as the ultimate quantum reality, as the ultimate background from which all the constructions of physics, such as space-time geometry, wavefunctions and wave functionals in the different regimes of quantum physics unfold as emergent aspects, as "explicate" orders.

Finally, in the light of the mathematical formalism analysed in chapters 3 and 4, an immediate and crucial link exists between the three-dimensional non-local quantum vacuum defined by **RS** processes and the symmetrized quantum potential. A parallelism may be made, on one hand, between the first, forward-time component of the symmetrized quantum potential and the **RS** processes of creation of quanta and, on the other hand, between the second, time-reverse component of the symmetrized quantum potential and the **RS** processes of annihilation of quanta of the three-dimensional quantum vacuum. Therefore, the interesting perspective is opened that the symmetrized quantum potential deriving from an opportune symmetrized quantum entropy, in each domain of quantum physics, represents the crucial physical aspect of the action of the quantum potential of the vacuum, at the most fundamental level. In other words, in the fundamental background of processes represented by the three-dimensional non-local quantum vacuum defined by **RS** processes corresponding to elementary fluctuations of the quantum vacuum energy density, the physical action of the quantum potential of the vacuum corresponds indeed to a symmetrized quantum potential which derives from an opportune symmetrized quantum entropy defining the geometrical properties of the fundamental background of processes.

References

1. G. Bacciagaluppi and A. Valentini, *Quantum Theory at the Crossroads: Reconsidering the 1927 Solvay Conference*, Cambridge University Press, Cambridge (2009); e-print arXiv:quant-ph/0609184.
2. D. Fiscaletti, *I fondamenti nella meccanica quantistica. Un'analisi critica dell'interpretazione ortodossa, della teoria di Bohm e della teoria GRW*, CLEUP, Padova (2003).
3. J. von Neumann, *Mathematische Grundlagen der Quantenmeckanik*, Springer, Berlin (1932); trad. it. *Mathematical Foundations of Quantum Mechanics*, Princeton University Press, Princeton (1955).
4. W. Heisenberg, *Physics and Philosophy*, Harper and Row, New York (1958).
5. M. Pavsic, *The landscape of theoretical physics: a global view*, Kluwer Academic Publishers, Boston/Dordrecht, London (2001).
6. C.F. Von Wezsacker, in *The Physicist's Conception of Nature*, edited by J. Mehra, Reidal Publishing Company, Boston (1973).
7. M. Born, "Zur Quantenmechanik der Stossvorgânge", *Zeitschrift für physik* **37**, 863–867 (1926).
8. N. Bohr, *Teoria dell'atomo e conoscenza umana*, Boringhieri, Torino (1961).
9. E. Schrödinger, *Naturwissenschaften* **23**, 48–50, 807–812, 823–828, 844–849 (1935).
10. E. Santamato, "Statistical interpretation of the Klein-Gordon equation in terms of the space-time Weyl curvature", *Journal of Mathematical Physics* **25**, 8, 2477ff (1984).
11. E. Santamato, "Gauge-invariant statistical mechanics and average action principle for the Klein-Gordon particle in geometric quantum mechanics", *Physical Review D* **32**, 10, 2615ff (1985).
12. J.T. Wheeler, "Quantum measurement and geometry", *Physical Review D* **41**, 2, 431–441 (1990).

13. W.R. Wood and G. Papini, "A geometric approach to the quantum mechanics of de-Broglie–Bohm and Vigier", in *The present status of quantum theory of light*, Proc. of Symposium in honor of J. P. Vigier (York University, Aug. 27–30 (1995)).
14. B.G. Sidhart, "Geometry and quantum mechanics", arXiv:physics/0211012 (2002).
15. R. Carroll, "Some remarks on Ricci flow and the quantum potential", arXiv:math-ph/0703065 (2007).
16. J.M. Isidro, J.L.G. Santander and P.F. de Cordoba, "Ricci flow, quantum mechanics and gravity", arXiv:0808.2351 [hep-th] (2008).
17. S. Abraham, P.F. de Cordoba, J.M. Isidro and J.L.G. Santander, "The Ricci flow on Riemann surfaces", arXiv:0810.2236 [hep-th] (2008).
18. S. Abraham, P.F. de Cordoba, J.M. Isidro and J.L.G. Santander, "A mechanics for the Ricci flow", *International Journal of Geometrical Methods in Modern Physics* **5**, 759–767 (2009); e-print arXiv:0810.2356 [hep-th].
19. J.M. Isidro, J.L.G. Santander and P.F. de Cordoba, "On the Ricci flow and the emergent quantum mechanics", *Journal of Physics Conference Series* **174**, 012033 (2009), e-print arXiv:0902.0143 [hep-th].
20. J.M. Isidro, J.L.G. Santander and P.F. de Cordoba, "Positive curvature can mimic a quantum", arXiv:0912.1535 [hep-th] (2009).
21. E. Gozzi and D. Mauro, "Quantization as a dimensional reduction phenomenon", *AIP Conference Proceedings* **844**, 158 (2006); e-print arXiv:quant-ph/0601209.
22. D. Dolce, "Compact time and determinism for bosons: foundations", *Foundations of Physics* **41**, 2, 178–203 (2011); e-print arXiv:0903.3680 [hep-th].
23. V. Guillemin and S. Sternberg, "Geometric Asymptotics", *Mathematics Surveys Monographs* **14**, American Mathematical Society, Providence (1978).
24. V. Guillemin and S. Sternberg, *Symplectic Techniques in Physics*, Cambridge University Press, Cambridge (1990).
25. M. De Gosson, *The Principles of Newtonian and Quantum Mechanics*, Imperial College Press, London (2001).
26. M. De Gosson, "The symplectic camel and quantum universal invariants: the angel of geometry versus the demon of algebra", *Quantum Matter* **3**, 3, 169–171 (2014).
27. L. De Broglie, in *Solvay Congress (1927), Electrons and photons: rapports et discussions du Cinquime Conseil de Physique tenu Bruxelles du 24 au Octobre 1927 sous les auspices de l'Istitut International de Physique Solvay*, Paris, Gauthier-Villars (1928).
28. L. de Broglie, *Une interpretation causale et non linéaire de la mécanique ondulatoire: la théorie de la doble solution*, Gauthier-Villars, Paris (1956).
29. L. de Broglie, "The reinterpretation of wave mechanics", *Foundations of Physics* **1**, 1, 5–15 (1970).

30. D. Bohm, "A new suggested interpretation of quantum theory in terms of hidden variables. Part I", *Physical Review* **85**, 166–179 (1952).
31. D. Bohm, "A new suggested interpretation of quantum theory in terms of hidden variables. Part II", *Physical Review* **85**, 180–193 (1952).
32. P.R. Holland, *The Quantum Theory of Motion*, Cambridge University Press, Cambridge (1993).
33. D. Fiscaletti, *I gatti di Schrödinger. Meccanica quantistica e visione del mondo*, Muzzio Editore, Roma (2007).
34. D. Fiscaletti, "The ontology of events: Bohmian mechanics versus GRW theory", *Quantum Biosystems* **2**, 93–101 (2007).
35. J.R. Croca, *Hyperphysics — The unification of physics, a new vision on physics, eurhythmy, emergence and non-linearity*, edited by J.R. Croca and J.E.F. Araüjo, Ed. Centro de Filosofia das Ciencias da universidade da Lisboa, Lisboa (2010), pp. 1–106.
36. J.R. Croca, *Towards a non-linear quantum physics*, World Scientific, London (2003).
37. J.R. Croca, *The principle of eurhythmy a key to the unity of physics, Special Sciences and the Unity of Sciences*, edited by O. Pombo, J.M. Torres, J. Symons and S. Rahman, Springer, New York (2012).
38. J.R. Croca, "Foundations of quantum physics at the beginning of 21st century physics", *Quantum Matter* **2**, 1, 1–8 (2013).
39. B.J. Hiley, "From the Heisenberg picture to Bohm: A new perspective on active information and its relation to Shannon information", in *Proc. Conf. Theory: reconsiderations of foundations*, edited by A. Khrennikov, Vaxjo University Press, pp. 141–162 (2002).
40. C. Philippidis, C. Dewdney and B. Hiley, "Quantum interference and the quantum potential", *Nuovo Cimento B* **52**, 1, 15–28 (1979).
41. B.G. Englert, M.O. Scully, G. Süssman e H. Walther, "Surrealistic Bohm trajectories", *Z. Naturforsch* **47a**, 1175–1186 (1992).
42. B.G. Englert, M.O. Scully, G. Süssman e H. Walther, "Reply to comment on surrealistic Bohm trajectories", *Z. Naturforsch* **48a**, 1263–1264 (1993).
43. H.D. Zeh, "Why Bohm's quantum theory", Foundations of Physics Letters 12, 2, 197–200 (1999); e-print arXiv:quant-ph/9812059v2.
44. D.H. Mahler, L. Rozema, K. Fisher, L. Vermeyden, K.J. Resch, H.W. Wiseman and A. Steinberg, "Experimental nonlocal and surreal Bohmian trajectories", *Science Advances* **2**, 2, e1501466 (2016).
45. C. Dewdney and B. Hiley, "A quantum potential description of one-dimensional time-dependent scattering from square barriers and square wells", *Foundations of Physics* **12**, 27–48 (1982).
46. *Quantum uncertainties — recent and future experiments and interpretations*, edited by W.M. Honig, D.W. Kraft and E. Panarella, Nato Asi Series, Plenum Press, New York (1987).
47. B. Hiley, "Some remarks on the evolution of Bohm' proposals for an alternative to standard quantum mechanics", preprint, www.bbk.ac.uk/tpru/BasilHiley/History_of_Bohm_s_QT.pdf (2010).

48. D. Dürr, S. Goldstein and N. Zanghi, "Quantum Equilibrium and the Origin of Absolute Uncertainty", *Journal of Statistical Physics* **67**, 843–907 (1992).
49. D. Bohm, *Quantum theory*, Routledge, London (1951).
50. D. Fiscaletti, "The geometrodynamic nature of the quantum potential", *Ukrainian Journal of Physics* **57**, 5, 560–572 (2012).
51. B.J. Hiley, "Non-commutative geometry, the Bohm interpretation and the mind-matter relationship", in *Proc. CASYS'2000*, Liege, Belgium, Aug. 7–12 (2000).
52. B.J. Hiley and M. Fernandes, in *Time, Temporality, and Now*, edited H. Atmanspacher and E. Ruhnau, Springer-Verlag (1997), pp. 365–382.
53. B.J. Hiley and N. Monk, "Quantum phase space and discrete Weyl algebra", *Modern Physics Letters A* **8**, 3225–3233 (1993).
54. S. Goldstein, S. Berndl, M. Daumer, D. Dürr and N. Zanghì, "A survey of Bohmian Mechanics", *Il Nuovo Cimento* **110B**, 737–750 (1995); e-print arXiv:quant-ph/9504010 (1995).
55. S. Goldstein, D. Dürr and N. Zanghì, "Bohmian Mechanics and Quantum Equilibrium", in *Stochastic Processes, Physics and Geometry II*, edited by S. Albeverio, U. Cattaneo and D. Merlini, World Scientific, Singapore (1995).
56. S. Goldstein, "Bohmian mechanics and the quantum revolution", *Synthese* **107**, 145–165 (1996).
57. S. Goldstein, D. Dürr and N. Zanghì, "Bohmian Mechanics as the Foundation of Quantum Mechanics", in *Bohmian Mechanics and Quantum Theory: An Appraisal*, edited by J.T. Cushing, A. Fine and S. Goldstein, Boston Studies in the Philosophy of Science **184**, 21–44, Kluwer, Dordrecth (1996); e-print arXiv:quant-ph/9511016 (1995).
58. D. Dürr, S. Goldstein and N. Zanghì, "Bohmian mechanics and meaning of the wave function", in *Experimental metaphysics. Quantum mechanical studies for Abner Shimony. Volume 1*, edited by R.S. Cohen, M. Horne and J. Stachel, Kluwer, Dordrecht (1997).
59. V. Allori and N. Zanghì, "What is Bohmian mechanics", arXiv:quant-ph/0112008v1 (2001).
60. S. Goldstein, D. Dürr, R. Tumulka and N. Zanghì, "Bohmian Mechanics", in *The Encyclopedia of Philosophy, Second Edition*, edited by D.M. Borchert, Macmillan Reference (2006).
61. S. Goldstein, D. Dürr, R. Tumulka and N. Zanghì, "Bohmian Mechanics", in *Compendium of Quantum Physics*, edited by F. Weinert, K. Hentschel and D. Greenberger, Springer-Verlag, Berlin (2009); e-print arXiv:0903.2601 [quant-ph] (2009).
62. S. Goldstein, R. Tumulka and N. Zanghì, "Bohmian Trajectories as the Foundation of Quantum Mechanics", in *Quantum Trajectories*, edited by P. Chattaraj, Taylor & Francis, Boca (2010); e-print arXiv:0912.2666 [quant-ph] (2009).
63. M. Atiq, M. Karamian and M. Golshani, "A quasi-Newtonian approach to Bohmian quantum mechanics", *Annales de la Fondation Louis de Broglie* **34**, 1, 67–81 (2009).

64. M. Abolhasani and M. Golshani, "The path integral approach in the frame work of causal interpretation", *Annales de la Fondation Louis de Broglie* **28**, 1, (2003).
65. D. Fiscaletti, "About the different approaches to Bohm's quantum potential in non-relativistic quantum mechanics", *Quantum Matter*, special issue "The quantum and the geometry" **3**, 3, 177–199 (2014).
66. D. Fiscaletti, "*Perspectives of Bohm's quantum potential towards a geometrodynamic interpretation of quantum physics. A critical survey*", Reviews in Theoretical Science **1**, 2, 103–144 (2013).
67. G. Grössing, "The vacuum fluctuation theorem: Exact Schrödinger equation via Nonequilibrium Thermodynamics", *Physics Letters A* **372**, 4556 (2008); e-print http://arxiv.org/abs/0711.4954 (2007).
68. G. Grössing, "On the thermodynamic origin of the quantum potential. On the thermodynamic origin of the quantum potential", *Physica A: Statistical Mechanics and its Applications* **388**, 6, 811–823 (2009); e-print arXiv: quant-ph 0808.35.39.pdf (2008).
69. D. Bohm and B. Hiley, *The Undivided Universe*, Routledge, London (1993).
70. V. Sbitnev, "Navier-Stokes equation describes the movement of a special superfluid medium", arXiv:1504.07497v1 [quant-ph] (2015).
71. V. Sbitnev, "Physical vacuum is a special superfluid medium", in *Selected Topics in Applications of Quantum Mechanics*, edited by M.R. Pahlavani, InTech, Rijeka, pp. 345–373 (2015).
72. V. Sbitnev, "Hydrodynamics of the physical vacuum: dark matter is an illusion", *Modern Physics Letters A* **30**, 35, 1550184 (2015).
73. V. Sbitnev, "Hydrodynamics of the physical vacuum. I: Scalar quantum sector", *Foundations of Physics* **46**, 5, 606–619 (2016); e-print arXiv:1504.07497.v2 [quant-ph] (2016).
74. V. Sbitnev, "Bohmian Trajectories and the Path Integral Paradigm — Complexified Lagrangian Mechanics", in *Theoretical Concepts of Quantum Mechanics*, edited by M.R. Pahlavani, InTech, Rijeka (2012), pp. 313–34; doi:10.5772/33064.
75. I. Licata and D. Fiscaletti, "Bell Length as Mutual Information in Quantum Interference", *Axioms* **3**, 153–165 (2014).
76. E. Nelson, "Derivation of the Schrödinger equation from Newtonian Mechanics", *Physical Review* **150**, 1079–1085 (1966).
77. A. Martins, "Fluidic Electrodynamics: On parallels between electromagnetic and fluidic inertia", http://arxiv.org/abs/1202.4611 (2012).
78. R.P. Feynman and A. Hibbs, *Quantum Mechanics and Path Integrals*, McGraw Hill, New York (1965).
79. H. Weyl, "Gravitation and Electrizidad", *Sitz. Ber. Preuss. Akad. Wiss.* **26**, 465–480 (1918).
80. H. Weyl, "Zur Gravitationstheorie", *Annalen der Physik* **54**, 117–145 (1918); H. Weyl, *Space, Time, Matter*, 4th edition, Dover Publications, Inc., New York (1952).
81. P.G. Bergmann, *Introduction to the Theory of Relativity*, Prentice-Hall, New Delhi (1969), 245ff.

82. P.A.M. Dirac, "Long range forces and broken symmetries", *Proceedings of Royal Society A* **333**, 403–418 (1973).
83. F. De Martini and E. Santamato, "Nonlocality, no-signalling and Bell's theorem investigated by Weyl's conformal differential geometry", *Physica Scripta T* **163**, 014015 (2014); e-print arXiv:1406.2970v1 [quant-ph].
84. F. De Martini and E. Santamato, "Solving the nonlocality riddle by conformal quantum geometrodynamics", *Journal of Physics: Conference Series* **442**, 012059 (2013); e-print arXiv:1203.0033v1 [quant-ph].
85. E. Madelung, "Quantentheorie in hydrodynamischer Form", *Zeitschrift für physik* **40**, 322–326 (1926).
86. E. Santamato, "Geometric derivation of the Schrödinger equation from classical mechanics in curved Weyl spaces", *Physical Review D* **29**, 2, 216ff (1984).
87. M. Inaba, "Quantum mechanics from a fluctuation in the cosmos", *International Journal of Modern Physics A* **16**, 17, 2965–2973 (2001).
88. B.G. Sidharth, "The universe of fluctuations", *International Journal of Modern Physics A* **13**, 15, 2599ff (1998).
89. B.G. Sidharth, "The Cosmology of Fluctuations", *Chaos, Solitons and Fractals* **16**, 4, 613–620 (2003).
90. F. De Martini and E. Santamato, "Ab initio derivation of the quantum Dirac equation by conformal differential geometry: the 'Affine Quantum Mechanics"', in *Quantum Foundations*, edited by A. Khrennikov, AIP, Melville (2012); e-print arXiv: quantph/1107.3168v1 (2011).
91. E. Santamato and F. De Martini, "Derivation of the Dirac equation by conformal differential geometry", *Foundations of Physics* **43**, 5, 631–641 (2013).
92. E. Santamato and F. De Martini, "A Conformal Geometric Approach to Quantum Entanglement for Spin-1/2 Particles", *European Physical Journal Web of Conferences* **58**, 010012 (2013).
93. M. Novello, J.M. Salim and F.T. Falciano, "On a Geometrical Description of Quantum Mechanics", *International Journal of Geometrical Methods in Modern Physics* **8**, 1, 87–98 (2011).
94. R.W. Carroll, *Fluctuations, Information, Gravity and the Quantum Potential*, Springer, Dordrecht (2006).
95. P.R. Holland, "Geometry of dislocated de Broglie waves", *Foundations of Physics* **17**, 4, 345–363 (1987).
96. A. Shojai and F. Shojai, "About some problems raised by the relativistic form of de Broglie-Bohm theory of pilot wave", *Physica Scripta* **64**, 5, 413–416 (2001).
97. F. Shojai and A. Shojai, "Understanding quantum theory in terms of geometry", in *New topics in quantum physics research*, edited by V. Krasnoholovets and F. Columbus, Nova Science Publishers, New York, pp. 59–98 (2006); e-print arXiv:gr-qc/0404102 v1 (2004).
98. G. Bertoldi, A. Faraggi and M. Matone, "Equivalence principle, higher dimensional Moebius group and the hidden antisymmetric tensor of quantum mechanics", arXiv:hep-th/9909201 (1999).

99. H. Brown, E. Sjöqvist and G. Bacciagaluppi, "Remarks on identical particles in de Broglie-Bohm theory", *Physics Letters A* 251, 229–235 (1999); e-print arXiv:quant-ph/9811054.
100. M. Brown, "The quantum potential: the breakdown of classical sympletic symmetry and the energy of localisation and dispersion", arXiv:quant-ph/9703007 (1997).
101. H. Brown and P. Holland, "Simple applications of Noether's first theorem in quantum mechanics and electromagnetism", *American Journal of Physics* **72**, 1, 34–39 (2004); e-print arXiv:quant-ph/0302062 (2003).
102. C. Dewdney and G. Horton, "A Relativistic Hidden-Variable Interpretation for the Massive Vector Field Based on Energy-Momentum Flows", *Foundations of Physics* **40**, 6, 658–678 (2010).
103. P.R. Holland, "New trajectory interpretation of quantum mechanics", *Foundations of Physics* **28**, 881–911 (1998).
104. P.R. Holland, "Hamiltonian theory of wave and particle in quantum mechanics I: Liouville's theorem and the interpretation of the de Broglie-Bohm theory", *Nuovo Cimento B* **116**, 1043–1070; "Hamiltonian theory of wave and particle in quantum mechanics II: Hamilton-Jacobi theory and particle back-reaction", *Nuovo Cimento B* **116**, 1143–1172 (2001).
105. P.R. Holland, "Causal interpretation of Fermi fields", *Physics Letters A* **128**, 1–2, 9–18 (1988).
106. P.R. Holland, "Uniqueness of conserved currents in quantum mechanics", arXiv:quant-ph/0305175 (2003).
107. P.R. Holland, "The de Broglie-Bohm theory of motion and quantum field theory", *Physics Reports* **224**, 95–150 (1993).
108. P.R. Holland, "Implications of Lorentz covariance for the guidance equation in two-slit quantum interference", arXiv:quant-ph/0302076 (2003).
109. P.R. Holland, "Computing the wavefunction from trajectories: particle and wave pictures in quantum mechanics and their relation", arXiv:quant-ph/0405145 (2004).
110. P.R. Holland, "Constructing the electromagnetic field from hydrodynamic trajectories", arXiv:quant-ph/0411141 (2004).
111. G. Horton and C. Dewdney, "A non-local, Lorentz-invariant, hidden variable interpretation of relativistic quantum mechanics based on particle trajectories", *Journal of Physics A: Mathematical and General* **34**, 46, 9871–9878 (2001); e-print arXiv:quant-ph/0110007 (2001).
112. G. Horton and C. Dewdney, "A relativistically covariant version of Bohm's quantum field theory for the scalar field", *Journal of Physics A: Mathematical and General* **37**, 49, 11935–11944; e-print arXiv:0407089 (2004).
113. C. Dewdney and G. Horton, "Relativistically invariant extension of the de Broglie Bohm theory of quantum mechanics", *Journal of Physics A: Mathematical and General* **35**, 47, 10117–10128 (2002).
114. G. Horton, C. Dewdney and U. Neeman, "De Broglie's pilot wave theory for the Klein-Gordon equation", arXiv:quant-ph/0109059 (2001).

115. G. Horton, C. Dewdney and H. Nesteruk, "Time-like flows of energy-momentum and particle trajectories for the Klein-Gordon equation", arXiv:quant-ph/0103114 (2001).
116. G. Horton, C. Dewdney and H. Nesteruk, "Time-like flows of energy-momentum and particle trajectories for the Klein-Gordon equation", *Journal of Physics A: Mathematical and General* **33**, 41, 7337–7352 (2000).
117. A. Mostafazadeh and F. Zamani, "Conserved current densities, localisation probabilities, and a new global gauge symmetry of Klein-Gordon fields", arXiv:quant-ph/0312078 (2003).
118. A. Mostafazadeh, "Quantum mechanics of Klein-Gordon-type fields and quantum cosmology", *Annals of Physics* **309**, 1–48; e-print arXiv:gr-qc/0306003 (2003).
119. R. Carroll, *Quantum theory, deformation, and integrability*, North-Holland, Amsterdam (2000).
120. R. Carroll, *Proceedings of the 10th International Conference on "Symmetry methods in physics"*, Kiev, Part I (2003), pp. 356–367.
121. R. Carroll, "(X, ψ) duality and enhanced dKdV on a Riemann surface", *Nuclear Physics B* **502**, 3, 561–593 (1997).
122. R. Carroll, "On the whitham equations and (X, ψ) duality", *Springer Lectures Notes in Physics* **502**, 33–56 (1998).
123. A. Faraggi and M. Matone, "The equivalence postulate of quantum mechanics", *Internationa Journal of Modern Physics A* **15**, 13, 1869–2017 (2000).
124. M. Matone, "The cocycle of quantum Hamilton-Jacobi equation and the stress tensor of CFT", *Brazilian Journal of Physics* **35**, 2A, 316–327; e-print arXiv: hep-th 0502134 (2005).
125. H. Nikolic, "Covariant canonical quantization of fields and Bohmian mechanics", *European Physical Journal C* **42**, 3, 365–374 (2005); e-print arXiv:hep-th/0407228.
126. H. Nikolic, "Bohmian particle trajectories in relativistic bosonic quantum field theory", *Foundations of Physics Letters* **17**, 4, 363–380 (2004); e-print arXiv:quant-ph/0208185.
127. H. Nikolic, "Relativistic quantum mechanics and the Bohmian interpretation", *Foundations of Physics Letters* **18**, 2, 123–138 and **18**, 6, 549–561 (2005); e-print arXiv:quant-ph/0406173.
128. H. Nikolic, "Quantum determinism from quantum general covariance", arXiv:hep-th/0601027 (2006).
129. D. Bohm and B.J. Hiley, "On the relativistic invariance of a quantum theory based on beables", *Foundations of Physics* **21**, 2, 243–250 (1991).
130. S. Gull, A. Lasenby and C. Doran, "Electron paths, tunnelling and diffraction in the spacetime algebra", *Foundations of Physics* **23**, 10, 1329–1356 (1993).
131. O. Chavoya-Aceves, "A de Broglie-Bohm like model for Dirac equation", arXiv:quant-ph/0304195v3 (2003).
132. H. Nikolic, "Time in relativistic and nonrelativistic quantum mechanics", *International Journal of Quantum Information* **7**, 3, 595–602 (2009).

133. H. Nikolic, "Relativistic quantum mechanics and the Bohmian interpretation", *Foundations of Physics Letters* **18**, 2, 123–138; e-print arXiv:quant-ph/0302152 (2003).
134. H. Nikolic, "There is no first quantization except in the de Broglie-Bohm interpretation", arXiv:quant-ph/0307179 (2003).
135. S. Hernandez-Zapata, "The Dirac equation from a Bohmian point of view", arXiv:1003.1558.pdf [quant-ph] (2010).
136. B.J. Hiley and R.E. Callaghan, "The Clifford algebra approach to quantum mechanics B: the Dirac particle and its relation to the Bohm approach", arXiv:1011.4033v1 [math-ph] (2010).
137. F. Sauter, "Lösung der Diracschen Gleichungen ohne Specializierung der Diracschen Opera-toren", *Zeitschrift für Physik* **63**, 803–814 (1930).
138. T. Takabayashi, "Relativistic hydrodynamics of the Dirac matter", *Supplement of the Progress in Theoretical Physics* **4**, 1–80 (1957).
139. R. Tommasi, *Milestones in Physics*, Aracne, Roma (2015).
140. E.G. Harris, *A Pedestrian Approach to Quantum Field Theory*, Wiley-Interscience, New York (1972).
141. I. Licata, "Visions of oneness. Space-time geometry and quantum physics", in *Vision of Oneness*, edited by I. Licata and A. Sakaji, Aracne Publishing, Rome (2011).
142. G. Preparata, in *An Introduction to a Realistic Quantum Physics*, World Scientific (2002).
143. I. Licata, "The keys and the door: for a unitary vision of the physicists' conception of nature", in *Vision of oneness*, edited by I. Licata and A.J. Sakaji, Aracne Publishing, Rome (2011).
144. I. Licata, *Osservando la sfinge*, Di Renzo, Roma, 3° edition (2009).
145. J.S. Bell, *Speakable and Unspeakable in Quantum Mechanics*, Cambridge University Press, Cambridge (1996).
146. P.N. Kaloyerou, "The causal interpretation of the electromagnetic field", *Physics Reports* **244**, 287–358 (1994).
147. A. Valentini, "Signal-locality, uncertainty, and the subquantum H-theorem", I. *Physics Letters A* **156**, 5–11 (1991).
148. H. Nikolic, "Quantum mechanics: myths and facts", arXiv:quant-ph/0609163 (2006).
149. H. Nikolic, "The general-covariant and gauge-invariant theory of quantum particles in classical backgrounds", *International Journal of Modern Physics D* **12**, 3, 407–444 (2003); e-print arXiv:hep-th/0202204.
150. H. Nikolic, "A general covariant concept of particles in curved background", *Physics Letters B* **527**, 1–2, 119–124 (2002); e-print arXiv:gr-qc/0111024.
151. H. Nikolic, Erratum to: "A general covariant concept of particles in curved background", *Physics Letters B* **529**, 3, 265 (2002).
152. H. Nikolic, "Covariant many-fingered time Bohmian interpretation of quantum field theory", *Physics Letters A* **348**, 166–171; e-print arXiv:hep-th/0501046 (2006).
153. H. Nikolic, "Bohmian particle trajectories in relativistic fermionic quantum field theory", *Foundations of Physics Letters* **18**, 2, 123–138; e-print arXiv:quant-ph/0302152 (2003).

154. H. Nikolic, "On the compatibility of Bohmian mechanics with standard quantum mechanics", arXiv:quant-ph/0305131 (2003).
155. I. Licata and D. Fiscaletti, *Quantum potential. Physics, geometry and algebra*, Springer, Heidelberg (2013).
156. A. Einstein, *Preussische Akademien der Wissenschaften. Sitzungsberichte (Berlin)* 688–696 (1916).
157. A. Shojai, "Quantum, gravity and geometry", *International Journal of Modern Physics A* **15**, 12, 1757–1771 (2000); e-print arXiv:gr-qc/0010013.
158. F. Shojai and A. Shojai, "Pure quantum solutions of Bohmian quantum gravity", arXiv:gr-qc/0105102 (2001).
159. F. Shojai and A. Shojai, "Quantum Einstein's equations and constraints algebra", *Pramana* **58**, 1, 13–19 (2002); e-print arXiv:gr-qc/0109052.
160. F. Shojai and A. Shojai, "On the relation of Weyl geometry and Bohmian quantum gravity", *Gravitation and Cosmology* **9**, 3, 163–168 (2003); e-print arXiv:gr-qc/0306099.
161. F. Shojai and A. Shojai, "Quantum Einstein's equations and constraints algebra", arXiv:gr-qc/0109052 (2004).
162. F. Shojai and A. Shojai, "Nonminimal scalar-tensor theories and quantum gravity", *International Journal of Modern Physics A* **15**, 13, 1859–1868 (2000); e-print arXiv:gr-qc/0010012.
163. A. Shojai and M. Golshani, "Direct particle interaction as the origin of the quantal behaviours", arXiv:quant-ph/9812019 (1998).
164. A. Shojai and M. Golshani, "On the relativistic quantum force", arXiv:quant-ph/9612023 (1996).
165. A. Shojai and M. Golshani, "Some observable results of the retarded's Bohm's theory", arXiv:quant-ph/9612020 (1996).
166. A. Shojai and M. Golshani, "Is superluminal motion in relativistic Bohm's theory observable?", arXiv:quant-ph/9612021 (1996).
167. F. Shojai and A. Shojai, "Constraint algebra and the equations of motion in the Bohmian interpretation of quantum gravity", *Classical and Quantum Gravity* **21**, 1–9 (2004); e-print arXiv:gr-qc/0311076.
168. F. Shojai and A. Shojai, "Causal loop quantum gravity and cosmological solutions", *Europhysics Letters* **71**, 6, 886–892 (2005); e-print arXiv:gr-qc/0409020.
169. F. Shojai and A. Shojai, "Constraint algebra in causal loop quantum gravity", arXiv:gr-qc/0409035 (2004).
170. L. de-Broglie, *Non-linear Wave Mechanics*, translated by A. J. Knodel, Elsevier Publishing Company, Amsterdam (1960).
171. D. Fiscaletti, "Bohm's quantum potential and the geometry of space", *Quantum Matter* **2**, 1, 45–53 (2013).
172. T. Horiguchi, "Quantum potential interpretation of the Wheeler-deWitt equation", *Modern Physics Letters A* **9**, 16, 1429–1443 (1994).
173. A. Blaut and J.K. Glikman, "Quantum potential approach to class of quantum cosmological models", *Classical and Quantum Gravity* **13**, 39–50 (1996).

174. S.P. Kim, "Quantum potential and cosmological singularities", *Physics Letters A* **236**, 11–15 (1997).
175. S.P. Kim, "Problem of unitarity and quantum corrections in semi-classical quantum gravity", *Physical Review D* **55**, 7511–7517 (1997).
176. J.T. Cushing, A. Fine and S. Goldstein eds., *Bohmiam mechanics and quantum theory: An appraisal*, Kluwer Academic Publishers, Boston (1996).
177. J.A. de Barros, N. Pinto-Neto and M.A. Sagioro-Leal, "The causal interpretation of dust and radiation fluid non-singular quantum cosmologies", *Physics Letters A* **241**, 229–239 (1998).
178. F. Shojai and M. Golshani, "On the geometrization of Bohmian mechanics: a new suggested approach to quantum gravity", *International Journal of Modern Physics A* **13**, 4, 677–693 (1998).
179. Jr. R. Colistete, J.C. Fabris and N. Pinto–Neto, "Singularities and Classical Limit in Quantum Cosmology with Scalar Fields", *Physical Review D* **57**, 8, 4707–4717 (1998).
180. P. Pinto-Neto and R. Colistete, "Graceful exit from inflation using quantum cosmology", *Physics Letters A* **290**, 5–6, 219–226 (2001).
181. M. Kenmoku, R. Sato and S. Uchida, "Classical and quantum solutions of (2+1)-dimensional gravity under the de Broglie-Bohm interpretation", *Classical and Quantum Gravity* **19**, 4, 799 (2002).
182. F. Shojai, A. Shojai and M. Golshani, "Conformal transformations and quantum gravity", *Modern Physics Letters A* **13**, 34, 2725–2729 (1998).
183. F. Shojai, A. Shojai and M. Golshani, "Scalar-tensor theories and quantum gravity", *Modern Physics Letters A* **13**, 36, 2915–2922 (1998).
184. A. Shojai, F. Shojai and M. Golshani, "Nonlocal effects in quantum gravity", *Modern Physics Letters A* **13**, 37, 2965–2969 (1998).
185. A. Shojai and F. Shojai, "Bohmian Quantum Gravity in the Linear Field Approximation", *Physica Scripta* **68**, 4, 207 (2003).
186. F. Shojai and M. Golshani, "On the general covariance in the Bohmian quantum gravity", arXiv:gr-qc/9903047v1 (1999).
187. F. Shojai and M. Golshani, "On Mach's principle in the Bohmian quantum cosmology", preprint (1997).
188. Y.M. Cho and D.H. Park, "Higher-dimensional unification and fifth force", *Nuovo Cimento B* **105**, 8–9, 817–829 (1990).
189. R. Omnès, *The Interpretation of Quantum Mechanics*, Princeton University Press, Princeton (1994).
190. M. Gell-Mann and J.B. Hartle, in *Complexity, Entropy and the Physics of Information*, edited W. H. Zurek, Addison Wesley (1990).
191. *The Many-Worlds Interpretation of Quantum Mechanics*, edited by B.S. DeWitt and N. Graham, Princeton University Press, Princeton (1973).
192. N. Pinto-Neto, "Quantum cosmology: how to interpret and obtain results", *Brazilian Journal of Physics* **30**, 2, 330–345 (2000).
193. A. Ashtekar and J. Lewandowski, "Quantum field theory of geometry", arXiv:hep-th/9603083 (1996).

194. A. Ashtekar, A. Corichi and J. Zapata, "Quantum theory of geometry III: non-commutativity of Riemannian structures", *Classical and Quantum Gravity* **15**, 2955–2972 (1998); e-print arXiv:gr-qc/9806041.
195. A. Ashtekar and J. Lewandowski, "Background independent quantum gravity: a status report", *Classical and Quantum Gravity* **21**, R53–R152 (2004); e-print arXiv:gr-qc/0404018.
196. A. Ashtekar and C. Isham, "Representations of holonomy algebras of gravity and non-abelian gauge theories", *Classical and Quantum Gravity* **9**, 1433–1467 (1992).
197. A. Ashtekar, "Gravity and the quantum", *New Journal of Physics* **7**, 200–232 (2005); e-print arXiv:gr-qc/0410054.
198. J. Baez (editor), *Knots and quantum gravity*, Oxford University Press, Oxford (1994).
199. S. Biswas, A. Shaw and D. Biswas, "Schrödinger Wheeler-DeWitt equation in multidimensional cosmology", *International Journal of Modern Physics D* **10**, 585–594 (2001); e-print arXiv:gr-qc/9906009.
200. B. Dittrich and T. Thiemann, "Testing the master constraint programme for loop quantum gravity I. General framework", *Classical and Quantum Gravity* **23**, 1025–1046 (2006); e-print arXiv:gr-qc/0411138 (2004).
201. B. Dittrich and T. Thiemann, "Testing the master constraint programme for loop quantum gravity II. Finite dimensional systems", *Classical and Quantum Gravity* **23**, 1067–1088 (2006); e-print arXiv: gr-qc 0411139 (2004).
202. B. Dittrich and T. Thiemann, "Testing the master constraint programme for loop quantum gravity III. SL(2R) models", *Classical and Quantum Gravity* **23**, 1089–1120 (2006); e-print arXiv:gr-qc/0411140 (2004).
203. B. Dittrich and T. Thiemann, "Testing the master constraint programme for loop quantum gravity IV. Free field theories", *Classical and Quantum Gravity* **23**, 1121–1142 (2006); e-print arXiv:gr-qc/0411141 (2004).
204. B. Dittrich and T. Thiemann, "Testing the master constraint programme for loop quantum gravity V. Interaction field theories", *Classical and Quantum Gravity* **23**, 1143–1162 (2006); e-print arXiv:gr-qc/0411142 (2004).
205. R. Gambini and J. Pullin, *Loops, knowts, gauge theories and quantum gravity*, Cambridge University Press, Cambridge (1996).
206. T. Horiguchi, K. Maeda and M. Sakamoto, "Analysis of the Wheeler-DeWitt equation beyond Planck scale and dimensional reduction", *Physics Letters B* **344**, 1, 105–109 (1995); e-print arXiv:hep-th/9409152.
207. M. Kenmoku, H. Kubortani, E. Takasugi and Y. Yamazaki, "De Broglie-Bohm interpretation for analytic solutions of the Wheeler-DeWitt equation in spherically symmetric space-time", arXiv:gr-qc/9906056 (1999).
208. C. Kiefer, *Quantum gravity*, Oxford University Press, Oxford (2004).
209. S. Kim, "Does Lorentz Boost Destroy Coherence?", arXiv:gr-qc/9703065 (1997).
210. S. Kim, "Quantum potential and cosmological singularities", *Physics Letters A* **236**, 11–15 (1997).

211. F. Markopoulou and L. Smolin, "Quantum theory from quantum gravity", *Physical Review D* **70**, 12, 124029–10 (2004); e-print arXiv:gr-qc/0311059 (2003).
212. I. Moss, *Quantum theory, black holes, and inflation*, Wiley, New York (1996).
213. C. Rovelli, *Quantum gravity*, Cambridge University Press, Cambridge (2004).
214. C. Rovelli, "Covariant hamiltonian formalism for field theory: Hamilton-Jacobi equation on the space G", *Lecture Notes in Physics* **633**, 36–62 (2003); e-print arXiv: gr-qc 0207043.
215. L. Smolin, *Three roads to quantum gravity*, Oxford University Press, Oxford (2000).
216. L. Smolin, "An invitation to loop quantum gravity", arXiv:hep-th/0408048 (2004).
217. L. Smolin, "Matrix models as non-local hidden variables theories", arXiv:hep-th/0201031 (2002).
218. T. Thiemann, "Introduction to modern canonical quantum general relativity", arXiv:gr-qc/0110034 (2001).
219. T. Thiemann, "Lectures on loop quantum gravity", *Lecture Notes Physics* **631**, 41–135 (2003).
220. T. Thiemann, "Anomaly-free formulation of non-perturbative, four-dimensional, Lorentzian quantum gravity", *Physics Letters B* **380**, 257–264 (1996).
221. T. Thiemann, "Quantum spin dynamics", *Classical and Quantum Gravity* **15**, 839–873 (1998).
222. T. Thiemann, "Quantum spin dynamics II", *Classical and Quantum Gravity* **15**, 875–905 (1998).
223. T. Thiemann, "QSD IV: 2+1 Euclidean quantum gravity as a model to test. 3+1 Lorentzian quantum gravity", *Classical and Quantum Gravity* **15**, 1249–1280 (1998).
224. T. Thiemann, "QSD V: quantum gravity as the natural regulator of matter quantum field theories", *Classical and Quantum Gravity* **15**, 1281–1314 (1998).
225. A. Shojai and F. Shojai, "Quantum effects and cluster formation", arXiv:astro-ph/0211272 (2002).
226. L. Smolin, "Could quantum mechanics be an approximation to another theory?", arXiv:quant-ph/0609109 (2006).
227. J. Vink, "Quantum mechanics in terms of discrete beables", *Physical Review A* **48**, 3, 1808–1818 (1993).
228. J. Vink, "Quantum potential interpretation of the wave function of the universe", *Nuclear Physics B* **369**, 707–728 (1992).
229. J.A. de Barros and N. Pinto-Neto, "The causal interpretation of quantum mechanics and the singularity problem and time issue in quantum cosmology", *International Journal of Modern Physics D* **7**, 201–213 (1998).

230. J. Kowalski-Glikman and J. C. Vink, "Gravity-matter mini-superspace: quantum regime, classical regime and in between", *Classical and Quantum Gravity* **7**, 901–918 (1990).
231. E.J. Squires, "A quantum solution to a cosmological mystery", *Physics Letters A* **162**, 1, 35–36 (1992).
232. N. Pinto-Neto, G. Santos and W. Struyve, "Quantum-to-classical transition of primordial cosmological perturbations in de Broglie Bohm quantum theory", *Physical Review D* **85**, 083506 (2012); e-print arXiv:gr-qc/0410001 (2004).
233. N. Pinto-Neto, "The Bohm interpretation of quantum cosmology", *Foundations of Physics* **35**, 4, 577–603 (2005); e-print arXiv:gr-qc/0410117.
234. N. Pinto-Neto, "Perturbations in bouncing cosmological models", *International Journal of Modern Physics D* **13**, 7, 1419–1424 (2004); e-print arXiv:gr-qc/0410225 (2004).
235. N. Pinto-Neto and E. Santini, "The consistency of causal quantum geometrodynamics and quantum field theory", *General Relativity and Gravitation* **34**, 4, 505–532 (2002); e-print arXiv:gr-qc/0009080.
236. N. Pinto-Neto and E. Santini, "The accelerated expansion of the universe as a quantum cosmological effect", *Physics Letters A* **315**, 36–50 (2003); e-print arXiv:gr-qc/0302112.
237. N. Pinto-Neto, F.T. Falciano, R. Pereira and E. Santini, "The Wheeler-deWitt quantization can solve the singularity problem", *Physical Review D* **86**, 063504 (2012).
238. Jr. R. Colistete, J.C. Fabris and N. Pinto-Neto, "Gaussian superpositions in scalar tensor quantum cosmological models", *Physical Review D* **62**, 083507 (2000).
239. N. Pinto-Neto and E. Santini, "Must quantum spacetimes be Euclidean?", *Physical Review D* **59**, 123517 (1999); e-print arXiv:gr-qc/9811067.
240. N. Pinto-Neto and J.C. Fabris, "Quantum cosmology from the de Broglie-Bohm perspective", arXiv:1306.0820v1 [gr-qc] (2013).
241. A. Farag Ali and S. Das, "Cosmology from quantum potential", *Physics Letters B* **741**, 276–279 (2015); arXiv:1404.3093v3 [gr-qc].
242. G. Leon, J. Saavedra and E.N. Saridakis, "Cosmological behaviour in extended nonlinear massive gravity", *Classical and Quantum Gravity* **30**, 135001 (2013); e-print arXiv:1301.7419.
243. R. Gannouji, M.W. Hossain, M. Sami and E.N. Saridakis, "Quasi-dilaton non linear massive gravity: investigation of background cosmological dynamics", *Physical Review D* **87**, 123536 (2013); e-print arXiv:1304.5095.
244. C. de Rham, M. Fasiello and A.J. Tolley, "Stable FLRW solutions in generalized massive gravity", arXiv:1410.0960 (2014).
245. S. Das and R.K. Bhaduri, "Dark energy and dark matter from Bose-Einstein condensate", arXiv:1411.0753 (2014).
246. C.M. Will, "Bounding the mass of the graviton using gravitational-wave observations of inspiralling compact binaries", *Physical Review D* **57**, 2061 (1998).

247. L.S. Finn and P.J. Sutton, "Bounding the mass of the graviton using binary pulsar observations", *Physical Review D* **65**, 044022 (2002); e-print arXiv:gr-qc/0109049.
248. A.S. Goldhaber and M.M. Nieto, "Photon and graviton mass limits", *Reviews in Modern Physics* **82**, 939 (2010); e-print arXiv:0809.1003.
249. E. Berti, J. Gair and A. Sesana, "Graviton mass bounds from space-based gravitational-wave observations of massive black hole populations", *Physical Review D* **84**, 101501(R) (2011).
250. F. Zwicky, "Cosmic and terretrial tests for the rest mass of gravitons", *Publications of the Astronomical Society of the Pacific* **73**, 434, 314 (1961).
251. J.R. Mureika and R.B. Mann, "Does entropic gravity bound the masses of the photon and graviton?", *Modern Physics Letters A* **26**, 171–181 (2011); e-print arXiv:1005.2214.
252. C. de Rham, G. Gabadadze, L. Heisenberg and D. Pirtskhalava, "Cosmology of the Galileon from massive gravity", *Physical Review D* **83**, 103516 (2011); e-print arXiv:1010.1780.
253. C. de Rham, G. Gabadadze and A.J. Tolley, "Cosmic acceleration and the helicity-0 graviton", *Physical Review Letters* **106**, 231101 (2011); e-print arXiv:1011.1232.
254. S. Majid, "Emergence of the Riemannian geometry and the massive graviton", arXiv:1401.0673.
255. C.H. Bennett, G. Brassard, C. Crépeau, R. Jozsa, A. Peres and W. Wootters, "Teleporting an unknown quantum state via dual classical and Einstein-Podolsky-Rosen channels", *Physical Review Letters* **70**, 13, 1895–1899 (1993).
256. A. Ekert, "Quantum cryptography based on Bell's theorem", *Physical Review Letters* **67**, 6, 661–663 (1991).
257. M.A. Nielsen and I.L. Chuang, *Quantum computation and quantum information*, Cambridge University Press, Cambridge (2000).
258. V. Vedral, *Introduction to quantum information science*, Oxford University Press, Oxford (2006).
259. L. Pezze and A. Smerzi, "Entanglement, Non-linear Dynamics and Heisenberg Limit", *Physical Review Letters* **102**, 100401 (2009).
260. D.Z. Albert and R. Galchen, "A quantum threat to special relativity", *Scientific American* **3**, 26–33 (2009).
261. A. Valentini, "Beyond the quantum", *Physics World* **22**, 11, 32–37 (2009); e-print arXiv:101.2758v1 [quant-ph].
262. P.G. Kwiat *et al.*, "New high-intensity source of polarization-entangled photon pairs", *Physical Review Letters* **75**, 24, 4337–4341 (1995).
263. Y. Makhlin, G. Schön and A. Shnirman, "Quantum-state engineering with Josephson-junction devices", *Reviews in Modern Physics* **73**, 2, 357–400 (2001).
264. A. Ramsak, J. Mravlje, R. Zitko and J. Bonca, "Spin qubits in double quantum dots — entanglement versus the Kondo effect", *Physical Review B* **74**, 241305(R) (2006).

265. T. Rejec, A. Ramsak and J.H. Jefferson, "Spin-dependent thermoelectric transport coefficients in near perfect quantum wires", *Physical Review B* **65**, 235301 (2002).
266. Q. Zhang *et al.*, "Experimental quantum teleportation of a two-qubit composite system", *Nature Physics* **2**, 678–682 (2006).
267. P.R. Holland, "Causal Interpretation of a System of Two Spin-1/2 Particles", *Physics Reports* **169**, 293–327 (1988).
268. A. Ramsak, "Geometrical view of quantum entanglement", arXiv:1109. 5537v2 [quant-ph] (2011).
269. A. Ramsak, "Spin-spin correlations of entangled qubit pairs in the Bohm interpretation of quantum mechanics", *Journal of Physics A: Mathematical and Theoretical* **45**, 115310 (2012); e-print arXiv:1202.1695v1 [quant-ph].
270. V.I. Sbitnev, "Bohmian split of the Schrödinger equation onto two equations describing evolution of real functions", *Kvantovaya Magiya* **5**, 1, 1101–1111 (2008); URL http://quantmagic.narod.ru/volumes/ VOL512008/p1101.html.
271. V.I. Sbitnev, "Bohmian trajectories and the path integral paradigm. Complexified lagrangian mechanics" *International Journal of Bifurcation and Chaos* **19**, 7, 2335–2346 (2009); e-print arXiv:0808.1245v1 [quant-ph].
272. D. Fiscaletti, "The quantum entropy as the ultimate visiting card of de Broglie-Bohm theory", *Ukrainian Journal of Physics* **57**, 9, 946–963 (2012).
273. D. Fiscaletti, "A geometrodynamic entropic approach to Bohm's quantum potential and the link with Feynman's path integrals formalism", *Quantum Matter* **2**, 2, 122–131 (2013).
274. D. Fiscaletti, "Perspectives about the quantum entropy in Bohm's approach to relativistic quantum mechanics and quantum field theory", *Quantum Matter* **4**, 5, 416–429 (2015).
275. L. Brillouin, *Science and information theory,* Courier Dover Publishing Inc., New York (2004).
276. D. Fiscaletti and I. Licata, "Bell length in the entanglement geometry", *International Journal of Theoretical Physics* **54**, 7, 2362–2381 (2015).
277. G. Resconi and I. Licata, "Quantum computing in non Euclidean geometry", arXiv:0911.0842 [quant-ph] (2009).
278. B. Poirier, "On flux continuity and probability conservation in complexified Bohmian mechanics", http://arXiv.org/abs/0803.0193 (2008).
279. P.V. Poluyan, "Nonclassical ontology and nonclassical movement", *Kvantovaya Magiya* **2**, 3, 3119–3134 (2005); available at http://quantmagic. narod.ru/volumes/VOL232005/p3119.html.
280. C. Grosche, *Path integrals, hyperbolic spaces, and Selberg trace formulae,* World Scientific, Singapore (1996).
281. D. Bohm, in *Symposium on the foundation of modern physics — 1987*, ontological foundation for the quantum theory, edited P. Lahti e P. Mittelstaedt, World Scientific, Singapore (1988).
282. D. Fiscaletti and I. Licata, "Weyl geometries, Fisher information and quantum entropy in quantum mechanics", *International Journal of Theoretical Physics* **51**, 11, 3587–3595 (2012).

283. G. Resconi, I. Licata and D. Fiscaletti, "Unification of quantum and gravity by non classical information entropy space", *Entropy* **15**, 3602–3619 (2013); e-print http://arxiv.org/abs/1110.5491 (2011).
284. I. Licata and D. Fiscaletti, "A Fisher-Bohm geometry for quantum information", *Electronic Journal of Theoretical Physics* **11**, 31, 71–88 (2014).
285. D. Fiscaletti and A. Sorli, "Non-locality and the symmetrized quantum potential", *Physics Essays* **21**, 4, 245–251 (2008).
286. D. Fiscaletti and A. Sorli, "Timeless space is a fundamental arena of quantum processes", *The IUP Journal of Physics* **3**, 4, 34–49 (2010).
287. D. Fiscaletti and A. Sorli, "Three-dimensional space as a medium of quantum entanglement", *Annales UMCS Sectio AAA: Physica* **57**, 47–72 (2012).
288. D. Fiscaletti, A. Sorli and D. Klinar, "The symmetryzed quantum potential and space as a direct information medium", *Annales de la Fondation Louis de Broglie* **37**, 41–71 (2012).
289. D. Fiscaletti and A. Sorli, "Toward an a-temporal interpretation of quantum potential", *Frontier Perspectives* **14**, 2, 43–54 (2005/2006).
290. D. Fiscaletti, "A-temporal physical space and quantum non-locality", *Electronic Journal of Theoretical Physics* **2**, 6 (2005).
291. J.S. Bell, "On the Einstein-Podolski-Rosen paradox", *Physics* **1**, 195–200 (1964).
292. A. Sorli and I.K. Sorli, "From space-time to a-temporal physical space", *Frontier Perspectives* **14**, 1, 38–40 (2005).
293. K.B. Wharton, "Time-symmetric quantum mechanics", *Foundations of Physics* **37**, 1, 159–168 (2007).
294. C. Dewdney, "Calculations in the causal interpretation of quantum mechanics", in *Quantum Uncertainties — Recent and Future Experiments and Interpretations*, edited by E. Panarella, Plenum Press, New York (1987), pp. 19–40.
295. J. Anandan, "Symmetries, quantum geometry and the fundamental interactions", *International Journal of Theoretical Physics* **41**, 2, 199–220 (2002); e-print arXiv:quant-ph/0012011v4 (2001).
296. D. Fiscaletti and A. Sorli, "Non-local quantum geometry and three-dimensional space as a direct information medium", *Quantum Matter* **3**, 3, 200–214 (2014).
297. D. Fiscaletti, *The timeless approach. Frontier perspectives in 21th century physics*, World Scientific, Singapore (2015).
298. A.D. Sakharov, "Vacuum quantum fluctuations in curved space and the theory of gravitation", *Doklady Akad. Nauk S.S.S.R.* **177**, 1, 70–71 (1967).
299. A. Rueda and B. Haisch, "Gravity and the quantum vacuum inertia hypothesis", *Annalen der Physik* **14**, 8, 479–498 (2005); e-print arXiv:gr-qc0504061v3 (2005).
300. H.E. Puthoff, "Polarizable-vacuum (PV) approach to general relativity", *Foundations of Physics* **32**, 6, 927–943 (2002).

301. M. Consoli, "Do potentials require massless particles?", *Physical Review Letters B* **672**, 3, 270–274 (2009).
302. M. Consoli, "On the low-energy spectrum of spontaneously broken phi4 theories", *Modern Physics Letters A* **26**, 531–542 (2011).
303. M. Consoli, "The vacuum condensates: a bridge between particle physics to gravity?", In *Vision of oneness*, edited by I. Licata and A. Sakaji, Aracne Editrice, Roma (2011).
304. P.W. Milonni, *The quantum vacuum — an introduction to Quantum Electrodynamics*, Academic Press, New York (1994).
305. J.W.M. Bush, "Pilot-Wave Hydrodynamics", *Annual Review of Fluid Mechanics* **47**, 1, 269–292 (2015).
306. M. Bhaumik, "Comprehending quantum theory from quantum fields", arXiv:13101251.pdf (2013).
307. S.E. Rugh and H. Zinkernagel, "The quantum vacuum and the cosmological constant problem", *Studies in History and Philosophy of Modern Physics* **33**, 4, 673–705 (2001).
308. S.F. Timashev, "Physical vacuum as a system manifesting itself on various scales — from nuclear physics to cosmology", arXiv:1107.pdf [gr-qc] (2011).
309. D. Fiscaletti, *I gatti di Schrödinger*, Editori Riuniti University Press, Roma, 2° edition (2015).
310. I. Licata, "Emergence and Computation at the Edge of Classical and Quantum Systems", in: *Physics of Emergence and Organization*, World Scientific, Singapore, pp. 1–25 (2008).
311. J.A. Wheeler, "Information, physics, quantum: The search for links", in W. Zurek, *Complexity, Entropy, and the Physics of Information*, Addison-Wesley, Redwood City, California (1990).
312. Reported in *Quanta and Reality*, Stephen Toulmin editor, Hutchinson, London (1971).
313. R.E. Kastner, "de Broglie waves as the 'bridge of becoming' between quantum theory and relativity", *Foundations of Science*, doi:10.1007/s10699–011–9273-4 (2011).
314. R.E. Kastner, "On delayed choice and contingent absorber experiments", *ISRN Mathematical Physics* **2012**, Article ID 617291, 9 pages (2012); doi:10.5402/2012/617291 (2012).
315. R.E. Kastner, "The broken symmetry of time", *AIP Conference Proceedings* **1408**, 7–21; doi:10.1063/1.3663714 (2011).
316. R. Kastner, *The new transactional interpretation of quantum theory: the reality of possibility*, Cambridge University Press, Cambridge (2012).
317. J.G. Cramer, "Generalized absorber theory and the Einstein–Podolsky–Rosen paradox", *Physical Review D* **22**, 2, 362–376 (1980).
318. J.G. Cramer, "The arrow of electromagnetic time and the generalized absorber theory", *Foundations of Physics* **13**, 9, 887–902 (1983).
319. J.G. Cramer, "The transactional interpretation of quantum mechanics", *Reviews of Modern Physics* **58**, 647–688 (1986).
320. J.G. Cramer, "An overview of the transactional interpretation", *International Journal of Theoretical Physics* **27**, 2, 227–236 (1988).

321. L. Chiatti, "The transaction as a quantum concept", in *Space-time geometry and quantum events*, edited by I. Licata, Nova Science Publishers, New York, pp. 11–44 (2014), e-print arXiv.org/pdf/1204.6636 (2012).
322. I. Licata, "Transaction and non-locality in quantum field theory", *European Physical Journal Web of Conferenes* **70**, 00039 (2014).
323. I. Licata and L. Chiatti, "The archaic universe: big bang, cosmological term and the quantum origin of time in projective cosmology", *International Journal of Theoretical Physics* **48**, 4, 1003–1018 (2009).
324. I. Licata and L. Chiatti, "Archaic universe and cosmological model: 'big-bang' as nucleation by vacuum", *International Journal of Theoretical Physics* **49**, 10, 2379–2402 (2010); e-print arXiv:genph/1004.1544.
325. J.D. Bjorken and S.D. Drell, *Relativistic Quantum Mechanics*, Mc Graw-Hill, New-York (1965).
326. D.M. Greenberger, M. Horne and A. Zeilinger, "Going Beyond Bell's Theorem", in *Quantum Theory and Conceptions of the Universe*, edited by M. Kafatos, Kluwer, Academic, Dordrecht (1989).
327. A. Aspect, P. Grangier and G. Roger, "Experimental realization of Einstein-Podolski-Rosen-Bohm gedankenexperiment: a new violation of Bell's inequalities", *Physical Review Letters* **49**, 2, 91–83 (1982).
328. A. Aspect, J. Dalibard e G. Roger, "Experimental tests of Bell's inequalities using time-varying analyzers", *Physical Review Letters* **49**, 25, 1804–1807 (1982).
329. C. Tittel, J. Breindel, H. Zbinden and N. Gisin, "Violation of Bell inequalities by photons more than 10 Km apart", *Physical Review Letters* **81**, 17, 3563–3566 (1998).
330. Z.Y. Ou, S.F. Pereira, H.J. Kimble and K.C. Peng, "Realization of the Einstein-Podolski-Rosen paradox for continuous variables", *Physical Review Letters* **68**, 25, 3663–3666 (1992).
331. E. Santos, "Quantum vacuum fluctuations and dark energy", <arXiv: 0812.4121v2 [gr-qc]>(2009).
332. E. Santos, "Space-time curvature induced by quantum vacuum fluctuations as an alternative to dark energy", *International Journal of Theoretical Physics* **50**, 7, 2125–2133 (2010).
333. D. Fiscaletti and A. Sorli, "Space-time curvature of general relativity and energy density of a three-dimensional quantum vacuum", *Annales UMCS Sectio AAA: Physica* **LXIX**, 55–81 (2014).
334. S. Ghao, "Why gravity is fundamental," arXiv:1001–3029v3 (2010).
335. Y.J. Ng, "Holographic Foam, Dark Energy and Infinite Statistics", *Physics Letters B* **657**, 1, 10–14 (2007).
336. Y.J. Ng, "Spacetime foam: from entropy and holography to infinite statistics and non-locality", *Entropy* **10**, 441–461 (2008).
337. Y.J. Ng, "Holographic quantum foam", arXiv:1001.0411v1 [gr-qc] (2010).
338. Y.J. Ng, "Various facets of spacetime foam", arXiv:1102.4109.v1 [gr-qc] (2011).

339. D. Fiscaletti and A. Sorli, "Perspectives about quantum mechanics in a model of a three-dimensional quantum vacuum where time is a mathematical dimension", *SOP Transactions on Theoretical Physics* **1**, 3, 11–38 (2014).
340. D. Fiscaletti and A. Sorli, "About a three-dimensional quantum vacuum as the ultimate origin of gravity, electromagnetic field, dark energy ... and quantum behaviour, *Ukrainian Journal of Physics* **61**, 5, 413–431 (2016).
341. D. Fiscaletti, "What is the actual behaviour of the electron? From Bohm's approach to the transactional interpretation to a three-dimensional timeless non-local quantum vacuum", *Electronic Journal of Theoretical Physics* **13**, 35, 13–38 (2016).
342. B. Haisch, A. Rueda and H.E. Puthoff, "Inertia as a zero-point field Lorentz force", *Physical Review A* **49**, 2, 678–694 (1994).
343. B. Haisch, A. Rueda and H.E. Puthoff, "Physics of the Zero-Point Field: Implications for Inertia, Gravitation and Mass", *Speculations in Science and Technology* **20**, 99–114 (1997).
344. A. Rueda and B. Haisch, "Contribution to the inertial mass by reaction of the vacuum to accelerated motion", *Foundations of Physics* **28**, 7, 1057–1108 (1998).
345. A. Rueda and B. Haisch, "Inertial mass as reaction of the vacuum to accelerated motion", *Physics Letters A* **240**, 3, 115–126 (1998).
346. B. Haisch, A. Rueda and Y. Dobyns, "Inertial mass and the quantum vacuum fields", *Annalen der Physik (Leipzig)* **10**, 5, 393–414 (2001).
347. D. Fiscaletti, "About dark energy and dark matter in a three-dimensional quantum vacuum model", *Foundations of Physics* **46**, 10, 1307–1340 (2016).
348. W.J.G. de Blok, S.S. McGaugh and V.C. Rubin, "High-resolution rotation curves of low surface brightness galaxies. II. Mass models", *The Astronomical Journal* **122**, 2396–2427 (2001); doi:10.1086/323450.
349. V.C. Rubin, "A brief history of dark matter", in *The Dark Universe: Matter, Energy and Gravity, Symposium Series: 15*, edited by M. Livio, pages 1–13, Cambridge University Press, Cambridge (2004).
350. D.S. Hajdukovic, "Is dark matter an illusion created by the gravitational polarization of the quantum vacuum?", *Astrophys. Space Sci.* **334**, 215–218 (2011).
351. J. Schaffer, "Monism: The priority of the Whole", *Philosophical Review* **119**, 31–76 (2010).
352. B. Hiley and R. Callaghan, "Clifford Algebras and the Dirac-Bohm Quantum Hamilton-Jacobi Equation", *Foundations of Physics* **42**, 1, 192–208 (2012).

Index

Printed in the United States
By Bookmasters